The Constitution of Algorithms

Inside Technology

Edited by Wiebe E. Bijker, Trevor J. Pinch, and Rebecca Slayton

A list of books in the series appears at the back of the book.

The Constitution of Algorithms

Ground-Truthing, Programming, Formulating

Florian Jaton

The MIT Press
Cambridge, Massachusetts
London, England

The open access edition of this book was made possible by generous funding from Arcadia—a charitable fund of Lisbet Rausing and Peter Baldwin.

This book was set in Stone Serif and Stone Sans by Westchester Publishing Services. Printed and bound in the United States of America.

Library of Congress Cataloging-in-Publication Data

Names: Jaton, Florian, author. | Bowker, Geoffrey C., writer of foreword.
Title: The constitution of algorithms : ground-truthing, programming, formulating / Florian Jaton ; foreword by Geoffrey C. Bowker.
Description: Cambridge, Massachusetts : The MIT Press, [2020] | Series: Inside technology | Includes bibliographical references and index.
Identifiers: LCCN 2020028166 | ISBN 9780262542142 (paperback)
Subjects: LCSH: Algorithms--Case studies. | Computer programming--Case studies. | Algorithms--Social aspects. | Mathematics--Philosophy.
Classification: LCC QA9.58 .J38 2020 | DDC 518/.1--dc23
LC record available at https://lccn.loc.gov/2020028166

10 9 8 7 6 5 4 3 2 1

To Fanny

Contents

Foreword

Geoffrey C. Bowker

Algorithms pervade our lives. They are political, cultural, and social facts that have become central to all parts of our existence over the past fifty years. Certainly, we had their forerunners before: endless checklists, safety protocols, and rules of conduct—each designed to take us out of ourselves and align our bodies, our selves with a bureaucratic or technical machine (in Foucault's better term, a set "dispositifs techniques"). Bureaucracy makes us act like machines, algorithms seek to make us into machines.

A corollary is that if we want to do fundamental social science and envision new forms of political life we need to go where the action is. We need to get to know algorithms from the inside. They did not parachute down from another planet to invade us (much as it may feel like this): they are human, fallible creations. The difficulties here are that social scientists and political actors often don't really understand the technical stakes, and symmetrically the computer scientists don't really get the social stakes.

This is precisely why this book is so important. It is a foundational text for exploring algorithms as a new form of social actor. How do algorithms get constructed to be effective actors; how do humans get constructed so that they create algorithms which surpass human understanding? Jaton's quest here has been fearless: go where the questions are, and locate the technical, social, and political issues on their home ground. As I read this book, I was constantly delighted as when reading a fine novel by not knowing what was going to come next (von Neumann architecture, tests for nascent computer engineers)—but by immediately feeling a sense of inevitability once the steps were taken.

I've been playing with a vision latterly of humans becoming progressively more irrelevant to the operation of our political economy: we do what we can but are increasingly interstitial. There is little doubt that we are creating machines that are more intelligent than we are and algorithms that know us better than we do ourselves. That's just fine. But how much richer and more beautiful a world we will create if we suffuse our algorithms with our own deeply held values created over thousands of years?

This book is not just for computer scientists or for social studies of science scholars: it speaks to some of the fundamental questions of human existence in this epoch. It provides tools and concepts for us to co-engineer our world (our planetary system, our species, our computers).

Chapeau! Florian. Happy reading all.

Acknowledgments

More than politeness, it is a matter of intellectual integrity to warmly thank those who helped me become the author of this book. To begin with, I would like to express my deepest gratitude to the members of the computer science laboratory who let me follow their day-to-day activities. Having an ethnographer around for more than two years must have been an odd experience. Yet I could not have wished for more comprehension toward my research topic and patience toward my clumsiness. It goes without saying that this inquiry could not have been written without the support of these brilliant computer scientists who quickly became my colleagues and friends.

If I enjoyed spending time in this computer science laboratory, it was also thanks to its director. By giving me an office, providing me with insightful feedback, and asking me to actively participate in the daily life of her laboratory, Sabine Süsstrunk of the Swiss Federal Institute of Technology Lausanne (EPFL) immensely facilitated my integration. I simply could never have dreamed of a better interdisciplinary collaboration.

My mentor Dominique Vinck has given me so many valuable tips, insights, and feedback throughout this inquiry that I wish I could have applied the following seal on the cover of this document: *Dominique Vinck Inside*®. It has been a privilege to be the student of such an inspiring professor.

This book also benefited from the insights of my colleagues at the Institute of Social Sciences of the University of Lausanne. Marc Audétat, Lola Auroy, Nicolas Baya Laffite, Boris Beaude, Luca Chiapperino, Laetitia Della Bianca, Olivier Glassey, Sara Guzmán, Anna Jobin, Nicky Lefeuvre, Pierre-Nicolas Oberhauser, Francesco Panese, Andréas Perret, Jessica Pidoux,

Margarita Rodriguez, Yohana Ruffiner, Marie Sautier, Romina Seminario, Tatiana Smirnova, Léa Stiefel, and Mylène Tanferri Machado: they all greatly contributed to my intellectual education. And I would like to extend a special thank you to Alexandre Camus, who, besides having given me great suggestions, has also stood for my fears, rants, and sudden bursts of joy (and despair).

To transform what was then a cumbersome thesis into an acceptable book, I benefited from a postdoctoral research stay at the Centre de Sociologie de l'Innovation of Mines Paristech, PSL Research University. And without the precise advice and comments of Félix Boilève, Jérôme Denis, Quentin Dufour, Liliana Doganova, Evan Fisher, Clément Gasull, Cornelius Heimstädt, Antoine Hennion, Brice Laurent, Fabian Muniesa, Émilie Perault, David Pontille, Mathieu Rajaoba, and Loïc Riom, this book would contain many more weaknesses than it has today. I also warmly thank Nassima Abdelghafour, Madeleine Akrich, Marie Alauzen, Mathieu Baudrin, Victoria Brun, Béatrice Cointe, Jean Danielou, Catherine Lucas, Alexandre Mallard, Morgan Meyer, Florence Paterson, Mathilde Pellizzari, Vololona Rabeharisoa, Roman Solé-Pomies, Sophie Tabouret, Félix Talvard, Carole-Anne Tisserand, Didier Torny, Frédéric Vergnaud, and Alexandre Violle for having welcomed me to their wonderful research center.

Easily distressed by administrative duties, I have been lucky to benefit from the help of amazing secretaries throughout my PhD and postdoctoral grants. To a great extent, it is thus thanks to Françoise Behn, Marianna Schismenou, Alba Brizzi, and Joëlle de Magalhaes that I could finally produce this document.

Funding is integral part of research. Thus I thank the Swiss National Science Foundation for its financial support throughout the completion of this work. Funding such a fundamental research project at the intersection of philosophy and computer science was for sure a risky investment. I cannot, of course, decide whether this work keeps the numerous promises I made to get both my PhD (POLAP1 148948) and postdoctoral grants (P2LAP1 184113). I can only assert that over the past few years, a great part of my vital energy was dedicated to the accomplishment of this project. I also wish to extend my thanks to the Société Académique Vaudoise for its generous support between October and December 2018.

From 2016 to 2017, I spent a year abroad at the EVOKE Lab and Studio of the University of California, Irvine (UCI) as part of my PhD program.

With regard to this formative experience, I must start by thanking Myles, Kyle, Dave, and Laura Jeffrey who never stopped considering me as part of their Californian family. I am also very grateful to my UCI colleagues at that time—Anja Bechmann, Roderic Crooks, Simon Penny, John Seberger, and Aubrey Slaughter—who greatly helped the completion of the book's second, third, and fourth chapters. And what can I say about the amazing collections of the University of California Libraries? Without the daily invisible work of University of California librarians, I could not have accessed the crucial references I needed to propose, I hope, innovative propositions. However, this Californian experience would have been impossible without the unconditional support of Geoffrey C. Bowker who believed in this project from the very beginning.

Obviously, this document benefited from the support of MIT Press, Inside Technology Series. In this regard, I want to thank the series' editorial staff for their kindness and unfailing availability throughout the publication process. I am also grateful to the anonymous reviewers and copyeditors who contributed to making this work better that it initially was. Of course, and this concerns all those who helped me to produce this book; all mistakes and low passes remain mine.

My close friends have helped, supported, and inspired me so much during my not-yet-really-started academic career that it will be unfair not to name them. Thus from the bottom of my heart, I want to thank Julien Bugnon, Gabriel Buser, Frédéric Clerc, Loïs de Goumoëns, Christophe Durant, Simon Duvoisin, Antoine Favre, Vincent Klaus, Nicolas and Vanessa Krieg, Naïke and Stéphane Lévy, Mathieu and Nancy Morier, Marco Picci, Coralie Pittet, Estelle and Vincent Rossire, Mathias Schild, Lucas Turrian, Nicolas Vautier, and Élise Vinckenbosch. It is a real privilege to be your friend.

As this work is the direct product of their unconditional affection, I finally wish to express my deepest gratitude to my mother, Katia; my father, Jean-Pierre; my sister, Laure; my brother, Damien; and my niece, Lina. And to Fanny, who lovingly supports me in the vicissitudes of intellectual life: Thank you for bringing infinite light.

Introduction

For critics and advocates alike, if we want to know algorithms, we may need to live with them.
—Seaver (2013, 11)

Let us start this introduction *in medias res*, in the middle of things:

Rearrangement 1

The election of Donald Trump in November 2016 was quite surprising: how could such a controversial figure reach the White House? The reasons, of course, are innumerous. But what if one of them was Facebook (Lapowsky 2016)? After all, Trump supporters never stopped using this platform to spread out disputed contents. What if voters were brainwashed by the "fake news" Facebook contributed to diffusing? What if this extensive interlinking participated in Trump's advertisement and fundraising? However harsh this claim might be, it seriously harms the image of the web application that would rather help to "connect people" than to build border walls (Isaac 2016). It seems then that monitoring needs to be increased, even though it may contradict some assumptions Mark Zuckerberg elevates as precepts (Zuckerberg 2016). The main target is the "News Feed," the spine of the application that displays stories posted by Facebook users. What about slightly modifying how News Feed automatically selects new stories to make it ignore "low quality posts"? This may help restore Facebook's public image, at least for a little bit, at least for a little while. And after several months of in-house research and testing, a new *algorithm* is made operational that—based on frequencies of posts and URLs of links—identifies spam users and automatically deprioritize the links they share (Isaac and Ember 2016). According to one of Facebook's vice presidents, this new method of computation should significantly reduce the diffusion of "low quality content such as clickbait, sensationalism, and misinformation" (Mosseri 2017).

Rearrangement 2

Planet Mars is a distant location. But hundreds of millions of kilometers did not dishearten the US National Aeronautics and Space Administration (NASA) from sending the robotic rover Curiosity to explore its surface. On May 6, 2012, the costly vehicle safely lands on Gale Crater. Quite a feat! Amazing high-resolution pictures are soon available on NASA's website, showing the world the jagged surface of this cold and arid planet. Of course, Curiosity is far more than a remote-controlled car taking exotic pictures. It is a genuine laboratory on wheels with many high-tech instruments: two cameras for true-color and multispectral imaging, two pairs of monochrome cameras for navigation, a robotic arm with an ultrahigh-definition camera, a laser-induced spectrometer, solar panels, two lithium-ion batteries, and so on (Jet Propulsion Laboratory 2015). Yet there is an obvious cost to this amazing remote-controlled laboratory: it needs to move its 350 kilograms (low gravity considered). The sharp, rocky surface of Mars does not alleviate the constant efforts of Curiosity's wheels, irremediably wearing down. And in January 2014, the situation becomes alarming (Webster 2015): Is there a way to extend the lifetime of Curiosity's wheels? After much research, a new driving *algorithm* becomes operational in June 2017 that uses real-time data from the navigation cameras to adjust Curiosity's speed when it comes to sharp Martian pebbles (Good 2017). By reducing the load of Curiosity's leading and middle wheels up to 20 percent, this new method of computation for navigation is considered a serious boost for the mission (Sharkey 2017).

Rearrangement 3

Israeli secret services in the West Bank are used to dismantling organizations they define as terrorist by means of preventive actions and intimidation. But what about individuals who commit attacks on a whim? Just like several police departments in the United States (Berg 2014), Israeli secret services are now supported by a security software whose *algorithm* generates profiles of potential attackers based on aggregated data posted on social media. Yet while several US civil courts are seriously considering the harmful bias of these new methods of computation (Angwin et al. 2016; Liptak 2017), Israeli military justice as applied to suspected Palestinian "attackers" prevents them from having any sort of legal protection. Thanks to the ability of the West Bank military commander to stamp administrative detentions, these "dangerous profiles" can be sentenced to a renewable six-month incarceration without any possibility of appeal. Many Palestinians targeted by this state-secret technology "have served long years without ever seeing a court" (Gurvitz 2017).

Rearrangement 4

How can people be made to eat more Nutella? It has not been easy these recent years for the Italian brand of chocolate spread. When palm oil production threatened remote orangutans, only a small fraction of citizens was eager to criticize its use in Nutella's recipe. But in May 2016, as soon as palm oil is suspected of speeding up the spread of cancer among Nutella consumers, there starts to be a worrying drop in sales (Landini and Navach 2017). For Nutella, something needs to be done to reconnect with the stomachs of its customers. What about a fresh new marketing campaign? In collaboration with advertising agency Ogilvy & Mather Italia, seven million uniquely designed Nutella jars are soon produced and sold in record time (Nudd 2017). At the heart of this successful marketing move lies an *algorithm* that computes a carefully selected set of colors and figures to generate unique pop patterns (Leadem 2017).

States of affairs change. In November 2016, News Feeds of Facebook users were subjected to spammers diffusing hoaxes and "fake news" that are presumed to have played a role in the election of Donald Trump. One month later, these News Feeds temporarily became monitored lists of stories worth being read. Similarly, Curiosity's weight together with sharp Martian pebbles first seriously affected the robot's wheels, thus compromising the initial duration of the mission. Yet a few years later, several changes in the locomotion system slowed down this unexpected wear. In another case, Israeli secret services were at first powerless against attacks that were not prepared within dismantable cell organizations. Yet these services soon were able to identify suspects and put them in jail without any kind of legal procedure. Finally, Nutella was first an old-fashioned chocolate spread whose recipe included orangutan-endangering and cancer-related palm oil. It then became, temporarily, a trendy pop product. For better or worse, collective configurations are rearranged, thus forming new states of affairs; relationships between humans and nonhumans are reconstituted, thus temporarily establishing new networks. According to this ontological position that is often called "process thought,"[1] the collective world is constantly reshaped in this way.[2]

That being said, we may wish to comprehend some of the dynamics of these messy *rearrangements* (RTs). After all, as we all have to coexist on the same planet, getting a clearer view of what is going on could not hurt; documenting a tiny set of the innumerous relationships that shape the world we inhabit may equip us with some kind of navigational instrument. Together, where do we go? What are we doing? What is going on? These are important, legitimate questions.

To address these questions, two approaches are generally used. Broadly speaking, the first approach consists in postulating the existence of aggregates capable of inducing states of affairs. Depending on academic traditions, such aggregates take different names: they are sometimes called "social forces," "fields and habitus," "economic rationality," or "social structures," among many other variations. These differently named yet a priori postulated aggregates are all pretenders to the definition of the social (or society), an influential yet evanescent matter that supposedly surrounds individuals and orientates their actions. The scientific study of this matter and the states of affairs it engenders is what I call the science of the social or, more succinctly, social science.

The second approach—the one this book embraces—consists in considering the social not as an evanescent matter surrounding individuals but as the small difference produced when two entities come into contact and temporarily *associate* with each other (Latour 2005).[3] This approach assumes that every new connection between two *actants*—humans (Bob, the president, Mark Zuckerberg) or nonhuman entities (a wheel, a virus, a document)— makes a small difference that can, sometimes, be accounted for. If we accept calling "social" the small difference produced when two actants temporally associate with each other, we may call "socio-logy" the activity that consists in producing specialized texts (*logos*) about these associations (*socius*).[4] Our initial four RTs are small examples of such an activity: Facebook, Curiosity, Israeli secret services, and Nutella temporarily associate themselves with new actants, and the blending of these new connections contributes to the formation of new configurations summarized within a text. Had I added several rearrangements and accounted for their constitutive associations a bit more thoroughly, I would have produced a genuine sociological work. On the contrary, had I invoked some hidden force to explain these reconfigurations; had I attributed the modifications of each state of affairs to some a priori postulated aggregate (e.g., economic rationality, society, culture), I would have produced a small work of social science. This distinction between sociology and social science will accompany us throughout this book. It is thus important to keep in mind that the present volume is—or, at least, is intended to be—a sociological work.

With these clarifications in mind, let us have a closer look on our four small sociological RTs. What do we see? We quickly notice that each RT is affected by an "algorithm," for now loosely defined as a *computerized method of calculation*. These four algorithms can be considered entities—or *actants*—as they all produce differences within specific configurations. In that sense, these algorithms are fundamentally not dissimilar to the other actants they, at some point, associate with. In RT1, *there is* Facebook, Donald Trump, spams, supporters, News Feed, a new algorithm, a Facebook vice president, and many other actants that, together, rearrange some state of affairs. In RT2, *there is* Mars, NASA, sharp pebbles, a navigational algorithm, lithium-ion batteries, and many other actants that, together, rearrange some state of affairs. The same is true of RT3 and RT4: algorithms are actants among many other actants.

Yet a closer look nonetheless suggests that the algorithms of our RTs possess characteristics that make them not completely akin to, say, sharp Martian pebbles or lithium-ions batteries. Contrary to such "firm" actants, the algorithms of our RTs appear more fluid; they seem to be able to move very quickly and make connections with other actants that were at first remote from each other. In RT1, Facebook's new algorithm can, in the end (and yet temporarily), associate itself with News Feeds of millions of users located all around the world almost instantaneously. In RT2, NASA's algorithm can reach Mars to make Curiosity's wheels cope with, potentially, all sharp Martian pebbles. In RT3, the algorithm used by Israeli secret services can classify thousands of social media texts sent by hundreds of thousand people located throughout a two-thousand square-mile territory. In RT4, Ogilvy & Mather Italia's algorithm can create millions of uniquely designed patterns instructing Nutella's packaging factories in Italy and France. It seems then that these algorithms can circulate and link up initially sparse actants in a very short amount of time. This is a nontrivial characteristic. To underline these algorithms' fluidity (they circulate), swiftness (they are fast), and distributivity (they are simultaneously scattered and united), let us temporarily categorize them as *devices*, a special category of actant that, according to philosopher Gilles Deleuze, is "tangled, multi-linear ensembles [that] trace processes that are always at disequilibrium, sometimes coming close to each other, sometimes getting distant from each other" (Deleuze 1989, 185).

If we continue considering our four RTs, we also quickly notice that each of these fluid, swift, and distributed devices called algorithms contributes to modifying a network of relationships. In every RT, one algorithm— well supported by many other entities (researchers, data, tests, computers, etc.)—participates in making Facebook less subject to the spread of hoaxes (RT1), Curiosity's wheels a bit more durable (RT2), Palestinians definitely more "jailable" (RT3), and Nutella temporarily more salable (RT4). Along with all the entities they are associated with, these methods of calculation seem then to participate in changing power dynamics: Facebook, Curiosity's wheels, Israeli security services, and Nutella become temporarily *stronger* than Trump-spamming supporters, sharp Martian pebbles, West Bank potential "terrorists," and palm oil scandals, respectively.

Scholars of *Science and Technology Studies* (STS)—a subfield of sociology and social science that aims to document the co-constitution of science, technology, and the collective world[5]—are nowadays prone to analyze algorithms' propensity to modify power dynamics in, for example, labor markets (Kushner 2013; Steiner 2012), surveillance strategies (Introna 2016; Introna and Wood 2002; Kraemer, van Overveld, and Peterson 2010), corporate finance (Lenglet 2011; MacKenzie 2014; Muniesa 2011a), cultural habits (Anderson 2011; Hallinan and Striphas 2014), or interpersonal relationships (Beer 2009; Bucher 2012). These scholars' works are of the most importance as they raise and maintain wakefulness with regard to what computerized methods of calculation do. Yet I must warn the reader right from the start: what algorithms do is not the main topic of this book.

However, as soon as one takes seriously into consideration the banal fact that objects and devices wear down and change, that "they break, malfunction and have to be constantly mended, retrofitted and repurposed" (Domínguez Rubio 2016, 60), thorough sociological studies of what algorithms do should be coupled with the studies of the maintenance and repair work required to keep them doing what they do. Whereas maintenance and repair work is currently receiving the attention of an increasing number of studies (e.g., de la Bellacasa 2011; Domínguez Rubio 2014, 2016; Denis and Pontille 2015; Lea and Pholeros 2010; Strebel, Bovet, and Sormani 2018), very few have specifically explored the work required to keep algorithms doing what they do (but see Crooks 2019). It is a shame since the differences algorithms produce should be, at least in principle, proportional to the work required to make them continue to produce such differences in constantly evolving situations. If we continue to draw upon our four initial RTs, we can for example imagine that to keep on protecting users from spammers, Facebook's new monitoring algorithm may need to be actualized to detect unexpected forms of trolling (RT1). Similarly, if Curiosity's balance of weight happens to change—such as if it loses a piece of equipment—the parameters of its driving algorithm will have to be modified (RT2). In a similar vein, due to the progressive accumulation of small differences in the computer equipment of Israeli secret services, the software package allowing the new security algorithm to effectively compute social media data and generate profiles will have to be slightly updated (RT3). Finally, for its algorithm to keep on supporting effective marketing *coups*, Ogilvy & Mather Italia will need to keep on convincing its clients that consumers are attached to singular products (RT4). In short, we can make the fair assumption that without constant efforts to make algorithms keep on fitting with constantly changing situations (and vice versa), these devices will not produce differences for very long. Although the work necessary to preserve the *agency* of algorithms (Introna 2016) is surely more and more common in contemporary economies, it remains poorly documented. Unfortunately, I will not contribute to filling in this gap; despite the need for such studies to better understand the collective world we live in, this book does not deal with the maintenance of algorithms.

What is this book's topic, then? We have quickly seen that, from a sociological standpoint, algorithms can be considered two kinds of entities:
devices that *do* things and devices that *need* things in order to keep on
doing what they do. Both views are, I believe, of great significance. Yet my
work follows a different path. Instead of starting from algorithms as devices
and studying their agency or need for maintenance, this book starts from
unrelated entities (e.g., documents, people, desires) and tries to account
for how they come into contact to form, in the end, devices we may call
"algorithms." In short, I am studying what is happening *before* algorithms
become fluid, swift, and distributed devices. Of course, things are not so
clear-cut; as we will see, projections on both agency and maintenance
requirements of future algorithms may impact on their constructions.
Moreover, already constructed algorithms participate in the formation of
new algorithms. But still, it is important for the reader to understand that
I will mainly inquire into the practical activities by which algorithms are
progressively assembled in assignable locations rather than what they may
suggest or require once they are assembled.

Negative Invisibilities

Already at this point, a question may arise: Why is it important to account
for the formation processes of algorithms? Why spending time and energy
writing—and reading—about their constitution? Are there not other things
to do than making the activities by which algorithms come into existence
visible?

Certainly. As Star and Strauss (1999) have suggested, some activities need
to remain provisionally invisible—that is, not accounted for—otherwise the
results of these activities may lose some of their capacities. The circus is one
example: making publicly visible the infrastructure and training practices
required to design and master, say, a Cirque du Soleil trapeze act may negatively affect the act itself. Wonder, surprise, or enchantment would potentially be counteracted by the down-to-earth and uncertain operations that
enabled the act. Here, a sociological account would take the risk of spoiling
the act; it may lower the act's *capacity to act*.[6] Following the distinction
made by Star and Strauss (1999, 23), the relative invisibility of the trapeze
act is, in that sense, *positive*: it helps the product of these circus practices
to be, by lack of a better term, adequate. The lack of any publicly available

account and the presence of secrecy help the act become an act, just as they help the public become the public of the act. In such a very specific situation, one may assume there is a mutual desire to believe in mastery.

But as soon as there are controversies about the products of some practices, the terms of their adequacy are disputed; when some capacity to act is put into question, disagreements about its formation need to be confronted. Let us, for example, imagine that the same Cirque du Soleil trapeze act leads to an accident. If disputes arise about this accident, there will be requests to make visible the practices that contributed to producing it. From being positively invisible, the practices required to do this trapeze act would become negatively invisible: for the different parties of the dispute to become able to *negotiate*, empirical accounts of how this act comes into existence will become necessary. What does the Cirque du Soleil need to perform this controversial act? Which elements could be changed to readjust this fragile assemblage? In short, in order to propose *compromises*, in order to better *compose*, disputants will benefit from empirical accounts of the practices of trapeze;[7] documenting what performers and entertainers cherish and fear and what they are attached to might allow constructive dissensions about the agency of what they produce to unfold.

Despite its obvious limits, this small imaginary example indicates that the request for visibility is somewhat correlated with the rise of controversies. When there are controversies over the products of practices, these products cannot be considered adequate anymore: positive invisibilities may thus switch to negative invisibilities that themselves call for empirical accounts—which can take the form of sociological investigations—on which disputes may arise *and* negotiations unfold. Of course, these accounts are very risky as they inherently speak in the name of individuals (Latour 2005, 121–140). To make visible what communities of practice need and cherish, and what they are attached to, the sociological account that may establish common grounds for further contentious negotiations would need to overcome many trials: Does the account make visible the actants that are crucial to the work of the practitioners? Do surprising but empirically supported connections unfold? Does the account propose new grips for collective composition? A single "no" to any one of these questions would make the sociological account fail to fulfill its initial commitment.

What about algorithms? Not so long ago, these devices attracted little attention. They were certainly involved in changing power relations, but these processes were not, or only to a limited extent, public issues. Things began to change in the late 1990s when sociologists started to question the discourse on empowerment and information accessibility put forward by the promoters of web technologies.[8] Hoffman and Novak (1998) showed, for example, that the accessibility and use of web technologies in the United States were largely function of racial differences. Lawrence and Giles (1999) stressed that, contrary to the promotional rhetoric of almost unlimited access, the search engines available in the late 1990s were only able to index a small and oriented fraction of the web. In the same vein, Introna and Nissenbaum (2000) underlined the underground—and potentially harmful— influence of the heuristics used for the classification of URLs by these same late-1990s search engines. The post-9/11 period that followed focused on criticisms of biases in programs and *algorithms*—the term appeared at that time in the critical literature[9]—for surveillance and preventive detection. In his study of the social implications of data mining technologies, Gandy (2002) warned, for example, that they are the gateway to rational discrimination, potentially strengthening correlative habits between social status and group membership. From a political economy perspective, Zureik and Hindle (2004) discussed biometric algorithms' propensity to trivialize social profiling, categorization, and exclusion of national groups. Another example is the work of Introna and Wood (2004): their analysis of facial recognition algorithms highlighted the potential biases of these devices, which were often, at that time, presented as impartial. This line of sociological research led, at the beginning of the 2010s, to numerous investigations on discriminations (e.g., Kraemer, van Overveld, and Peterson 2010; Gillepsie 2014 Steiner 2012) and invisibilizations (Bucher 2012; Bozdag 2013) induced by the use of algorithms.

This research direction has continued in recent years, with increasingly comprehensive works revealing the contrasting, and often questionable, effects of algorithms on contemporary societies (e.g., Crawford and Calo 2016; Noble 2018; O'Neil 2016; Pasquale 2015). These awareness-raising efforts were also reported in the press, further making algorithms matters of public concern (e.g., Mazzotti 2017; Risen and Poitras 2017; Smith 2018). This dynamic—too complex to be thoroughly dealt with in this introduction[10]— has led to the current situation where the collective world is steadily affected by controversies over algorithms. A quick look at the news, at the time of writing, suffices to remind us of it. UK police is about to use a new algorithm to identify online hate crime on social media (Roberts 2017)? This soon triggers hostile reactions from the nonprofit organization "Big Brother Watch," ready to "fight any attempt to curb free speech online" (Parker 2018). A new algorithm is published in an academic journal that can presumably deduce people's sexuality from photographs of faces (Levin 2017)? The Gay & Lesbian Alliance Against Defamation soon condemns such a "dangerous and flawed research that could cause harm to LGBTQ people around the world" (Anderson 2017).[11] Facebook's algorithm continues to bombard a grieved woman by parenting ads after the stillbirth of her son (Brockell 2018)? Thousands of tweets soon denounce gender bias from tech companies (Mahdawi 2018). Every week, a new dispute arises regarding the consequences—actual or potential—of new algorithms, often preceded by changing attributive nouns such as big data, machine learning, or more recently, artificial intelligence.

The intended relevance of this book should be considered in the light of the current controversies over the agency of algorithms. Following in the footsteps of authors such as Bechman and Bowker (2019), Barocas and Selbst (2016), and Grosman and Reigeluth (2019)—to whom I shall return later in the book—my aim here is to propose intellectual tools to prepare the elaboration of compromises. The invisibility of the practices underlying the development of algorithms can indeed no longer be considered positive: as they are the object of repeated disputes, it is now certainly important, or at least interesting, to document the practical processes that enable them to come into existence. Roughly put, if sociology has looked, with a certain success, at the effects of algorithms, it is now time for it to inquire into the *causes of these effects*, however distributed and multiple they may be. A gap needs to be filled in; by means of empirical accounts of how computer scientists and engineers nurture algorithms, some risky yet refreshing grounds for constructive disputes may be provided.[12] The needs, attachments, and values of those who design algorithms—as documented by my limited sociological account—may contradict other needs, attachments, and values. But at least, in these days of controversies, parties in dispute may slowly start to negotiate, as Walter Lippmann says, "under their own colors" (1982, 91). Yet before considering how I intend to effectively run this inquiry into the practical formation of algorithms, I quickly need to further specify its political dimension. To do so, I shall now make a quick detour by discussing the unconventional term "constitution" I use here to qualify my venture.

Why "Constitution" (And Not Simply "Construction")?

At the beginning of this introduction, I asserted that the collective world is constantly rearranged: heterogeneous entities never stop associating with each other, the blending of these associations temporarily establishing new states of affairs. From this (debatable) ontological position, it follows that the world is not "out there," ready to be grasped from some outside stand-point. Instead, according to this processual ontology, the world is always *becoming*; it is the active product of associations between human and non-human actants.

Yet one may rightly argue that everything is not always reinvented. While some associations bring about ephemeral actants (e.g., a cry of joy, tears of sadness, laughs at some joke), some other associations bring about actants that are more enduring. Many entities that populate/generate the collective world are of this sort: Mark Zuckerberg, the planet Mars, West Bank jails, Nutella jars—just to mention some entities we encountered in our small initial RTs—are quite enduring entities. Such actants, thanks to their ability to live on beyond the here and now of their instantiation, may in turn associate themselves with other actants, thus contributing to the continuous generation of the collective world. Such relatively stable actants possess some *durability* that allows them to bring about and orient what is becoming.

If we continue considering differences among actants, we quickly notice that some durable actants can *move* from one place to another more or less easily. Let us keep on using familiar entities to illustrate this point. If we consider the planet Mars and West Bank jails, these entities appear rather static. It is difficult for them to associate with actants capable of making them deviate from their initial trajectories: without important mobilization efforts, the planet Mars and West Bank jails will just stay where they are. This is not quite the case for Mark Zuckerberg who, once associated with actants such as "shoes," "cars," or "roads," can markedly change his initial trajectory and, in turn, associate himself with other actants that were at first distant from him. Yet, largely due to his body envelope, Mark Zuckerberg's relative mobility is rather costly: in order for him to somehow keep on being Mark Zuckerberg, in order for him to maintain most of his durability while he is moving, he would need to associate with many other actants (e.g., oxygen, food, space for his legs, coffee breaks) protecting him from being too much altered. In the case of Nutella jars, the story is a bit different. They too need to associate with other actants to deviate from their initial trajectories (e.g., supply chain managers, railway lines, sale contracts, delivery people). But contrary to Mark Zuckerberg, one can make the fair assumption that Nutella jars' alteration is slower: due to their proper materiality, due to their own *medium*, they can, for example, be stored, piled up, and handled without being significantly transformed. Among our exemplary durable entities, Nutella jars seem then the most durable *and* mobile: when compared to the planet Mars, West Bank jails, or even Mark Zuckerberg—and when provided adequate associations—these jars can move from one place to another without being too much altered.

When cumulated, durability and mobility are nontrivial characteristics: entities that combine both abilities are more likely to associate with other entities, thus actively contributing to the generation of the collective world. But a very special category of entities cumulates another ability that makes them certainly the most world-generative of all. These entities go by different names: Jack Goody calls them "graphical objects" (1977); Bruno Latour and Steve Woolgar call them "inscriptions" (1986, 43–91); Dorothy Smith calls them "accounts" or "documents" (1974). But no matter how these are labeled, sociologists have long emphasized on these actants' fascinating capacity to be durable and mobile *and* to carry with them some characteristics of other actants—or of other associations between actants. This is essentially what texts, tables, graphs, or drawings do: thanks to the presence and constant maintenance of specific habits, rules, and technologies—what Jérôme Denis (2018) calls *scriptural infrastructures*—these often durable and mobile inscriptions can host some aspects of actants and associations and present them *again* (*re*-present) somewhere else. This scriptural transport of (part of) actants—that itself necessitates many other actants to unfold—may in turn create a link between what has happened and what is to become. This sounds like an odd statement, but such a phenomenon is in fact very common: Every time I read a *New York Times* article, a connection is made between what has happened in the past (some events) and what is happening now (me, considering this event and, eventually, reacting to it). Of course, this connection, this link has been formatted in order to be hosted in the specific materiality of the inscription I am considering (here, the newspaper article). Such a link is thus always a partial, but potentially faithful, in-formed version of what has happened. When I'm reading the *New York Times*, I don't see migrants struggling to reach

Europe in horrendous conditions; I see a flat surface with words that re-present me those migrants; this re-presentation triggering in me feelings of helplessness, shame, and despair, evanescent actants that will, in turn, con-tribute to the continuous generation of the collective world (though quite insignificantly). To qualify inscriptions' capacity to carry some properties of actants-associations and establish formatted yet generative connections between times and locations, I shall use the term "re-presentability." More than just being durable and mobile actants, inscriptions are thus also re-present*able*: they can—together with suitable infrastructures—carry, trans-port, and display properties that are not only theirs.

Durability, mobility, re-presentability: these are capacities not to be under-estimated. Inscriptions, despite their often-modest appearances (lists of num-bers, drawings, articles, tables, graphs), greatly participate in the shaping of our world. A new molecule appears that revolutionizes our understanding of the human hypothalamus? As well documented by Latour and Woolgar (1986), such an association-prone actant derives, to a large extent, from inscriptions assembled, accumulated, compiled, and compared within and between laboratories. A new management technique starts to align corpo-rate activities to a single arbitrary standard? As proposed by Thévenot (1984) and Yates (1989), such Taylorist normalization—and its consequences—heavily relies on measures, coding, and equity methods whose scriptural circulation allows the centralization of control over the workers. A new algorithm is published that may ignite original avenues of research in digi-tal image processing? As I will try to show throughout this book, the for-mation of such an actant owes a great deal to the production, circulation, transformation, and compilation of many different types of inscriptions. We will more thoroughly examine the world-generative capacity of inscrip-tions in due time (especially in chapters 4, 5, and 6). For now, suffice it to say that these durable, mobile, and re-presentable actants contribute *a lot* to what is constantly happening.

But whatever their generative power, "inscriptions" do not exist by themselves: they obviously need to be produced before they start to circu-late. In that sense, every inscription needs to be *inscribed*. Extracting some aspects of associations (or "events"; at this point, both terms are equivalent) and re-presenting them on flat, durable media is not at all evident: What part of the event shall be kept and written down? What language shall be used? What protocol shall be followed to later compare this inscription with some others and produce, in turn, new compiled inscriptions? Considering the world-generative potential of inscriptions, these are major issues, most of time supported by organizational and professional practices with their own goals, rules, and principles that every day engage hundreds of millions of people and instruments. This oriented work consisting in producing inscriptions and, eventually, capitalizing on their world-generative potential is what Dorothy Smith (1974) calls "the fabric of documentary reality."[13] And this fabric is highly political.

To illustrate her point, Smith takes the a priori mundane example of birth certificates. Inscribing a birth on a report is, in fact, not evident nor neutral. It is the product of an organizational and professional practice that shapes births and their accounts in very peculiar terms, very different from, say, how mothers and fathers may want to remember it. As she put it:

> "Jessie Franck was born on July 9th, 1963" appears maximally unequivocal in this respect. But as we examine how it has been fabricated it becomes apparent that its character as merely a record is part of how it has been contrived. Everything that a mother and a father might want to have remembered as how the birth of Jessie Franck was for them is stored elsewhere and is specifically discarded as irrelevant in the practices of the recording agency. The latter is concerned only to set up a certified and permanent link between the birth of a particular individual—an actual event, and a name and certain social coordinates essential to locating that individual—the names of her parents, where she was born, etc. (Smith 1974, 264)

Birth certificates are very selective—they only keep a very small part of birth events—and this selection is oriented toward the potential of such concise inscriptions—their features can, in turn, be used for identification purposes or government statistics. Moreover, as being inscriptions that can be remobilized in other spaces, birth certificates and their desired purposes make a specific version of births that will, in many cases, impose on other concurrent versions. Despite their very partial and partisan origins, these circulating inscriptions will form a fulcrum for other inscriptions, progressively establishing formal, factual, and so-called "neutral" versions of births.

This political aspect of inscription practices which aim to make partial partisan versions of events does not only concern administration. The power of Smith's argument lies in that it is also applicable to *any* inscription as it is materially impossible to fully inscribe an event in all its subtleties: choices need to be made regarding what will be kept (and formatted) and what will be ignored. What inscriptions gain as world-generators also lose as world-betrayers, the latter being even a condition to the former.[14]

With these elements in mind, let us now come back to this present book. Have I not said it intends to be a sociological work? Have I not said it intends to account for associations that progressively form devices we call algorithms? At this point, these assertions can be further specified. Sociology, as a professional activity that consists in producing specialized texts (*logos*) about associations (*socius*), does not escape what I shall now call "Dorothy Smith's law": however descriptive it is, sociology brings into being—by means of inscriptions—partial realities to the detriment of other realities. What is true for administrators (Desrosières 2010), economists (MacKenzie, Muniesa, and Siu 2007), or scientists (Latour 1987) is also true for sociologists: while describing realities by means of texts, they also enact these realities.

As Law and Urry (2004, 396) well summarized it, *there is no innocence*:[15] a text, however faithful—and some texts are definitely more faithful than others—is also a wishful accomplishment. I must then admit that what I intend to do in this book is not only describing what happens in particular, algorithm-related, situations: due to this book's very existence as a textual inscription, it is also an attempt at enacting a world to the detriment of other enacted worlds. My gesture is thus analytical and political: it aims to produce a descriptive account of how algorithms come into existence—we can keep that—*but also*, and in the same movement, to propose a new version of their realities. The motivation behind this analytico-political move were presented in the previous section: in these days of controversies over the agency of algorithms, a refined—yet formatted and thus intrinsically limited—account of their inner components may establish grounds for constructive disputes about and with algorithms.

To come back to the title of this section, I assume the classical notion of "construction" does not well express such a venture. Construction has been for sure a useful term for sociology as it has equipped many valuable critiques of naturalized matters: studies on the construction of gender (Lorber and Farrell 1991), patriarchy (Lerner 1986), or maternity (Badinter 1981), just to mention some classics, have all been wonderfully liberating. But considering recent developments in STS and sociology in general, it appears that construction suffers from being two-faced: while it well expresses its descriptive aspirations—showing how results have been produced—it also tends to hide its political claims—generating realities to the detriment of others.[16] Due to its propensity to hide "Dorothy Smith's law" under the cover of analytical ambitions, I consider it wiser to renounce using the term "construction" to qualify my overall gesture.

I am not the first sociologist to dismiss construction. It is in fact quite a popular move, motivated by more or less the same arguments as presented above. Law and Urry (2004) prefer to use "enactment" as it better expresses the performativity of descriptive ventures. Latour (2013), inspired by Souriau ([1943] 2015), has recourse to "instauration" as it underlines the fragility of practical, succeeding assemblages. Ingold (2014), in the wake of Rorty (1980), gives priority to "edification" as it stresses the continuous and never fully achieved aspect of what is about to happen. All these notions are surely interesting alternatives to construction. But at the risk of feeding in a sociological jargon already well supplied, I choose here to use the notion of "constitution" as it has the significant advantage of containing natively a double signification: a *process* by which something occurs as well as a *document* advocating for rights and prerogatives. Here lies an interesting tension that may recall the assumed ambivalence of my gesture: describing and contesting. Moreover, as a constitution is never fixed once and for all (it can be amended, completed, abolished), the notion forces us to recognize the necessary incompleteness of my venture, the three activities that I try to put into existence here—ground-truthing, programming, and formulating (more on this later, obviously)—must be considered partial and temporary. Many more gerund articles, as long as they are supported by empirical materials, can be potentially added to the present constituent act of algorithms.

For all these reasons, this book's title *The Constitution of Algorithms* should be understood as the putting into text and existence—simultaneously empirical and activist—of what algorithms *shall be*. At the very end of the inquiry, in light of the accounted elements, I will come back to the implications of this analytical/insurrectional gesture in a section borrowing from Antonio Negri's (1999) work on "constituent power." For now, let us just note and accept this ambivalence by using the term constitution, a constant reminder of this inquiry's bipolarity.

A Laboratory Study

At this point, I have no other choice than to ask the reader to follow me—at least temporarily—in assuming that in these days of controversies over the agency of algorithms, the invisibility of the work required to design, shape, and diffuse them is negative as it prevents disputing parties from having common grounds for negotiations. Let us also assume that one way to propose such grounds, and thus to suggest constructive disputes and composition attempts, could be to conduct sociological inquiries in order to make visible the work practices required to make algorithms come into existence. Let us finally assume that this volume is an attempt at such an inquiry that, in its capacity as a world-generative inscription, cannot but be a partial, partisan, and open-ended (while also faithful and empirical) constitution of algorithms. If we accept these debatable assumptions, the next question could be: How can I effectively run such a partial, empirical, and activist inquiry? On what materials can I ground it?

It would be tempting to use readily available sources, such as the many academic papers and manuals describing the internal workings of algorithms. This is in fact what several STS scholars have done in some very interesting works.[17] However, I have reasons to believe that the sole use of these sources surreptitiously contributes to the perpetuation of the negative invisibility of algorithms' components. Regarding computer science papers published in academic journals, it would, of course, be incorrect to say that this literature is erroneous: on the contrary, it attests to what is about to, perhaps, become scientifically true.[18] But as many important science studies have shown, these scientific publications tend to report the results of processes, not the practical activities that led to those results. Under these conditions, it is problematic to solely use academic publications to make the formation of algorithms visible since these documents are themselves supported and framed by unstated elements. Michael Lynch (1985) well summarized this problem inherent in the analysis of scientific publications:

> [Methods sections of scientific research papers] supply step-by-step maxims of conduct for the already competent practitioner to assimilate within an indefinite mix of common sense and unformulated, but specifically scientific, practices of inquiry. These unformulated practices are necessarily omitted from the domain of study when science studies rely upon the literary residues of laboratory inquiry as the observable and analyzable presence of scientific work. (Lynch 1985, 3)

Moreover, for entangled reasons we will cover throughout this book, authors of academic papers tend also to *defend* their algorithms against concurrent algorithms. A claim published in a scientific journal is indeed directed against other claims and is intended to obtain the reader's support. Hence the importance of captation techniques that aim "to lay out the text so that wherever the reader is there is only one way to go" (Latour 1987, 57). These conviction habits and the additional necessity they provide—essential elements to establish objective constructions—tend to *purify* the scientific accounts of algorithms of the many disparate elements that have contributed to their textual existence. When relying on these documents to analyze computerized methods of calculation, it is therefore the hesitations, doubts, and "infra-ordinary" equipment and writings that tend to escape the analyst's gaze.[19]

But what about the numerous manuals that teach us how to design algorithms?[20] Do they not provide descriptions of how to assemble computerized methods of calculation? Are they not, in that sense, *connectors* between algorithms and the collective world they contribute to shaping? These pedagogical resources are certainly crucial to inculcate students and newcomers with the basic components of computerized methods of calculation, which are essential to their sociological analysis. Yet, as Lucy Suchman (1995) reminded us, these resources are, by definition, normative accounts of how work should be done, not of how work is effectively done. This is a crucial but often forgotten precision: "[These] normative accounts represent idealization and typifications. As such, they depend for their writing on the deletion of contingencies and differences" (Suchman 1995, 61). Instead of accounting for what it is being done during mundane situations, manuals account for what ought to be done. They are (important) peremptory recipes, not empirically grounded accounts of practices.[21] This is, I believe, the main limitation of contemporary studies that rely mainly upon textbooks and classes on algorithmic design: they inform about how contemporary pedagogues want algorithms to be constructed, not on how these algorithms are constructed on a day-to-day basis. Instead of getting closer to computer scientists by accounting for their work, these studies, otherwise very interesting, tend to move them further away.[22]

Academic papers and manuals are therefore sources that should be handled with precautions. But how to reach what these sources, which remain useful and important, contribute to keeping out of sight? How to get a higher definition, yet still intrinsically limited, picture of the work required to assemble algorithms? Fortunately, for this very specific purpose, I can rely on a proven STS analytical genre often labeled "laboratory study." The first such studies appeared in the 1970s, mostly in the United States. In a sense, the collective (Western) world was at that time not so dissimilar to the one we are experiencing today: controversies about types of agencies were arising continuously. But instead of algorithms, these controversies mostly concerned *scientific facts* often developed in life science, physics, and neurology. For many reasons that are too entangled to be discussed in this introduction,[23] several scholars felt the need to deflate the delusive aspect of scientific facts by sociologically accounting for mundane practices of natural scientists trying to *manufacture* certified knowledge (Collins 1975; Knorr-Cetina 1981; Lynch 1985; Latour and Woolgar 1986). The method of these scholars was quite radical: in reaction to the authoritative precepts of epistemology, these authors borrowed from ethnography its *in situ* analytical perspective to document "the soft underbelly of science" (Edge 1976). As Latour and Woolgar put it:

> We envisaged a research procedure analogous with that of an intrepid explorer of the Ivory Coast, who, having studied the belief system or material production of "savage minds" by living with tribesmen, sharing their hardship and almost becoming one of them, eventually returns with a body of observations which he can present as a preliminary research report. ... We attach particular importance to the collection and description of observations of scientific activity obtained in *a particular setting*. (1986, 28; emphasis in the original)

Instead of starting from scientific theories, minds, or "laws of Reason," these laboratory ethnographers—who actively participated in the launching of *Science and Technology Studies*—decided to start from mundane actions and work practices to document and make visible how scientific facts were progressively assembled. Several other monographs accounting for the practices of physicists (Traweek 1992; Sormani 2014) and design engineers (Vinck 2003) followed the seminal 1980s laboratory studies, each time providing insightful new results. We will cover some of these results in due time. For now, suffice it to say that the present sociological inquiry is based almost entirely on these works. But what does that concretely imply?

It first implies locating places where individuals work daily to assemble algorithms. For my case, this localization exercise was not very difficult as I was institutionally close to a European technical institute with about twenty computer science laboratories working every day to propose new algorithms and to make them circulate in broader academic and industrial networks. A more arduous task was to convince the director of one these laboratories to let me describe the practical shaping of algorithms as an "intrepid explorer." Fortunately, institutional movements related to the establishment of a new institute of digital humanities enabled me to share my research ambitions with a computer science professor open to inter-disciplinarity.[24] And after several trials, I could be part of her laboratory of digital image processing for two and half years, from November 2013 to March 2016. These were no passive moments: as required by the analytical genre of laboratory studies and also by the rules of the laboratory to which I was affiliated as full member, I had to participate in the life of the labora-tory and thus become somewhat competent. Although the skills I progres-sively acquired certainly did not make me become a computer scientist, they were nonetheless crucial for speaking adequately about issues that mattered to my new colleagues. But participating and discussing were not enough: I also had to write down, collect, and compile what I did, saw, and discussed. Very concretely, this implied taking *a lot* of notes. Discussions, meetings, presentations, actions: everything I experienced had, ideally, to be written down, referenced in notebooks and computer documents to be later retrieved, compared, sampled, and analyzed. This full-time data com-pilation work implied one last move: after my stay within the computer sci-ence laboratory—during which I participated in projects, held discussions with colleagues, observed what they did, wrote down as much as I could, and made presentations about my preliminary results (processes that have deeply transformed me and the sociology I now do)—I had to return to my own community of research to more thoroughly work on the collected materials and write an investigation report that, progressively, has become the present book.

But these all-too-basic elements—that will be more thoroughly presented in chapter 1—elude one important question: How to effectively account for, and thus write down and analyze, what computer scientists do as they try to shape new algorithms within their laboratory? How to experience, capture, and analyze their *actions*?

Courses of Action

As soon as one is convinced of, and enabled to, undertake a laboratory study to document—in a partial yet faithful way—the constitution of algo-rithms, one quickly lands in uncharted territory. If there are laboratory studies of life sciences, physics, medicine, or brain sciences, very little has been published on computer science work.[25] The cost of entry and the time required to carry out this type of investigation certainly contributed to this situation. But it is also possible that a peculiar habit of thought participated in this disinterest. Indeed, for entangled reasons I will try to tackle in chapters 3 and 5, the fair assumption that computer code and mathematics actively contribute to the shaping of computerized methods of calculation is often doubled with the not-so-fair assumption that both code and mathematics have no, or little, empirical thickness. This assumed evanescence of the ingredients of algorithms contributes, in turn, to making them appear inscrutable. This common habit—that Ziewitz (2016) associated with an "algorithmic drama"[26]—may have discouraged sociologists from entering sites where algorithms are shaped, diffused, and maintained: Why bother trying to inquire into these places since everything happens in the heads of those who work there?

But like any ethnographer involved in the daily work of a scientific laboratory—trying to participate, talk adequately, and compile empirical materials—I quickly realized that very few things could be attributed to the brains of my colleagues, however clever they were. Of course, they never stopped doing things—writing on scratch paper, comparing graphs, typing on keyboards, inspecting databases, moving their mouse cursors, taking coffee breaks—that at first appeared unrelated. But as I stubbornly accounted for these things in my logbooks, I soon realized that the succession of these small elementary "blocks" of action sometimes ended up forming bigger accomplishments: a database, a script, a complete program, an algorithm. By remaining continuously with my new colleagues in their laboratory, conscientiously writing down observations and even recording some work sequences (with their prior authorization), I was soon forced to admit that what we call "practice" is in fact *a term without opposite* (Latour 1996). In the artificial setting of my laboratory study, accounting for as many associations as possible, I soon realized that the much-debated distinction between "theory" and "practice" was an artifact. In the laboratory, there were only practices whose *successions* ended up sometimes forming "databases," "computer programs," "mathematical models," or "algorithms." A little-equipped retrospective look on these trajectories could easily ignore their importance. But once I managed to slow these trajectories down and patiently account for them—sometimes with the help of those who were realizing them—I realized that I could almost do without any internal "abstract" cognitive mechanisms.

Following the seminal work of Jacques Theureau (2003), I shall use the term *courses of action* for these accountable chronological sequences of gestures, looks, speeches, movements, and interactions between humans and nonhumans whose articulations may end up producing *something* (a piece of steel, a plank, a court decision, an algorithm, etc.).[27] Sticking to this generic definition is crucial as it will help us resist the supposed abstraction of computer science work: what ends up being called a "mathematical model," "code," or even "algorithm" must be, one way or another, the product of accountable courses of action unfolding within specific situations and carried out by assignable actants. Moreover, I shall include under the generic term "activity" courses of action unfolding in different times and locations that yet lead to related achievements. In this volume, an activity will then be understood as *a set of intertwining courses of actions sharing common finalities*. The three parts of this volume are all adventurous attempts to present activities taking part to the formation of algorithms; hence their respective titles ending with *ing*: ground-truth*ing*, programm*ing*, formulat*ing*.

This leads to one potential limitation of courses of action as laboratory studies allow them to be accounted for. I mentioned earlier that trajectories must often be slowed down to identify the courses of action whose articulation may lead to the formation of something. This slowing down is salutary as it allows many crucial shaping actions to unfold. But it also has one flaw: it forces one to proceed *very* slowly. As a consequence, any small a priori mundane course of action may unfold on a dozen pages, thus limiting the number of cases.[28]

Three Gerund Parts (But Potentially More)

I hope the reader has gotten a sense of why I decided to make this inquiry, how I tried to conduct it, and where it may eventually lead. But before diving in this exploratory study, I shall briefly present the three parts of this book that, following my action-oriented methodology, are all gerunds: ground-truthing, programming, formulating.

Part I mainly deals with the work required to define problems capable of being solved computationally. In chapter 1, I present the overall setting of the inquiry and introduce basic notions in digital image processing and standard algorithmic study. In chapter 2, I go directly to the heart of the matter and follow a group of young computer scientists trying to publish one of their algorithms. During this first case study of image processing in the making, we will encounter what computer scientists call "ground truths": referential repositories that work as material bases for algorithms. The centrality of ground truths and of the work required to build them make me assert that, to a certain extent, *we get the algorithms of our ground truths*.

Part II tries something that has rarely been attempted: considering computer programming as a practical, situated activity. In chapter 3, I propose historical and conceptual reasons why programming has resisted—and still resists—ethnographic scrutiny. At the end of the chapter, I focus on the computational metaphor of the mind, the main conceptual stumbling stone preventing any close analysis of computer programming practices. In chapter 4, building on notions and concepts introduced in the previous chapters, I carefully describe computer programming courses of action I attended during my laboratory study. Besides opening new avenues of research, this second case study leads, inter alia, to the following proposition: *a programmer may never solve any problem*.

In part III, I consider the role of mathematics in the formation of algorithms. In chapter 5, I first build on STS-inspired inquiries into mathematics to present mathematical practices as stakeholders of scientific activity. I then use this unconventional view on mathematics to define formulating as the activity of translating entities until they acquire the same form as previously-defined mathematical objects. In chapter 6, I build on these theoretical arguments to account for courses of action that successfully formulated some of the relationships among the data of a ground-truth database. This third and last case study will also make us appreciate some of the numerous links between ground-truthing, programming, and formulating activities, entangled processes that, sometimes, leads to the shaping of algorithms. These elements will finally allow me to touch on the topic of machine learning and artificial intelligence, here considered audacious yet costly attempts at automating formulating practices. In the conclusion, I develop some corollaries of the empirical and theoretical elements this inquiry unfolded.

Although ground-truthing, programming, and formulating activities follow each other in the present volume, they do not necessarily do so in the "real" life of action. In places such as the computer science laboratory we will soon get to know, these activities form a whirlwind process whose elements influence each other in a *dance of agency* (Pickering 1995). Moreover, even though this book's narrative thread is sequential—with subsequent chapters sometimes referring to previous ones—one may browse through it in different ways. Readers interested in ethnographic accounts may, for example, jump from one case study to another before eventually coming back to more theoretical pieces such as chapters 3 and 5. Readers who favor conceptual ventures may wish to go the other way round, starting with intellectual matters before coming back to down-to-earth accounts of practices. Of course, curious readers without specific expectations may also follow the book's thread, starting from chapter 1 and ending with the conclusion.

As mentioned earlier, it is important to keep in mind—almost like a mantra—that these three activities forming an empirical and partisan version of what algorithms shall be are not fixed nor exclusive. Even though they form, I believe, a refreshing and faithful conception of how algorithms come into existence, the precise ecology of algorithms would clearly benefit from further investigations. There are surely many more activities contributing to the formation of algorithms that future ethnographies and case studies will, hopefully, unfold. In that sense, although this volume does intend to bring about an alternative action-oriented constitution of algorithms, my arguments should also be considered preliminary propositions asking for further considerations.

At any rate, inscriptions make worlds only when read: at this point, my main concern is that readers—sociologists interested in the constitutive relationships of algorithms; computer scientists curious about an alternative action-oriented account of their work; or in fact, anyone concerned about the power, and beauty, of algorithms—are intrigued enough to come with me to explore some of the things that are happening in a computer science laboratory.

I   Ground-Truthing

The fact that techniques mediate advances suggests a way in which mathematical problems that arise in society are ultimately in some relationships with the techniques which that society has forged. This, in turn, suggests that mathematicians, like societies, can only pose those questions to which a potentiality of a response exists.

—Ritter (1995, 72)

The introduction presented the rationale of this inquiry. Now, obviously, the hard work begins: effectively doing it! We will start smoothly though, with two straightforward chapters. Chapter 1 specifies the overall setting of the inquiry: a well-respected computer science laboratory that specializes in digital image processing; I shall call it "the Lab." I start by presenting its environment and some aspects of its organization as well as its place, modest but substantive, in the heterogeneous ecosystem of computer science industry. I will also consider methodological matters and discuss the notion of algorithm as it is generally presented in the specialized literature. Chapter 2 starts in the middle of things at the Lab's cafeteria during a working session where the Group—three young computer scientists—tries to coordinate the development of a new algorithm. After a quick parenthesis where I present the basic issues at stake, we will closely follow this project, meeting along the way entities called "ground truths" whose importance in the constitution of algorithms we will learn to appreciate. The last section of chapter 2 will be a brief summary.

1 Studying Computer Scientists

This inquiry took place in a European technical institute (ETI) between November 2013 and February 2016. This public school was integral part of the global academic landscape and hosted more than five thousand undergraduate and twenty-five hundred graduate students in five faculties: basic sciences, engineering, life sciences, architecture, and computer science. In this investigation, I will mainly focus on the computer science faculty (CSF), one of the most renowned within the ETI for its ability to attract foreign students and professors, to raise important research funds, and to engage in numerous partnerships with the industry.

Over the time of this inquiry, the CSF employed nearly forty professors supervising the training of more than 780 undergraduate and 550 graduate students. The CSF professors were supported in their teaching activities by around 250 doctoral students who were also working on the completion of their PhD theses, generally over four years. Research among CSF members was extremely varied, ranging from theoretical computer science and hardware architecture to machine learning and signal processing. Significant human and material resources were invested to gird the whole domain of computer science and take active part to its development.

Teaching, research, and administrative activities of the CSF were mainly located in six buildings linked to each other by a system of paths, footbridges, and underground passages. Within this complex, the most recent building (inaugurated in 2004) served as a nerve center, housing most of the laboratories, the best equipped conference rooms, and the faculty's cafeteria, highly prized for its breathtaking view of the surroundings (figure 1.1). Opposite the CSF's main building, on the other side of a small road, was another complex of buildings housing around one hundred start-ups and

Figure 1.1
The CSF main building. On the left and right sides of the central patio, lines of offices and seminar rooms. In the center of the image, in air-conditioned rooms with glazed windows, three server farms store local programs, experiments, and databases. On the top floor, illuminated, one can discern the entrance to the faculty cafeteria.

spin-offs as well as several offices of large companies and service providers. Created in the 1990s, this innovation area had the explicit purpose of bringing fundamental research outputs closer to the industry, according to dynamics of scientific valorization close to those analyzed by Liliana Doganova (2012). Members of this innovation area often interacted with members of the CSF during both formal and informal events, many of which took place in the CSF main building.

However, the vast majority of CSF students did not launch start-ups at the end of their training programs. Rather, they tended to be hired by large national and international technology companies. This was particularly true for doctoral students whose research funds were frequently supported by large companies such as Google, IBM, NEC, or Facebook following calls for projects, thus creating multiple and regular professional connections. Visiting trips and internships were also routinely organized within technology companies as part of master's and doctoral programs. This was another distinctive feature of CSF: within the ETI, CSF students had the greatest employability.

But public money nonetheless constituted the main financial resource for ongoing research projects. Here, too, the CSF seemed to have a strategic advantage within the ETI, heavily capitalizing on and participating in public speeches reporting the advent of a new industrial revolution around big data, machine learning, and artificial intelligence. In addition, thanks to the CSF's reputation as a potential trainer of a new generation of digital entrepreneurs (with several iconic precedents participating in this reputation), its financing requests could play the renewal of industry card, a goal explicitly put forward by national research funding agencies. Relative to its size within the ETI, the CSF was thus one of the faculties to which the most public research funds were allocated.

Although the CSF hosted cutting-edge computer equipment, its premises remained open most of the time. From 7 a.m. to 7 p.m., apart from inconspicuous surveillance cameras placed in sensitive areas such as server farms, no special security procedures were in place. Unlike, for example, Vincent-Antonin Lépinay's (2011) analysis of General Bank's trading rooms, my ethnographic inquiry was largely conducted in an open environment with no explicit surveillance mechanisms. For example, it was common to meet tourists who came to visit and photograph the high-tech architecture of the CSF premises. From 7 p.m. to 7 a.m., the security system was complemented by two night watchmen and locked entrance doors (with alarms) for those without an access card.

Nevertheless, while the CSF premises remained open most of the time, I of course needed institutional support to collaborate with computer scientists and document their courses of action. Without an e-mail address and an account within the administrative system, it was, for example, impossible to connect to the CSF servers or use advanced software, both constituting the basic infrastructure of most ongoing projects. Moreover, given the deliberately small size of most of the CSF laboratories (around twenty collaborators under the supervision of a professor), it was impossible to blend into the mass and investigate in a hidden way.

As a Science and Technology Studies (STS) sociologist without any formal training in computer science, I first had difficulty raising the interest of the CSF professors as my research questions appeared too abstract and their impact too uncertain. Fortunately, at some point I had the opportunity to surf on a broader institutional movement seeking to bring the CSF closer to the faculty of human sciences (FHS) of a neighboring university to which I was then affiliated. In early 2013, with the stated desire to penetrate cultural spheres, the ETI's management started to invest in the establishment of a center for digital humanities. As this movement involved the recruitment of new teaching and research staff, it quickly created links between humanity scholars of FHS—some of them STS-inspired—and computer scientists of ETI, and it was in this context of disciplinary rapprochement that I met the director of a laboratory that specialized in digital image processing. After several furtive yet decisive exchanges, I obtained her support to apply for a national fellowship promoting interdisciplinary research. Following several selection rounds, my application was finally retained in September 2013, therefore committing me to run a four-year FHS-CSF doctoral project with the stated ambition of carrying out an ethnographic inquiry into the formation of algorithms.[1] This dual institutional affiliation allowed me to be officially accredited as full member of CSF's image-processing laboratory for a period of two-and-a-half years. From November 2013 to March 2016, I had not only the same rights as any laboratory member, notably in terms of research infrastructure, but also the same prerogatives, notably in terms of presentation of results. While these conditions of investigation were at first quite tough—after all, I had initially no experience in computer science—they gave me the unique opportunity to stay, observe, and work for what I will from now on call "the Lab."

The Lab

The Lab was located on the third floor of the CSF main building. Typical of the organization of the CSF, it was centered upon the tutelary figure of a full professor, the director of the Lab. The director was assisted by a secretary dealing with administrative issues that were often complex due to the high proportion of collaborators who came from abroad (especially from Persia, India, and China).[2] Among these collaborators, one postdoc student stayed at the Lab for one-and-a-half years. An invited scholar also had a desk and took active part in teaching and research activities. Members of spin-offs, sometimes related to the innovation area mentioned earlier, also stayed within the Lab for the duration of their fund raising, ranging from one to two years. It was not uncommon for these spin-off collaborators to make presentations at Lab seminars (more on this later), though in these situations the other collaborators were required to respect an unofficial "nondisclosure arrangement." Some collaborators in between two research contracts were also sometimes hired as "scientists," a temporary position allowing them to pursue their ongoing work in decent conditions. However, most of the Lab's members were PhD students aged from twenty-three to thirty years old and generally holders of four-year employment contracts, at the end of which they were asked to submit doctoral theses allowing them to become doctors of computer science. During my time in the Lab, the number of PhD students varied from six to ten and depended on the number of submitted theses and awarded research contracts. In parallel to their research activities, these students also had to work as teaching assistants for bachelor's and master's classes, including those given by the Lab's director. All in all, for the two-and-a-half years of my collaboration, the Lab hosted between ten and sixteen people, including myself.

Like many CSF professors, the director continuously tried to establish community dynamics within her Lab. This involved, for example, bringing cakes and biscuits to encourage informal chatting at the end of the weekly Lab meetings, during which one or two collaborators presented their work in progress. Two Lab dinners at nearby restaurants were also organized each year; one around Christmas, the other at the end of June. Echoing a corporate outing, a two-day excursion was organized during the summer as well. The Lab's PhD students also contributed to this dynamic by frequently organizing "after-work" outings to the school pub on their own initiative. All these facilitation efforts effectively created and maintained relationships among collaborators, many of whom had initially arrived in the Lab without knowing anyone in the area.

To some extent, the architectural organization of the Lab also participated in these community dynamics as the seven offices, generally occupied by two researchers facing each other, were each aligned along the same hall (see figures 1.2 and 1.3). The Lab also had a private cafeteria that provided tables, chairs, fridges, and coffee machines. As we will see later, this cafeteria was often used as a meeting point, even though the Lab had its own seminar room.

If these community dynamics, greatly encouraged by the Lab's director, did contribute to creating an enriching work environment, then they also went along with managerial aspects. For example, attendance and contribution to Lab meetings were mandatory, with each collaborator being required to make at least one presentation per semester. In addition, similar to corporate settings, collaborators were required to inform the secretary in the event of illness or incapacity, thus suggesting they should be at the Lab

Figure 1.2
The Lab's hall. On the left, behind closed doors, the Lab's cafeteria and seminar room. On the right, seven offices most of the time occupied by two researchers.

Figure 1.3
Inside one of the Lab's offices. Two researchers were generally facing each other, though they were behind one to three large monitors.

every working day unless otherwise specified. Moreover, scientific collaborators were asked to meet with the director at least once every two weeks to inform her of their research progress. This allowed the director to have an actualized view on the ongoing projects while committing collaborators to sharing results, questions, problems, or doubts with her.

This leads us to one central element penetrating many aspects of the Lab: researchers were asked to produce outputs. This incentive to produce tangible results derived from a broader dynamic, now common to research institutions desiring to achieve, and maintain, the heights of the academic rankings of world universities (Espeland and Sauder 2016). Although most of the CSF laboratory directors held stable academic positions, they nonetheless had to be accountable for the performance of their research teams as the category of output having the greatest impact on these evaluations were articles published in peer-reviewed journals and conferences. Most of the research efforts I attended and participated in were then directed toward this very specific goal: publishing peer-reviewed articles. Despite its close relations with the tech industry and its effective support for the launch of spin-offs, the Lab was, in that sense, mainly academic-paper oriented.

But what was the content of the peer-reviewed articles that members of the Lab sought to publish in academic journals and conference proceedings? What was the Lab working on? The research field of the Lab was existentially linked to the advent of a piece of equipment called the charge-coupled device (CCD). The history of the CCD's development, from its patented concept at Bell Labs in the late 1960s to the many norms and standards that supported its industrialization during the 1990s, is a long and tortuous story.[3] In addition, a precise understanding of its now-stabilized internal functioning would require foundations in solid-state physics.[4] For what interests us here—superficially understanding the main topic of the Lab's academic papers—we can just focus on what CCDs and their different variations such as complementary metal-oxide semiconductors (CMOSs)[5] allowed the Lab to do (i.e., the potentialities these devices suggest).

In a nutshell, through the translation of electromagnetic photons into electron charges as well as their amplification and digitalization, CCDs and CMOSs—as industrially produced devices supported by many standards—enable the production of digital images constituted of discrete square elements called pixels.[6] Organized according to a coordinate system allowing the identification of their locations within a grid, these discrete pixels—assigned

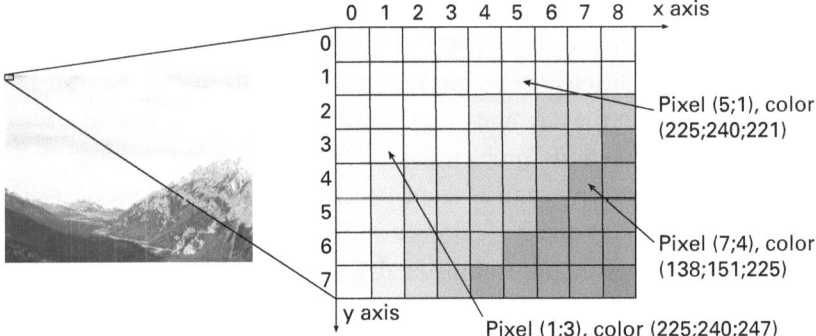

Figure 1.4
Schematic of the pixel organization of a digital photograph as enabled by industrially produced and standardized CCDs and CMOSs. The schematic on the right is an
imaginary zoom of the digital photograph on the left. Every pixel is identified by its
location within a coordinate system (*x/y*). Moreover, assuming the image on the left
is a color image, each pixel is described by three complementary values, commonly
referred to as a red, green, and blue (RGB) color scheme. As most standard computers
now express RGB values as eight-bit memory addresses (e.g., one byte), these triplets
can vary from zero to 255 or, in hexadecimal writing, from 00 to FF.

eight-bit red, green, and blue values in the case of color images (see figure 1.4)—
have the ability to be processed by computer programs that are themselves,
most of time, inspired by certified mathematical statements. Many terms of
the former sentence will be discussed at length in the following chapters.
For now, it is enough to comprehend that in each of the seven offices of the
Lab as well as in many other scientific and industrial locations, pictures of
buildings, shadows, mountains, smiles, or elephants—as produced by standardized CCDs and CMOSs—were also considered *two-dimensional signals*
that could be processed by means of computerized methods of calculation.[7]
The design and shaping of these methods, their presentation within academic papers, and their expression as computer programs able to automatically compute the constitutive elements of digital photographs (often called
"natural images") was the main research focus of the Lab.[8] This specific area
of practice was and is generally called "two-dimensional digital signal processing" or, more succinctly, "image processing" or "image recognition" (when
it deals with recognition tasks).

Even though spending time and energy assembling computerized methods of calculation capable of processing CDD- and CMOS-derived pixels in

meaningful ways might at first sound esoteric, such an activity plays an important role in contemporary economies.[9] This is to be related with the unprecedented production, circulation, and accessibility of digital photographs:[10] thanks to image-processing algorithms, these numerous two-dimensional signals have become *traces* potentially indicating habits, attributes, preferences, and desires. Instead of a noisy, expansive stream of inscrutable data, the many digital photographs produced and shared every day have turned into *valuable assets* (Birch and Muniesa 2020) with the advent of image processing and recognition. This is a phenomenon whose magnitude must be grasped. Giant technology services companies such as Facebook, Google, Amazon, Apple, IBM, or Microsoft all have laboratories whose members work every day to manufacture new algorithms to commercially exploit the infinite potential of digital photographs, tangible expressions of what users, clients, and partners are assumedly attached to.[11] Nation-states are not to be left out either; powerful public agencies also massively invest in image processing to make use of the capabilities of digital photographs for security, control, and disciplinary purposes.[12] In recent years, similar to what Hine (2008) described for the case of biological systematics, image processing has been seen as a resource in control and planning and, to this end, has increasingly become the object of strategic policy concern and support.

All this may sound gloomy. However, image processing is inextricably a fascinating research area with many dedicated academic journals[13] and conferences.[14] The research issue is indeed appealing: how to make box-like computing machines *see* and possibly use their formalist ecology to make them detect, recognize, and reveal things that we, as bipedal mammals, cannot grasp with our organic senses? Huge academic efforts are invested every day in the development of algorithms capable of manipulating CCD- and CMOS-enabled pixels to make computers become genuine visual equipment. It is important to note, however, that a clear-cut boundary among image-processing groups cannot be easily drawn: academic researchers are funded by public agencies but also by private companies that themselves are sometimes solicited by public agencies that then take part in the development of industrial products. For better or worse, these heterogeneous actants associate with each other and cooperatively participate in the development and worldwide diffusion of image-processing algorithms through computing devices. And at its own level, the Lab was participating in this highly collective endeavor.

Yet one may rightly object that a sixteen-person academic laboratory for image processing such as the Lab is not akin to, say, a giant technology services company such as Google or a powerful state agency such as the National Security Agency. How dare I treat on the same level a small yet respected academic institution welcoming an ethnographer interested in the manufacture of algorithms and gigantic actors attached to secrecy and daily contributing to the progressive establishment of a "black box society" (Pasquale 2015)? It is true that important differences exist between an algorithm as an academic proposition and an algorithm as a commercial product or an actual control device (notably in terms of optimization and software implementation). Nevertheless, it is crucial to specify that academic contributions such as those of the Lab do irrigate the work of large industrial and state actors. These connections are often made visible during in-house talks where alumni working in the industry are invited to discuss their ongoing projects in academic settings. During my stay at the Lab, I attended many such talks and was at first surprised to find that behind a priori impressive affiliations such as Google Brain or IBM Watson lay a computer scientist not so dissimilar to the ones I daily interacted with, saying more or less the same things, and working in teams of similar proportions (though for a significantly different salary). For example, in November 2015, the director of the Lab invited an Instagram employee—an alumnus of the Lab—to talk about their new browsing system whose main components derived from a paper published in the *Proceedings of the 2014 IEEE Conference on Computer Vision and Pattern Recognition*. In June 2014, a former Lab member working for NEC in a five-person team also presented her ongoing algorithmic project as deriving from a series of papers presented at the 2013 European Conference on Computer Vision in which she participated. Other people—mostly from IBM and Google—also took part in these "invited talks" organized by the Lab and neighboring CSF signal-processing laboratories, most of the time mentioning and using state-of-the-art publications.[15] Actors who were officially part of the industry appeared then closely connected to the academic community, working in teams of similar size, participating in the same events, and sharing the same references. Better still, this continuous interaction between academic laboratories such as the Lab and the gigantic tech industry was a two-way street: companies like Google, Facebook, and Microsoft also organized academic events, sponsored international conferences, and published papers in the best-ranked journals (see figure 1.5).[16]

Nonetheless it remains true that academic publications are not commercial products; if university and industrial laboratories both publish papers presenting new image-processing algorithms, then these methods are rarely workable as they are. To become genuine goods capable of making important differences in the collective world, they must take part in wider *passivation* and *valuation* processes that will significantly modify their initial properties (Callon 2017; Muniesa 2011b). Depending on their circulation within differentiated networks, some computerized methods of calculation initially designed by industrial or academic image-processing laboratories can thus remain very specialized and intended for ad hoc purposes (e.g., superpixel segmentation algorithms), whereas others can become widespread and industrially implemented in broader assemblages such as digital cameras (e.g., red-eye-removal algorithms), expensive software, and large information systems (e.g., text-recognition algorithms, compression schemes, or feature clustering). However, before they may circulate in broader networks and hybridize to the point of becoming parts of larger systems, image-processing algorithms first need to be designed, discussed, and shared among a heterogeneous research community in which the Lab played an active role. Whether widespread or specialized, image-processing algorithms—also sometimes just called "models" within the computer science community—first need to be nurtured, trained, evaluated, and compared in places like the Lab.

Developing image-processing algorithms and publishing them in peer-reviewed academic journals and conferences was thus a central activity within the Lab, and it was this activity that I intended to account for. Yet I still had to find a way to document the courses of action that took place there.

Collecting Materials

Thanks to my interdisciplinary research contract, I was part of the Lab for two-and-a-half years. Just as any other collaborator, I had a desk, an e-mail address, and an account within the administrative system. Yet despite these optimal conditions for ethnographic investigation, it would be an understatement to claim that the first days were difficult: everything happening around me seemed at first out of reach. Fortunately, the rules of the Lab that I had to observe quickly allowed me to experience assignable situations. I divided these situations progressively into seven different yet interrelated *types* whose systematic account and referencing ended up constituting my corpus of field data.

The first type of situation I experienced was the Lab meetings I mentioned earlier. During these weekly meetings, the Lab's members gathered in a small conference room to attend and react to presentations of works in progress. Every PhD student (me included), postdoc, spin-off member, or invited scholar were asked to make at least one presentation each semester. These meetings turned out to be crucial to my inquiry for at least three reasons. First, they helped me identify the research topics of my new colleagues. I could then use this information to initiate discussions with them in more informal settings. Second, Lab meetings allowed me to present my research project as well as some of its preliminary propositions in front of the whole Lab. These mandatory exercises thus forced me to put my exploratory intuitions to the test and, often, retrofit them. Third, these situations gave me opportunities to share doubts and needs as in September 2015 when I used this tribune to publicly ask for help in my attempts to better document computer programming practices (more on this in chapter 4). Yet although these Lab meetings were essential to the advancement of my inquiry, most of the data I will use in the following chapters were not collected during these situations. Indeed, as these meetings mostly dealt with results of ongoing research projects within the Lab, the empirical processes and courses of action that led to these results were generally not at the center of the discussions.

The second type of situation was conferences organized by the Lab and neighbored signal-processing laboratories. As mentioned earlier, some of these conferences were invited talks where alumni working in the industry came to discuss ongoing projects. Other conferences were closer to traditional keynotes and gave the floor to prominent researchers, mainly from academic institutions. Though, again, I do not directly use data collected from these conferences in the empirical chapters, these events were nonetheless crucial situations to experience and account for as they allowed me to identify current debates in computer science and better appreciate some of the relationships between research and industry.

A third type of situation I experienced was the so-called Group meetings in which I participated between November 2013 and June 2014. These Group meetings were part of an image-processing project to which the Lab's director had assigned me, and they were precious for my ethnographic inquiry as they made me encounter what computer scientists call ground truths—inconspicuous entities that are yet central to the formation of algorithms. These entities will be introduced in chapter 2 and will accompany us throughout the rest of the book.

A fourth type of situation took place at the office desks of the Lab. Finding appropriate ways to account for these "desk situations" was an important felicity condition of this inquiry as it was at these precise moments and locations that courses of action crucial to the actual construction of algorithms often took place. I had the chance to follow and account for such desk situations during a small part of the image-processing project to which I was assigned between November 2013 and June 2014 (more on this in chapter 6) as well as during several computer programming episodes that took place between September 2015 and February 2016 (more on this in chapter 4).

A fifth type of situation was the numerous classes and tutorials in which I participated throughout my time at the Lab. From basic signal-processing classes to advanced Python programming tutorials, a significant part of my time and energy was dedicated to learning the language of computer science. Even if I do not directly use elements I saw in classes or during tutorials in the following case studies, these situations nonetheless greatly helped me speak with my computer scientist colleagues. Though quite time consuming—again, I had initially no experience in computer science— these learning activities were crucial prerequisites to interact adequately with my fellow workers about issues that mattered to them.

A sixth type of situation was the semi-structured interviews I conducted throughout my stay at the Lab. These interviews were initially exploratory in nature and aimed to give me a better understanding of how my colleagues saw their work. However, as the investigation progressed, I instead used interviews as retroactive tools to revisit with Lab members the events for which I could only partially account. This helped me fill in some of the many gaps in my data.

Finally, a seventh generic type of situation was the informal discussions I had daily with the Lab's members. Although I conducted twenty-five semi-structured interviews, these were clearly not as valuable as the numerous conversations I had during coffee breaks, lunches, Christmas parties, corporate outings, or after-work sessions at the pub. Besides facilitating my integration within the Lab, these situations helped me share what I was experiencing and documenting. During these informal moments, I could, for example, discuss past presentations, recently published papers, ongoing projects, forthcoming programming operations, or unclear elements I had seen in class.

From November 2013 to April 2016, I spent most of my working time in and around the Lab, switching among these seven types of situations and trying to account for them in my logbooks the best I could. At the end of the day, sometimes until late in the evening, I used a text editor to clean up these notes, classify them according to an increasingly consistent taxonomy, and reference them to the paper pages from which they derived (see figure 1.6). This collecting and referencing system was at first very messy as the number of situational categories increased to the point of no longer being relevant and my single initial Word document became increasingly cumbersome. However, after a couple of months, I could identify the seven different yet interrelated situational categories I have just presented, and thanks to the computer programming skills I progressively acquired through classes and tutorials, I decided to stick to individual .txt files whose content could be browsed by simple yet powerful Python programs I started to draft (see figure 1.7). Once systematized, this ad hoc data management plan more or less nimbly allowed me to juggle my digitized data while maintaining access to the original paper notes.

In April 2016, after a small farewell party, I left the Lab with around one thousand pages of handwritten notes; two thousand .txt files; a dozen modulable Python scripts; and hundreds of audio, image, and movie recordings as well as numerous half-finished analytical propositions. And with all these empirical materials literally under my arm, I (temporarily) exited my field site, asking myself serious questions about the significance of all this.

A Torturous Interlude

Ethnography is a transformative experience. Encountering worlds and writing about them—what is the point of even trying such an odd exercise? Computer science now gives me comfort. And as for my former sociologist peers, what will they think of this new me? I cannot talk anymore. Hell of a journey, significant metamorphosis: "I understand, and since I cannot express myself except in pagan terms, I would rather keep quiet," someone said a long time ago. Yet words shall be written, promises kept, and something not forgotten: my new "new" colleagues (the former ones) have all gone through similar journeys. After all, we are in the same shaky

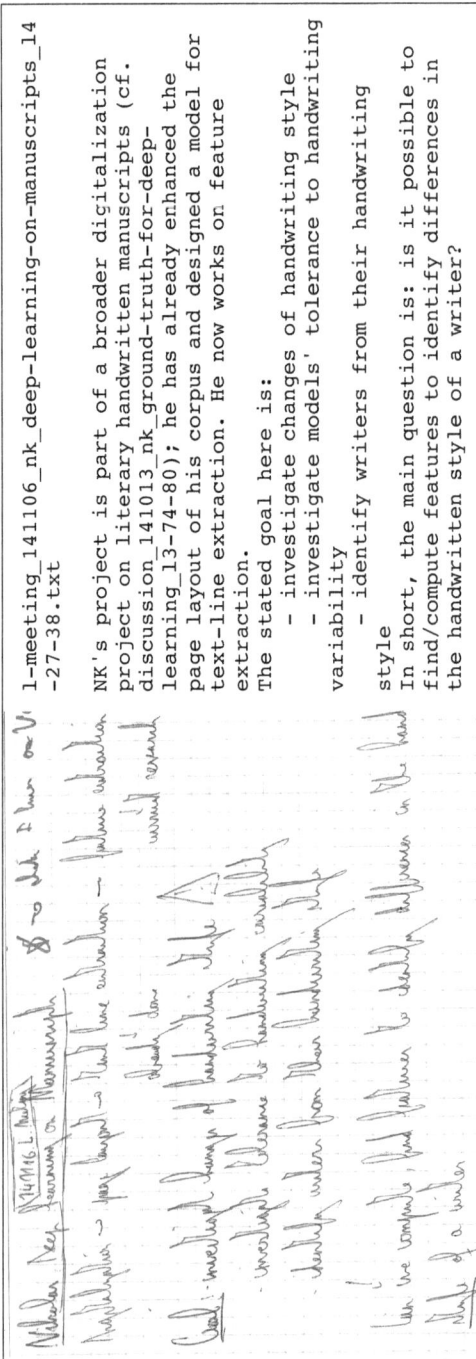

```
l-meeting_141106_nk_deep-learning-on-manuscripts_14
-27-38.txt

NK's project is part of a broader digitalization
project on literary handwritten manuscripts (cf.
discussion_141013_nk_ground-truth-for-deep-
learning_13-74-80); he has already enhanced the
page layout of his corpus and designed a model for
text-line extraction. He now works on feature
extraction.
The stated goal here is:
     - investigate changes of handwriting style
     - investigate models' tolerance to handwriting
variability
     - identify writers from their handwriting
style
In short, the main question is: is it possible to
find/compute features to identify differences in
the handwritten style of a writer?
```

Figure 1.6

Excerpt from one of my logbooks and its translation into a .txt file. On the left, notes taken during a Lab meeting on November 16, 2014. On the right, the translation of these notes into a .txt file. The name of the file starts with "l-meeting," thus indicating it refers to a Lab meeting. The second section, "141106," refers to the date of the logbook entry. The third section, "nk," refers to the initials of the collaborator the note concerns. The fourth section, "deep-learning-on-manuscripts," refers to the title of the presentation. The fifth and last section (14–27–38) indicates the location of the original note, here in logbook number 4, from page 27 to page 38.

```
1.  import OS
2.  import mmap
3.
4.  for i in os.listdir("/Users/florianjaton/logbook"):
5.      if i.endswith("txt"):
6.          f = open(i)
7.          s = mmap.(f.fileno(), 0, access=mmap.ACCESS_READ)
8.          if s.find("ground truth" and "NK") != -1:
9.              file = open("0_list-entries", "a")
10.             file.write(i)
11.             file.write("\n")
```

Figure 1.7
Example of a small Python script used to browse the content of the .txt files. This script, working as a small computer program, makes the computer list the names of the .txt files whose content include the keywords "ground truth" and "NK" in a new document named "0_list-entries."

boat, trying to write faithful sociological documents from scattered empirical data. But how can I do justice to my limited yet empirical materials, distorted voices of those for whom I proposed to become the spokesperson (without any mandate)? I lack everything: a history, a medium, a language. Where do I start? Maybe in the middle of things, as always. Back to fundamentals, to practices, to *courses of action*. Read and reread classics; dive again and again into my materials while sharing them with my colleagues who are gradually becoming pairs again (how could I have forgotten that?). Half-relevant things start to emerge—almost-analytical propositions. What data can make them bloom in a written document? Not even a fraction, an infinitesimal quantity: tiny snapshot of an enlightened world. Accountable activities start taking shape on text pages. But are they still readable? Inscriptions only make worlds when read. Conceptual shortage: both computer science and sociology may not have the means to confront the manufacture of algorithms. The slightest little programming sequence soon suggests the rewriting of computers' history; any small formula demands an alternative philosophy of mathematics (what a cluttered topic!). We walk around with eyes wide shut. Gradually, though, patterns emerge: courses of action become vectors tracing genuine, accountable activities; an impressionist draft from which adversarial lines appear: they may be powerful but not inscrutable. How could we start composing *with* algorithms? The hope is so dim, and the means so limited. "A voice cries out in the desert," and so on and so on. Enough laments: the whole thing is driven by issues more important than my small personal troubles. And I guess I must now validate my return ticket to propose a partial-yet-empirical constitution of algorithms, somehow.

Algorithm, You Say?

Going through the previous, unusual section, I hope the reader could appreciate that writing an ethnographic document about the shaping of algorithms can somewhat be tortuous—even more so when one realizes that in computer science the notion of algorithm is rarely problematic! As a sociologist and ethnographer interested in the manufacture of algorithms, I indeed landed in an academic field whose most illustrious figures have dedicated—and still dedicate—their lives to the study of algorithms. To many computer science professionals then, the fuss about "what an algorithm is" is overhyped; as one colleague suggested me on my first week in the Lab, taking the local undergraduate course in "algorithmic study" may allow me to complete my research in record time... In order to specify my analytical gesture, it is thus important to look at this well-established computer-science-oriented take on algorithms to consider the present work as an original complement to it.

When browsing through the numerous—yet not infinite—computer science manuals on algorithmic study, one notices algorithms are defined in quite a homogeneous way. Authors typically start with a short history of the term[17] before quickly shifting to its general contemporary acceptation as a *systematic method composed of different steps*.[18] Authors then specify that the rules of an algorithm's steps should be univocal enough to be implemented in computing devices, thus differentiating algorithms from other a priori systematic methods such as cooking recipes or installation guides. In the same movement, it is also specified that these step-by-step computer-implementable methods always refer to a problem they are designed to solve.[19] This second definitional element assigns algorithms a *function*, allowing computers to provide answers that are correct relative to specific problems at hand.

Right after these opening statements, computer science manuals tend to organize these functional step-by-step computer-implementable problem-solving methods around "inputs" and "outputs." The functional activity of algorithms is thus further specified: the way algorithms may provide right answers to defined problems is by *transforming* inputs into outputs. This third definitional movement leads to the standard well-accepted conception of algorithm as "a procedure that takes any of the possible input instances and transforms it to the desired output" (Skiena 2008, 3).[20]

These a priori all-too-basic elements are, in fact, not trivial as they push ahead with an evaluation stance and frame algorithms in a very oriented way. Indeed, by endowing itself with problems-inputs and solutions-outputs, this take on algorithms can emphasize on the *adequacy relation* between these two poles. The study of algorithms becomes then the study of their *effectiveness*. This overlooking position is fundamental and penetrates the entire field of algorithmic study whose scientific agenda is well summarized by Knuth: "We often are faced with several algorithms for the same problem and we must decide *which is best*" (1997a, 7; italics added).[21] From this point, algorithmic analyses can focus on the elaboration of meta-methods that allow the systematization of the formal evaluation of algorithms.

Borrowing from a wide variety of mathematical branches (e.g., set theory, complexity theory), methods for analyzing algorithms as proposed by algorithmic students can be extremely elegant and powerful. Moreover, in the light of the significant advances in terms of implementation, data structuration, optimization, and theoretical understanding, this standard conception of algorithms as more or less functional interfaces between inputs and outputs—themselves defined by specific problems—certainly deserves its high respectability. However, I believe this standard conception has some limits that, in these days of controversies over algorithms, are important enough to suggest complementary alternatives that yet still need to be submitted.

First, the standard conception of algorithms overlooks the definition of the problems that algorithms are intended to solve. According to this view, problems and their potential solutions are already made, and the role of algorithmic studies is to evaluate the effectiveness of the steps leading to the transformation of inputs into outputs. Yet it is fair to assume that problems and the terms that define them do not exist by themselves. As it is shown in chapter 2 of this book, for example, problems are delicately irrigated products of *problematization processes* engaging habits, desires, skills, and values. And these collective processes greatly participate in the way algorithms—as problem-solving devices—will further be designed.

The second limit is linked to the first one: if one considers problematization as part of algorithmic design, the nature of the competition among algorithms changes. The best algorithms are not only the ones whose formal characteristics certify their superiority but also the ones that managed to associate with their problems' definitions the procedures capable of evaluating their results. By concentrating on formal criterions—without taking into account how these formalisms participated in the initial shaping of the problems at hand—the standard conception of algorithms tends to cover up the evaluation infrastructure and politics of algorithms. As shown in chapter 2, for example, evaluative procedures do not necessarily follow the design of algorithms; they also, sometimes, precede and influence it.

Third, the actual computerization of the iterative methods is not considered. Even though the standard conception of algorithms rightly insists on the centrality of computer code for the optimal execution of algorithms, this insistence takes the shape of programming methodologies that do not consider the experience of programming as it is lived at computer terminals. According to this standard conception of algorithms, writing numbered lists of instructions capable of triggering electric pulses in desired ways is mainly considered a means to an end. But as it is shown in chapters 4 and 6 of this book, programming practices—by virtue of the collective processes they require in order to unfold—also sometimes influence the way algorithms come into existence.

Fourth, little is said about how mathematical statements end up being enrolled for the transformation of inputs into outputs and how this enrollment affects the considered algorithms. To the standard conception of algorithms, mathematical statements appear out of the blue, ready to be scrutinized by means of other mathematical statements capable of evaluating their effectiveness. Yet as the chapter 6 of this book indicates, enrolling mathematical statements in order to operate the transformation of inputs into outputs is a problematic process in its own right, and again, this impacts the nature of algorithms. The initial conception of the dataset and its progressive problematization, reorganization, and reduction engage expectations and anticipations that fully participate in the ecology of algorithms in the wild.

The present work therefore intends to open up algorithms and extend them to processes that they are attached to but whose standard conception prevents from appreciating. If this venture does not, of course, aim to contest the results of algorithmic studies, it intends to enrich it with grounded sociological considerations.

2 A First Case Study

Let us start this ethnographic inquiry into the constitution of algorithms with a first dive into the life of the Lab. More precisely, let us start on November 7, 2013, at the Lab's cafeteria. At that time, I had only been at the Lab for a few days. During my first Lab meeting, I introduced myself as an ethnographer who had four years to submit a PhD thesis on the practical shaping of algorithms. Reactions had been courteous, although tinged with some indifference. Attention went up a notch when the director told the invited postdoc CL, the third-year PhD student GY, and the first-year PhD student BJ that I would take part to their ongoing project. It is this project we will follow in this first case study centered around several Group meetings, collective working sessions where CL, GY, and BJ (and myself) tried to coordinate the submission of a paper on a new algorithm.[1]

Entering the Lab's Cafeteria

Around 3 p.m. on November 7, 2013, I (FJ) entered the Lab's cafeteria for the first Group meeting. By that time, the Group and the topic of the project had already been defined: previous discussions among the Lab associates agreed that a new collective publication in *saliency detection* was relevant regarding the state of the art as well as the expertise of CL, GY, and BJ. Naturally, as any ethnographer freshly landed on his field site, I was terribly anxious: Would I live up to the expectations? Would they help me understand what they do? My participation in the project was clearly a top-down decision as the Lab's director had assigned me to the project to help me properly start my inquiry. Would the Group welcome me? I tried to read some papers on saliency detection that CL previously sent me but

I was confused by their tacit postulates. How would it be possible to detect this strange thing called "saliency" since what is important in a digital image certainly varies from person to person? And what is this odd notion of "ground truth" that the papers' algorithms seem to rely on? "Ground" and "truth": for an STS scholar, such a conjunction sounded highly problematic. As soon as I entered the Lab's cafeteria though, the members of the Group presented me with the ambitions of the project and how they intended to run it:[2]

Group meeting, the Lab's cafeteria, November 7, 2013

CL: "So you heard about saliency, right?"

FJ: "Well, I've read some stuff."

CL: "Huge topic, but basically, when you look at an image, not everything is important usually, and you focus only on some elements. ... What we try to do basically, it's like a model that detects elements in an image that should attract attention. ... GY's worked on a model that uses contrasts to segment objects and BJ has a model that detects faces. We'll use them as a base. ... For now, most saliency models only detect objects and don't pay attention to faces. There's no ground truth for that. But what we say is that faces are also important and usually attract directly the attention. ... And that's the point: we want to include faces to saliency, basically."

GY: "And segment faces. Because face detectors output only rectangles. ... There can be many applications [for the model], like in display or compression for example."

Many questions immediately arose. How and why is it important to focus on "elements that should attract attention"? Why is it problematic not to have a "ground truth" to detect "multiple objects and faces"? And what is a ground truth anyway? Why is it related to "saliency" and its potential industrial applications? Already at this early stage of the inquiry, the meandering flows of ethnography somewhat deprive us from our landmarks. To follow the Group and become able to fully explore these materials, some more equipment is obviously needed. I will thus temporally "pause" the account of the Group's project and consider for a while the sociohistorical background of saliency detection that underlies the Group's framing of its project. Once these introductory elements are acquired, I will be come back to this first Group meeting.

Backstage Elements: Saliency Detection and Digital Image Processing

"Saliency" for computer scientists in image processing is a blurry term with a history that is difficult to track. The term "saliency" was gradually created by straddling different—yet closely related—research areas. One point of departure could be the 1970s when explicative models developed in cognitive psychology and neurobiology[3] started to schematize how the human brain could quickly handle an amount of visual data that is far larger than its estimated processing capabilities (Eason, Harter, and White 1969; Lappin and Uttal 1976; Shiffrin and Gardner 1972).[4] After many disputes and controversies, a rough agreement about the overall process of humans' "selective visual attention method" had progressively emerged that distinguishes between two neuronal processes of selecting and gating visual information (Itti and Koch 2001; Heinke and Humphreys 2004).[5] On the one hand, there is a task-independent and rapid "bottom-up visual attention process" that selects conspicuous stimuli such as color contrasts, feature orientations, or spatial frequency. On the other hand, there is a slower "top-down visual attention process" that operates selectively based on tasks to accomplish. The term "saliency map" was proposed by Koch and Ullman (1985) to define the final result of the brain's bottom-up visual attention process.

In the 1980s, the way that cognitive psychologists and neurobiologists theorized two different "paths" for the brain to process light signals—one fast and generic, the other slower and task-specific—inspired scientists whose machines face a similar problem in computer vision: the stream of sampled digital signals that emanated from CCDs were too large to be processed all at once. From this point, two different classes of image-processing detection algorithms have progressively been shaped. The first class was inspired by the assumed bottom-up schematic process of visual attention and tried to detect "low-level features" inscribed within the pixels of a given image, such as intensity, color, orientation, and texture.[6] Through the academic efforts of Laurent Itti and Christof Koch in the 2000s (Itti, Koch, and Niebur 1998; Itti, Koch, and Braun 2000; Itti and Koch 2001; Elazary and Itti 2008; Zhao and Koch 2011), the term "saliency" was progressively assimilated into this first class of algorithms that became labeled *saliency-detection algorithms*. The second class of image-processing detection algorithms was inspired by the assumed top-down schematic process of visual attention and is based on "high-level features" that have to be learned by machines according to specific metrics (e.g., face or car detection). This often involves automated learning procedures and the management of increasingly large databases (Grimson and Lozano-Perez 1983; Lowe 1999).

Despite differences in terms of substratum, both high-level and low-level detection algorithms were, and are, bound to the same construction workflow that consists of five interrelated and problematic steps:

1. The acquisition of a finite dataset.

2. On the data of this dataset, the manual labeling of clear *targets*, defined here as the elements (faces, cars, salient regions) the desired algorithm will be asked to detect.

3. The construction of a database gathering the unlabeled data and their manually labeled counterparts. This database is usually called "ground truth" by the research community.

4. The design of the algorithm's calculating properties and parameters based on a representative part of the ground-truth database.

5. The evaluation of the algorithm's performances based on the rest of the ground-truth database.

To illustrate this schematic workflow, let us hypothesize the existence of ϕ, a standard detection algorithm in image processing. The very existence of ϕ depends upon a finite set of digital images for which human workers have previously labeled targets (e.g., faces, cars, salient regions). The unlabeled images and their manually labeled counterparts are then gathered together within a database to form the *ground truth* of ϕ. To design and code ϕ, the ground truth is randomly split into two parts: the "training set" and the "evaluation set." The designers of ϕ would use the training set to extract formal information about the targets, often with help of mathematical expressions. Once formulated and translated into machine-readable code, the algorithm ϕ is tested on the evaluation set to see how well it detects targets that were not used to design its properties. From its confrontation with the evaluation set, ϕ produces a precise number of outputs that can be qualified either as "true positives," "false negatives," or "false positives," thanks to the previous human-labeling work. Out of this comparison between manually designed targets and automatically produced outputs, statistical measures such as precision (the fraction of detected items that were previously defined as targets) and recall (the fraction of targets among the detected items) can be obtained to compare and rank competing algorithms (see figure 2.1).

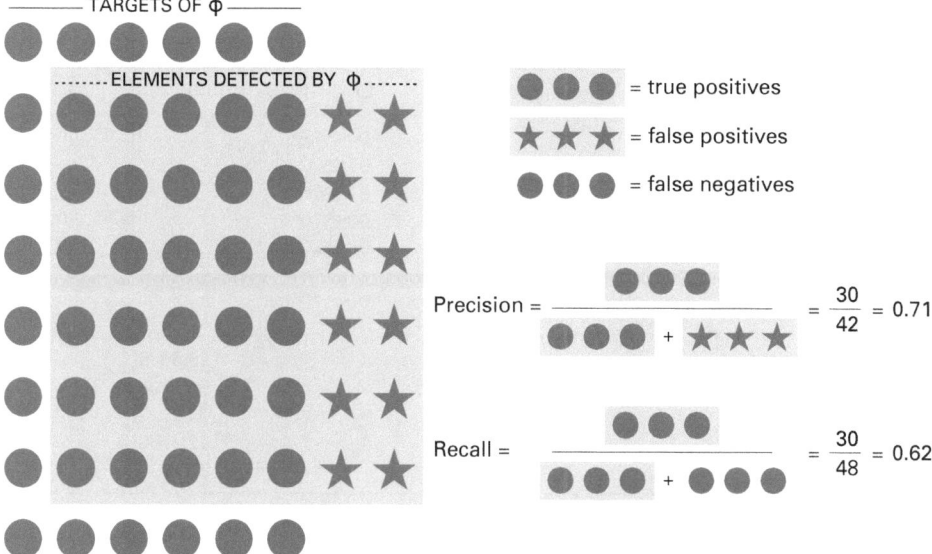

Figure 2.1
Schematic of precision and recall measures on ϕ. In this hypothetical example, ϕ (grey background) detected thirty targets (true positives) but missed eighteen of them (false negatives). This performance means that ϕ has a recall score of 0.62. The algorithm ϕ also detected twelve elements that are not targets (false positives), and this makes it have a precision score of 0.71. From this point, other algorithms intended to detect the same targets can be tested on the same ground truth and may have better or worse precision and recall scores than ϕ.

One drawback of high-level detection algorithms is that they are task-specific and cannot by themselves detect different types of targets: a face-detection algorithm will detect faces, a car-detection algorithm will detect cars, a plane-detection algorithm will detect planes, and so on.[7] Yet, one of the benefits of such high-level detection algorithms is that the definition of their targets (faces, cars, planes) often involves minor ambiguities for those who design them: cars, faces, or planes have rather unambiguous characteristics that facilitate agreement. Targets and ground truths can then be manually shaped by computer scientists in order to train high-level detection algorithms. Moreover, these ground truths can also serve as referees among competing high-level detection algorithms as they provide precision and recall metrics. The subfield of face detection with its numerous ground truths and algorithmic propositions provides a paradigmatic example of a highly

Results Reported in Terms of Percentage Correct Detection (CD) and Number of False Positives (FP), CD/FP, on the CMU and MIT Datasets

Face detection system	CMU-130	CMU-125	MIT-23	MIT-20
Schneiderman & Kanade—E[a] [170]		94.4%/65		
Schneiderman & Kanade—W[b] [170]		90.2%/110		
Yang et al.—FA [217]		92.3%/8		89.4%/3
Yang et al.—LDA [217]		93.6%/7		91.5%/1
Roth et al. [157]		94.8%/7		94.1%/3
Rowley et al. [158]	86.2%/23		84.5%/8	
Feraud et al. [42]	86%/8			
Colmenarez & Huang [22]	93.9%/8122			
Sung & Poggio [182]			79.9%/5	
Lew & Huijsmans [107]			94.1%/64	
Osuna et al. [140]			74.2%/20	
Lin et al. [113]			72.3%/6	
Guand Li [54]			87.1%/0	

[a]Eigenvector coefficients.
[b]Wavelet coefficients.

Figure 2.2
An exemplary comparison table among high-level face-detection algorithms. Two ground truths are used for this comparison table from Carnegie Mellon University (CMU) and the Massachusetts Institute of Technology (MIT). On the left, a list of algorithms named according to the papers in which they were proposed. In this table, the 'Percentage of Correct Detection' (CD) indicates the recall values and the 'Number of False Positives' (FP) suggests the precision values. *Source:* Hjelmås and Low (2001, 262). Reproduced with permission from Elsevier.

developed and competitive topic in image processing since at least the 2000s (see figure 2.2).

In the 2000s, unlike research in high-level detection, low-level saliency detection had no "natural" ground truth allowing the design and evaluation of computational models.[8] At that time, if the task-independent and adaptive character of saliency detection was theoretically interesting for automatic image cropping (Santella et al. 2006), adaptive display on small devices (Chen et al. 2003), advertising design, and image compression (Itti 2000), the absence of any ground truth that could allow the training and evaluation of computational models prevented saliency detection from being an active topic in digital image processing. As Itti, Koch, and Niebur (1998) confessed when they tested the very first saliency-detection algorithm on natural images:

> With many such [natural] images, it is difficult to objectively evaluate the model, because *no objective reference is available for comparison*, and observers may disagree on which locations are the most salient. (Itti, Koch, and Niebur 1998, 1258; italics added)

Saliency detection in natural images is an equivocal topic not easily expressed in a ground truth. Whereas it is usually straightforward (and yet time consuming) to define univocal targets for training and evaluating high-level face-detection or car-detection algorithms, it is far more complex to do so for saliency-detection algorithms because what is considered as salient in a natural image tends to change from person to person. While in the 2000s saliency-detection algorithms might have been promising for many industrial applications, no one in the field of image processing had found a way to design a ground truth for natural images.

In 2007, Liu et al. proposed an innovative solution to this issue and created the very first ground truth for saliency detection in natural images. Their shift was smart, costly, and contributed greatly to framing and establishing the subfield of saliency detection in the image-processing literature. Liu et al.'s first move was to propose one possible scope of saliency detection by incorporating concepts from high-level detection. According to them, instead of trying to highlight salient areas within digital images, computational models for saliency should detect *the most salient object* within a given digital image. They thus framed the saliency problem as being binary and one-off object related. According to them, to get around the impasse of saliency detection, saliency-detection algorithms should distinguish one salient object from the rest of the image:

> We incorporate the high-level concept of salient object into the process of visual attention in each respective image. We call them salient objects, or foreground objects that we are familiar with. ... We formulate salient object detection as a binary labelling problem that separates a salient object from the background. Like face detection, we detect a familiar object; unlike face detection, we detect a familiar yet unknown object in an image. (Liu et al. 2007, 1–2)

Thanks to this refinement of the concept of saliency (from "anything that first attracts attention" to "the one object in a picture that first attracts attention"), Liu et al. could organize an experiment in order to construct legitimate targets to be retrieved by computational models. They first randomly collected 130,099 high-quality natural images from internet forums and search engines. Then they manually selected 20,840 images that fit

Figure 2.3
Samples from Liu et al.'s dataset. Pictures contain one centered and contrastive element. *Source:* Microsoft Research Asia (MSRA) public dataset, Liu et al. (2007).

with their definition of the saliency problem: images that, according to them, contained only one salient object. This initial selection operation was crucial as it excluded images with several potential salient objects. The result was an initial dataset of no complex pictures with mixed features (see figure 2.3).

They then proceeded in two steps. First, they asked three human workers to manually draw a rectangle on what they thought was the most salient object in each image. For each image, Liu et al. then obtained three different rectangles whose consistencies could be measured by the percentage of shared pixels. For a given image, if its three rectangles were more consistent than a chosen threshold (here, 80 percent of pixels in common), the image was considered as containing a "highly consistent salient object" (Liu et al. 2007, 2). After this first selection step, their dataset called α contained around thirteen thousand images.

For the second step, Liu et al. randomly selected five thousand highly consistent salient-object images from α to create a second dataset called β. They then asked nine other human workers to label the salient object of every image in β with a rectangle. This time, Liu et al. obtained for every image nine different yet highly consistent rectangles whose average surface was considered their "saliency probability map" (Liu et al. 2007, 3). Thanks to this constructed social agreement, the five thousand saliency probability maps—in a computer science perspective, tangible *matrices* constituted of specific numerical values—could then be considered the best solutions to the saliency problem as they framed it. The whole ground truth—the database gathering the natural images and their corresponding saliency probability maps—became the material base on which the desired algorithm could be developed. By constructing this ground truth, Liu et al. defined the terms of a new problem whose solutions could be retrieved by means of calculating methods.

The shift here was not trivial. Indeed, by organizing this survey, inviting people into their laboratory, welcoming them, explaining the topic to them, writing the appropriate computer programs to make them label the images, and gathering the results in a proper database in order to statistically process them, Liu et al. transformed their initial reduced conception of saliency detection into workable and unambiguous targets with specific numerical values. At the end of this laborious process, Liu et al. could randomly select two thousand images from set α and one thousand images from set β to construct a training set (Liu et al. 2007, 5–6) to analyze the shared features of their constructed-yet-sound-by-virtue-of-agreement targets. Once the adequate numerical features were extracted from the targets of the training set and implemented in machine-readable language, they used the four thousand remaining images from set β to statistically measure the performances of their algorithm. Further, and for the very first time, they also could compare the detection performances of their algorithm with two competing algorithms that had already been proposed by other laboratories but that could not have been evaluated on natural images before due to the lack of any "natural" targets related to saliency. Besides the actual completion of their saliency-detection algorithm, the great innovation of Liu et al. was then to redefine the saliency problem so that it could allow performance evaluations (see figure 2.4).

By publishing their paper and also publicly providing their ground truth online, it is not an exaggeration to say that Liu et al. established a newly assessable research direction in image processing. A costly infrastructure had been put together, ready to be reused to support other competing algorithmic propositions with perhaps better performances according to Liu et al's ground truth and the definition of saliency it encapsulates. Their publication was more than a paper: it was a paper that allowed other papers to be published as they provided a ground truth that could be used by other researchers as long as they properly quote the seminal paper and accept the ground truth's restricted—yet operational—definition of saliency.[9]

Another important paper for saliency detection—and therefore also for the Group's project that we shall soon continue to follow—was published

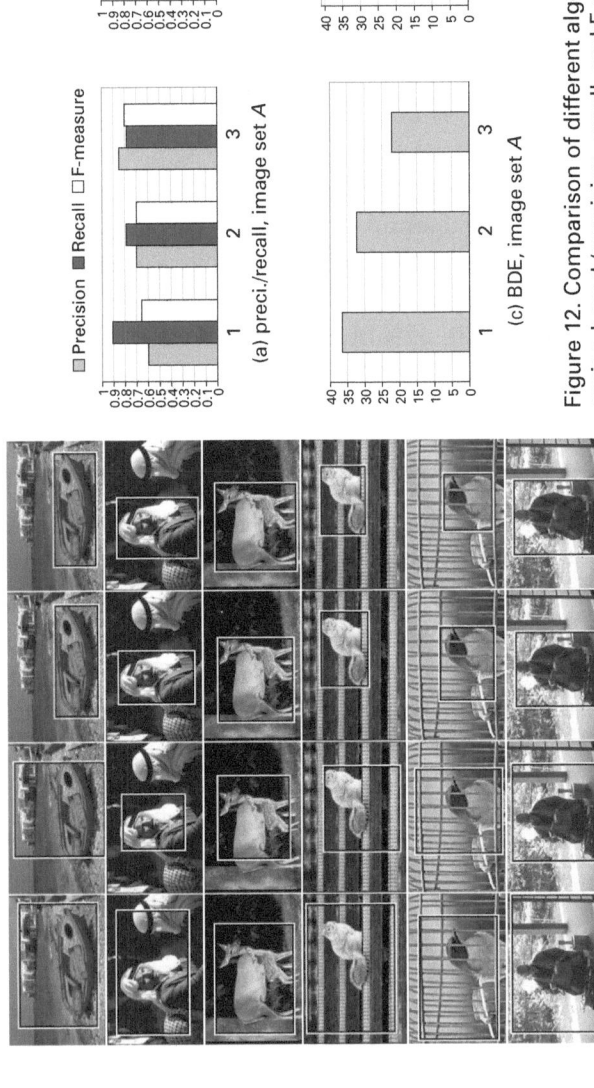

Precision ■ Recall □ F-measure

(a) preci./recall, image set A

Precision ■ Recall □ F-measure

(b) preci./recall, image set B

(c) BDE, image set A

(d) BDE, image set B

Figure 12. Comparison of different algorithms. (a-b) and (c-d) are region-based (precision, recall, and F-measure) and boundary-based (BDE—boundary displacement error) evaluations. 1. FG. 2. SM. 3. our approach.

Figure 14. Comparison of different algorithms. From left to right: FG, SM, our approach, and ground-truth.

Figure 2.4

Performance evaluations on Liu et al.'s ground truth. On the left, a visual comparison among three different saliency-detection algorithms according to the ground truth. On the right, histograms that summarize the statistical performances of the three algorithms. In these histograms, the ground truth corresponds to the y axis, the best possible saliency-detection performance that enables the evaluation. *Source:* Liu et al. (2007, 7). Reproduced with permission from IEEE.

(a)   (b)   (c)   (d)   (e)

Figure 2.5
Image (a) is an unlabeled image of Liu et al.'s ground truth; image (b) is the result of Wang & Li's saliency-detection algorithm; image (c) is the imaginary result of some other saliency-detection algorithm on (a); and image (d) is the bounding-box target as provided by Liu et al.'s ground truth. Even though (b) is more accurate than (c), it will obtain a lower statistical evaluation if compared to (d). This is why Wang & Li propose (e), a binary target that matches the contours of the already defined salient object. *Source:* Wang and Li (2008, 968). Reproduced with permission from IEEE.

in 2008 by Wang and Li. To them, even though Liu et al. (2007) were right to frame the saliency problem as a binary problem, their bounding-box ground truth remained unsatisfactory as it could well evaluate inaccurate results (see figure 2.5). To refine the measures of Liu et al.'s very first ground truth for saliency detection, Wang and Li randomly selected three hundred images from β dataset and used a segmentation tool to manually label the contours of each of the three hundred salient objects. What they proposed and evaluated then was a saliency-detection algorithm that "not only captures the rough location and region of the salient objects, but also roughly keeps the contours right" (Wang and Li 2008, 965).

From this point, saliency detection in image-processing was almost set: even though many algorithms exploiting different low-level pixel information were later proposed (Achanta et al. 2009; Chang et al. 2011; Cheng et al. 2011; Goferman, Zelnik-Manor, and Tal 2012; Shen and Wu 2012; Wang et al. 2010), they were all bound to the saliency problem as defined by Liu et al. in 2007. And even though other ground truths have later been proposed in published papers (Judd, Durand, and Torralba 2012; Movahedi and Elder 2010) to widen the scope of saliency detection (notably by proposing images with two objects that could be decentered), Liu et al.'s seminal framing of saliency detection as a binary object-related problem remained unchallenged. And when the Group started their project in November 2013,

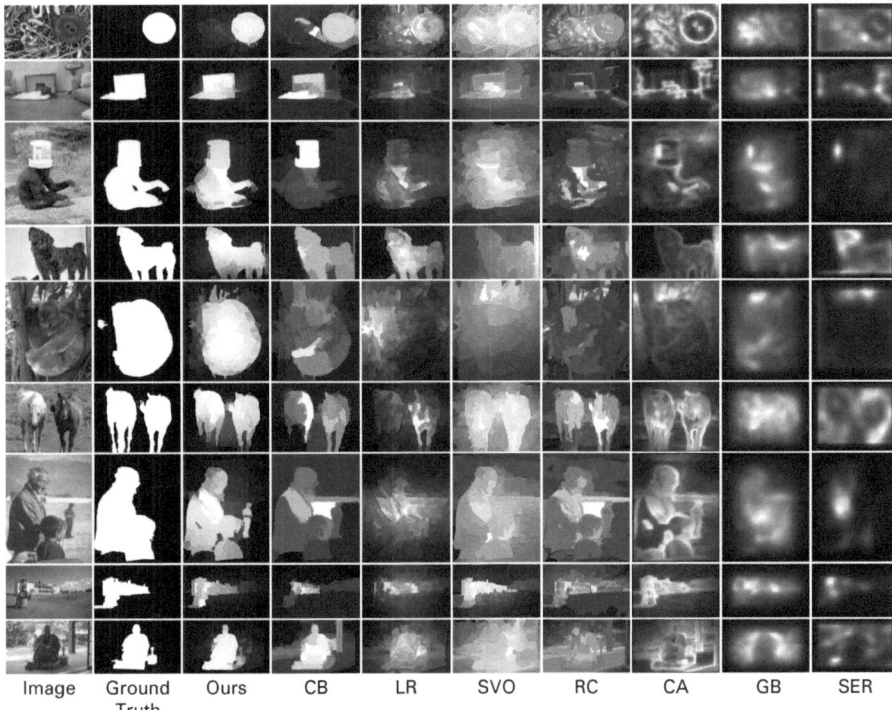

| Image | Ground Truth | Ours | CB | LR | SVO | RC | CA | GB | SER |

Figure 9. Comparison of different methods on the ASD, SED and SOD datasets. The first three rows are from the ASD dataset, the middle three rows are from the SED dataset, the last three rows are from the SOD dataset.

Table 1. Comparison of average execution time (seconds per image).

Method	Ours	CB	SVO	RC	LR	CA	GB	SER	FT	LC	SR	IT
Time(s)	0.105	1.179	40.33	0.106	11.92	36.05	0.418	25.19	0.016	0.002	0.002	0.165
Code	Matlab	Matlab	Matlab	C++	Matlab	Matlab	Matlab	C++	C++	C++	C++	Matlab

Figure 2.6

2013 comparison table between different saliency-detection algorithms. The number of competing algorithms has increased since 2007. Here, three ground truths are used for performance evaluations: ASD (Achanta et al. 2009), SED (Alpert et al. 2007), and SOD (Movahedi and Elder 2010). Below the figure, a table compares the execution time of each implemented algorithm. *Source:* Jiang et al. (2013, 1672). Reproduced with permission from IEEE.

Liu et al.'s problematization of the saliency problem was continuing to support a competition among algorithms that differentiated themselves by speed and accuracy (see figure 2.6).

With this brief history of saliency in image processing, we are better equipped to follow the Group as it tries to construct its own innovative saliency-detection algorithm. Social surveys, salient objects whose contours define the targets of competing algorithms, ground truths bound to a binary problematization of saliency, promising industrial applications: the stage we are about to explore is supported by all of these elements, constraining the members of the Group in the shaping of their project as well as providing them opportunities for further reconfigurations.

Reframing Saliency

If, at the beginning of the chapter, the Group's explanations appeared quite cryptic, the previous introductory review should now enable us to understand them critically. Let us thus look at the same excerpt once again:

> **Group meeting, the Lab's cafeteria, November 7, 2013**
>
> **CL:** "So, you heard about saliency, right?"
>
> **FJ:** "Well, I've read some stuff."
>
> **CL:** "Huge topic, but basically, when you look at an image, not everything is important usually, and you focus only on some elements. … What we try to do basically, it's like a model that detects elements in an image that should attract attention. … GY's worked on a model that uses contrasts to segment objects and BJ has a model that detects faces. We'll use them as a base. … For now, most saliency models only detect objects and don't pay attention to faces. There's no ground truth for that. But what we say is that faces are also important and usually attract directly the attention. … And that's the point: we want to include faces to saliency, basically."
>
> **GY:** "And segment faces. Because face detectors output only rectangles. … There can be many applications [for the model], like in display or compression for example."

According to the Group, saliency-detection models should also take human faces into account as faces are important in human attention mechanisms. Moreover, investing this interstice within saliency detection would be a good opportunity to merge some of the Group's recent researches on both low-level segmentation and high-level face detection. The idea to combine high-level face detection with low-level saliency detection derived from previous image-processing papers (Borji 2012; Karthikeyan, Jagadeesh, and Manjunath 2013) inspired themselves by studies in gaze prediction (Cerf, Frady, and Koch 2009), cognitive psychology (Little, Jones, and DeBruine 2011), and neurobiology (Dekowska, Kuniecki, and Jaśkowski 2008). But the

Group's ambition here was to go further in the saliency direction as framed by Wang and Li (2008), after Liu et al. (2007), by proposing an algorithm capable of detecting and segmenting the *contours* of faces. In order to accomplish such subtle results, the previous work done by GY on segmentation and BJ on face detection would constitute a precious resource to work on.

The Group also wanted to construct a saliency-detection model that could effectively process a larger range of natural images:

Group meeting, the Lab's cafeteria, November 7, 2013

GY: "But you know [to FJ], we hope the algorithm could detect multiple objects and faces. Because in saliency detection, models can only detect like one or two objects on simple images. They don't detect multiple salient objects in complex images. ... But the problem is that there's no ground truth for that. There's only ground truth with like one or two objects, and not that many faces."

In many cases, natural images not only capture one or two objects distinguished from a clear background; pictures produced by users of digital cameras—according to the Group—are generally more cluttered than those used to train and evaluate saliency-detection algorithms in the wake of Liu et al. (2007). Indeed, at least in November 2013, saliency detection was becoming a research area where algorithms were more and more efficient *only* on those—rare—natural images with clear and untangled features. But the Group also knew that this issue was intimately related to the then available ground truths for saliency detection that were all bound to Liu et al's restricted initial definition of saliency that only fit simple images. From this point, as the Group wanted to propose a model that could detect a different and more subtle saliency, it had to construct the targets of such saliency; as it wanted to propose a model that could calculate and detect multiple salient features (objects and faces) in more complex and realistic images, it had to construct a new ground truth that would gather complex images and their corresponding multiple salient features.

The Group's desire to redefine the terms of the saliency problem did not come ex nihilo. When Liu et al. did their research on saliency in 2007, it was difficult for computer scientists to organize large social surveys on complex images. But in November 2013, the growing availability of crowd-sourcing services enabled new potentialities:

Group meeting, the Lab's cafeteria, November 7, 2013

GY: "But we want to use crowdsourcing to do a new ground truth and ask people to label features they think are salient. … And then we could use that for our model and compare the results, you see?"

In broad strokes, crowdsourcing—a contraction of "crowd" and "outsourcing" initially coined by journalist Howe (2006)—is "a type of participative online activity in which an individual, an institution, a non-profit organization, or a company proposes to a group of individuals of varying knowledge, heterogeneity, and number, via a flexible open call, the voluntary undertaking of a task" (Estellés-Arolas and González-Ladrón-de-Guevara 2012, 195). In November 2013, this service was offered by several companies such as Amazon (*via* Amazon Mechanical Turk), ClickWorker, or Employment Crossing (*via* ShortTask), whose own application programming interfaces (APIs)[10] recommended surveys to registered online contingent workers mainly located in the United States and India. Once a worker submits their completed task—which can vary greatly in time and complexity—the organization that designed the survey (e.g., a research institution, a company, an individual) can decide on its validity. If the task is considered valid, the worker receives from the crowdsourcing company the amount of money initially indicated in the open call. If the task is considered not valid, the worker receives nothing and has, most of the time, no possibility of appeal. As the moral economy of crowdsourcing has recently been the object of critical sociological studies, it is necessary to devote a short sidebar to it.

Contingent work has long supported industrial efforts. As, for example, documented by Pennington and Westover (1989), the textile industry as it developed in England in the 1850s relied heavily on off-site manufacturing operations, often referred to as "industrial homework." Women and children living in the countryside, operating as proto-on-demand workers, were asked to make crucial finishing touches too fine for the machines of the time. Almost simultaneously, a similar phenomenon was taking place in the United States, particularly in the Pittsburg, Pennsylvania, area: even though it was often seen as a reminiscence of a preindustrial era that was doomed to disappear, "piecework" organized on a commission basis in partnership with rural households was a necessary lever for the scaling up of mass manufacturing (Albrecht 1982). And if trade unions did later manage, through painful struggles, to somewhat improve the working conditions of employees (e.g., US Fair Labor Standards Act in 1938, French *Accords de Matignon* in 1936), these improvements mostly concerned full-time work carried out on designated production sites that was mostly reserved for white male adults. The concessions made to salaried workers during the first half of the twentieth century thus mostly concerned those who benefited from visibility and proximity: contingent work, which was scattered, not very visible, little valued, and considered unskilled, continued to pass under the radar. To this—and to many other things that are beyond the scope of this sidebar[11]—was later added a more or less explicit corporate strategy of circumventing unionization and work regulations (which were already reserved for specific trades) based notably on the growing availability of information and communication technologies. This strategy of "fissuration of the workplace" (Weil 2014), well in line with the financialization of Western economies,[12] helped to further promote outsourcing: instead of depending on employees benefiting from statutory logic, it has become preferable and valued to depend on remote worldwide networks of contingent staff. And crowdsourcing, as distributed computer-supported on-demand low-valued work, can be seen as the continuation of contingent work's support to and modification of industrial capitalism. As Gray and Suri (2019, 58) noted: "Those on-demand jobs today are the latest iteration of expendable ghost work. They are, on the one hand, necessary in the moment, but they are too easily devalued because the tasks that they do are typically dismissed as mundane or rote and the people often employed to do them carry no cultural clout."[13]

Let us come back to the Lab. In November 2013, like most people, the Group was not aware of the dynamics underlying generalized outsourcing and devaluation of contingent labor as supported by contemporary crowdsourcing processes. An indication of this unawareness could be found in the term "users" the Group often employed to refer to the anonymous *workers* engaged in this new form of precariat.[14] For the Group, at that moment, the estimated benefits of crowdsourcing were huge: once the desired web application was coded and set with an instruction, such as "please highlight the features that directly attract your attention," the Group would be able to pay a crowdsourcing company whose API would take charge of linking the survey to dozens of low paid "users" of the Group's web application. In turn, these "users"—that I will from now on call "workers"—would feed the

Group's server with labeling coordinates that could be processed on software packages such as Matlab.[15] For our story, crowdsourcing—as a rather easily available paid service—created a difference: the gathering of many manually labeled salient features became more manageable for the Group than it had been for Liu et al. in 2007, and an extension of the notion of saliency to multiple features became—at least in November 2013—doable.

Another difference effected by crowdsourcing was a potential redefinition of the saliency problem as being *continuous*:

Group meeting, the Lab's cafeteria, November 7, 2013

FJ: "So, basically you want many labels?"

GY: "Yes because you know, in the state-of-the-art face detection or saliency models only detect things in a binary way, like face/no face, salient/not salient. What we also try to do is a model that evaluates the importance of faces and objects and segments them. Like 'this face is more important than this other face which is more important than that object' and so on. ... But anyways, to do that [a ground truth based on the results of a crowdsourcing task], we first need a dataset with many images with different contents."

CL: "Yes, we thought about something like 1,000 image at least, to train and evaluate. But it has to be images with different objects and faces with different sizes."

GY: "And we have to select the images; good images to run the survey. ... We'll try to propose a paper in [the] spring so it would be good to have finished crowdsourcing in January, I guess."

If the images used to construct the ground truth contained only one or two objects and were labeled only by several individuals, no relational values among the labeled features could be calculated. From this point, defining saliency as a binary problem in the manner of Liu et al. (2007) would make complete sense. Yet as the Group could afford to launch a social survey that asked for many labels on a dataset with complex images containing many features, it would become methodologically possible to assign *relative importance values* to the different labeled features. This was a question of arithmetic values: if one feature were manually labeled as salient, the Group could only obtain a binary value (foreground and background). But if several features were labeled as more or less salient by many workers, the Group could obtain a continuous subset of results. In short, for the Group, crowdsourcing once again created a difference by making it possible to create new types of targets with relatively continuous values. It was difficult at this point to predict if the Group's algorithm would effectively be able to approach these subtle results. Nevertheless, the ground truth the Group wanted to constitute would enable the development of such an algorithm by providing the targets that the model should try to retrieve in the best possible way.

Even though the Group had managed to build on previous works in saliency detection and other related fields to reframe the problem of saliency, it still lacked the ground truth that could numerically establish the terms of this new problem: both the inputs the desired algorithm should work on and the outputs (the "targets") it should try to retrieve still needed to be constructed. In that sense, the Group was only at the beginning of the problematization process that may lead to a new computational model: its new definition of the saliency problem still needed to be equipped (Vinck 2011) with tangible elements (a new set of complex images, a crowdsourcing task, continuous values, segmented faces) to form a referential database that would, in turn, constitute the material base of the new computerized method of calculation. Borrowing from Michel Callon (1986), we might say that, for the members of the Group, the new ground truth appeared as an *obligatory passage point* that could make them become—perhaps—indispensable for the research community in saliency detection. Without a new ground truth, saliency-detection models would still operate on unrealistic images; they would still be one-off object related; they would still ignore the detection and segmentation of faces; and they would still, therefore, be irrelevant for real-world applications. With the help of a new ground truth, these shortcomings that the Group attributed to saliency detection may be overcome. In a similar vein—this time borrowing from Joan Fujimura (1987)—we might say that, at this point, the Group's saliency problem was doable *only* at the level of its laboratory. The Group had indeed been given time and money to conduct the project and had insights on how to run it. But without any ground truth, the Group had no tangible means to articulate this "laboratory level" with both the research communities in image processing *and* the specific tasks required to effectively define a working model of computation. It is only by constructing a database gathering "input-data" and "output-targets" that the Group would be able to propose and, eventually, publish an algorithm capable of solving the saliency problem as the Group reframed it.

Constructing a New Ground Truth

We have now a better sense of some of the pitfalls that sometimes get in the way of computer scientists trying to shape a new algorithm. As we were following the Group in the beginning of its saliency-detection project, we realized that the constitution of an image-processing algorithm capable of establishing a new research direction goes along with the shaping of a new ground truth that should precisely support and equip the constitution of the algorithm. Yet for now, we only considered the reasons why the Group needed to design a new ground truth. But how did it actually *make* it?

In addition to working on the coding of the crowdsourcing web application, the Group also dedicated November and December 2013 to the selection of images that echo the algorithm's three expected performances: (1) detecting and segmenting the contours of salient features, including faces; (2) detecting and segmenting these salient features in complex images; and (3) evaluating the relative importance of the detected and segmented salient features. These specifications led to several Group meetings specifically organized to discuss the content and distribution of the selected images:

Group meeting, the Lab's cafeteria, November 21, 2013

BJ: "Well, we may avoid this kind of basketball photo because these players may be famous-like. They are good because the ball contrasts with faces, but at least I know some of the players. And if I know, we include other features like 'I know this face,' so I label it."

CL: "I think maybe if you have somebody that is famous, the importance of the face increases and then we just want to avoid modeling that in our method."

...

CL: "OK. And the distributions are looking better?"

FJ: "Yes definitely. BJ just showed me what to improve."

CL: "OK. So what other variables do we consider?"

GY: "Like frontal and so on. But equalizing them is real pain."

CL: "But we can cover some of them; maybe not equalize. So there should be like the front face with images of just the front of the face and then there is the side face, and a mixture in between."

The selection process took time because a wide variety of image contents (e.g., sport, portraits, side faces) had to be gathered to cover more natural situations than the other ground truths. Also, no famous features (e.g., buildings, comedians, athletes) that could influence attention processes should be part of the content. As we can see, the Group's anticipated capabilities for the algorithm oriented this manual selection process: similarly to Liu et al. (2007) but in a manner that made the Group include more complex "natural situations," the assembling of a dataset was driven by the algorithm's future tasks.[16] By December 2013, eight hundred high-resolution images were gathered— mostly from Flickr—and stored in the Lab's server. Since the Group considered the inclusion of faces within saliency detection as the most significant contribution of the project, 632 of the selected images included human faces.

In parallel to this problem-oriented selection of images, organizational work on the selected images had to be defined in order not to be overloaded by the increasing number of files and by the huge amount of labeled results to be gathered throughout the crowdsourcing task. This kind of organizational procedure was very close to data management and implied the realization of a whole new database for which information could be easily retrieved and anticipated. Moreover, the shaping of the crowdsourcing survey also required coordination and adjustments: What question would be asked? How would answers be collected and processed in order to fulfill the ambitions of the project? Those were crucial issues as the "raw" labeled answers obtained via crowdsourcing could only be rectangles and not precise contours:

Group meeting, the Lab's cafeteria, December 12, 2013

CL: "But for the database, do we rename the images so that we have a consistency?"

BJ: "Hum. … I don't think so because now we can track the files back to the website with their ID. And with Matlab you can like store the jpg files in one folder and retrieve all of them automatically"

…

CL: "What do you think, GY? Can we ask people to select a region of the image or to do something like segmenting directly on it?"

GY: "I don't think you can get pixel-precision answers with crowdsourcing. We'll need to do the pixel-precision [in the Lab] because if we ask them, it's gonna be a very sloppy job. Or too slow and expensive anyway."

CL: "So what do you want? There is your Matlab code to segment features, right?"

GY: "Yes, but that's low-level stuff, pixel-precision [segmentation]. It's gonna be for later, after we collect the coordinates, I guess. I still need to finish the scripts [to collect the coordinates] anyway. Real pain. ... But what I thought was just like ask people to draw rectangles on the salient things, then collect the coordinates with their ID and then use this information to deduce which feature is more salient than the other on each image. Location of the salient feature is a really fuzzy decision, but cutting up the edges is not that dependent. ... You know where the tree ends, and that's what we want. Nobody will come and say 'No! The tree ends here!' There is not so many variances between people I guess in most of the cases."

CL: "OK, let's code for rectangles then. If that's easy for the users, let's just do that."

The IDs of the selected images allowed the Group to put the images in a Matlab database rather easily. But within the images, the salient features labeled by the crowdworkers were more difficult to handle since GY's interactive tool to get the precise boundaries of image contents was based on low-level information. As a consequence, segmenting the boundaries of low-contrasted features such as faces could take several minutes, whereas *affordable* crowdsourcing was about small and quick tasks. The Group could not take the risk of either collecting "sloppy" tasks or spending an infeasible amount of money to do so.[17] The labeled features would thus have to be post-processed within the Lab to obtain precise contours.

Moreover, another potential point of failure of the project resided in the development of the crowdsourcing web application. Indeed, asking people to draw rectangles around features, translating these rectangles into coordinates, and storing them into files to process them statistically required nontrivial programming skills. By January 2014, when the crowdsourcing web application was made fully operational, it comprised seven different scripts (around seven hundred lines of code) written in html, PHP, and JavaScript that responded to each other depending on the workers' inputs (see figure 2.7). Yet, if the Lab's computer scientists were at ease with numerical computing and programming languages such as Matlab, C, or C++, web designing and social pooling were not competencies for which they were necessarily trained.

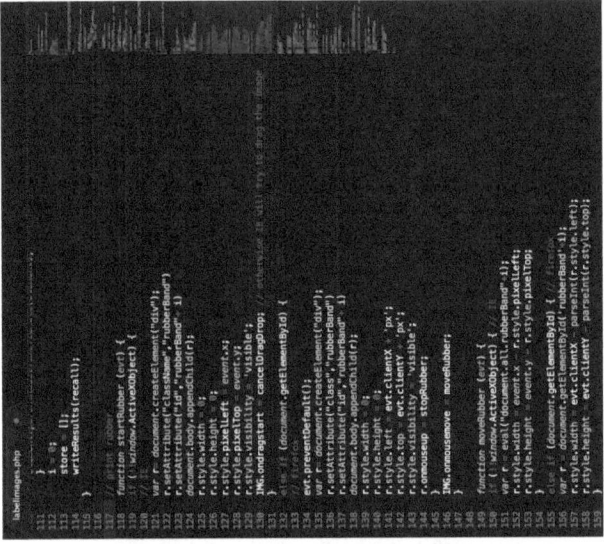

Figure 2.7

Screen captures of the web application designed by the Group for its crowdsourcing task. On the left, the application when ran by a web browser. Once workers created a username, they could start the experiment and draw rectangles. When workers clicked on "Next Image" button, the coordinates of the rectangles were stored in .txt files on the Lab's server. On the right, one excerpt of one of the seven scripts required to realize such interactive labels and data storage.

Once coded and debugged—a delicate process in its own right (see chapter 4)—the different scripts were stored in one section of the Lab's server whose address was made available in January 2014 to the now-defunct company ShortTask whose API offered the best-rated contingent workers. By February 2014, thirty workers' tasks qua tens of thousands of rectangles' coordinates were stored in the Group's database as .txt files, ready to be processed thanks to the previous preparatory steps. At this point, each image of the previously collected dataset was linked with many different rectangles drawn by the workers. By superimposing all the coordinates of the different rectangles on Matlab, the Group created for each image a "weight map" with varying intensities that indicated the relative consensus on salient regions (see figure 2.8). The Group then applied to each image a widely used threshold taken from Otsu (1979)—part of Matlab's internal library—to keep only weighty regions that had been considered salient by the workers. In a third step that took two entire weeks, the Group—in fact, BJ and me—manually segmented the contours of the salient elements within the salient regions to obtain "salient features." Finally, the Group assigned the mean value of the salient regions' map to the corresponding salient features to obtain the final targets capable of defining and evaluating new kinds of saliency-detection algorithms. This laborious process took place between February and March 2014; almost a month was dedicated to the processing of the coordinates produced by the workers and then collected by the html-JavaScript-PHP scripts and database.

By March 2014, the Group successfully managed to create targets with relative saliency values. The selected images and their corresponding targets could then be organized as a single database that finally constituted the ground truth. From this point, one could consider that the Group effectively managed to redefine the terms of the saliency problem: the transformations the desired algorithm should conduct were—finally—numerically defined. Thanks to the definition of inputs (the selected images) and the definition of outputs (the targets), the Group finally possessed a problem that numerical computing could take care of.

Of course, establishing the terms of a problem by means of a new ground truth was not enough: to propose an actual algorithm, the Group also had to design and code lists of instructions that could effectively transform input-data into output-targets according to the problem they had just established. To design and code these lists of instructions, the Group randomly

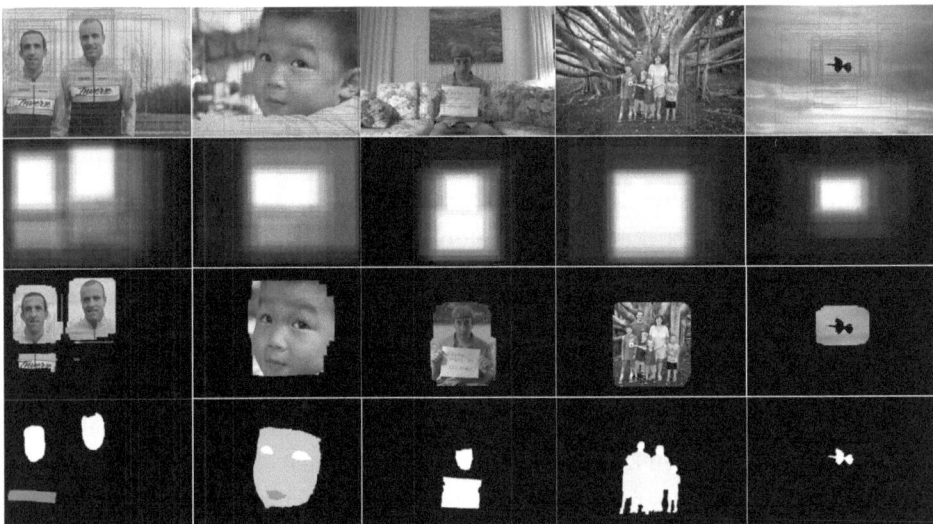

Figure 2.8
Matlab table summarizing the different steps required for the processing of the coordinates produced by the workers who accomplished the crowdsourcing task. The first row shows examples of images and rectangular labels collected from the crowdsourcing task. The second row shows the weight maps obtained from the superposition of the labels. The third row shows the salient regions produced by using Otsu's (1979) threshold. The last row presents the final targets with relative saliency values. The first three steps could be automated, but the last segmentation step had to be done manually. At the end of this process, the images (first row, without the labels) and their corresponding targets (last row) were gathered in a single database that constituted the Group's ground truth.

selected two hundred images out of the ground truth to form a training set. After formal analysis of the relationships between the inputs and the targets of this training set, the Group extracted several numerical features that expressed—though not completely—these input-target relationships.[18] The whole process of extracting and verifying numerical features and parameters from the training set and translating them sequentially into Matlab programming language took almost a month. But at the end of this process, the Group possessed a list of Matlab instructions that was able to transform the input values of the training set into values relatively close to those of the targets.

By the end of March 2014, the Group used the remainder of its ground-truth database to evaluate the algorithm and compare it with already available saliency-detection algorithms in terms of precision and recall measures (see figure 2.9). The results of this confrontation being satisfactory, the features and performances of the Group's algorithm were finally summarized in a draft paper and submitted to an important European Conference on image processing.

As these Group meetings and documents show, the Group's algorithm could only be made operational once the newly defined problem of saliency had been solved by human workers and expressed in a ground-truth database. In that sense, the finalization of Matlab lists of instructions capable of solving the newly defined problem of saliency *followed* the problematization process in which the Group was engaged. The theoretical reframing of saliency, the selection of specific images on Flickr, the coding of a web application, the creation of a Matlab database, the processing of the

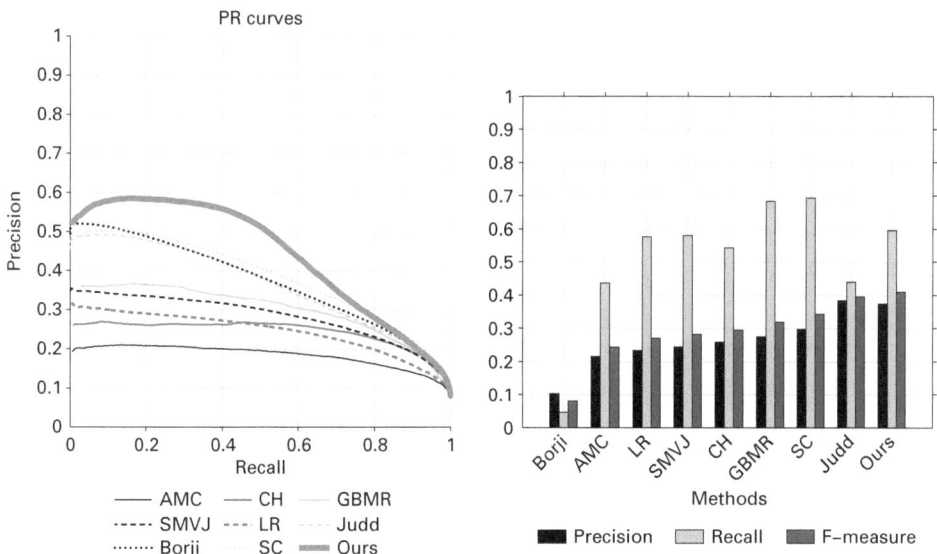

Figure 2.9
Two Matlab-generated graphs comparing the performances of the Group's algorithm ("Ours") with already published ones ("AMC," "CH," etc.). The new ground truth enabled both graphs. In the graph on the left, the curves represented the variation of precision ("*y*" axis) and recall ("*x*" axis) scores for all the images in the ground truth when processed by each algorithm. In the graph on the right, histograms measured the same data while also including F-Measure values, the weighted average of precision and recall values. Both graphs indicated that, according to the new ground truth, the Group's algorithm significantly outperformed all state-of-the-art algorithms.

workers' coordinates: all these practices were required to design the ground truth that ended up allowing the extraction of the relevant numerical features of the algorithm as well as its evaluation. Of course, the mundane work required for the construction of the ground truth was not *sufficient* to complete the complex lists of Matlab instructions that ended up effectively processing the pixels of the images: critical certified mathematical claims also needed to be articulated and expressed into machine-readable format. Yet, by providing the training set to extract the numerical features of the algorithm and by providing the evaluation set to measure the algorithm's performances, the ground truth greatly participated in the completion of the algorithm.

The above elements are not so trivial, and some deeper reflections are required before moving forward. In November 2013, the Group had only few elements at its disposal. It had desires (e.g., contesting previous papers), skills (e.g., mathematical and programming abilities), means (e.g., access to academic journals, powerful computers), and hopes (e.g., make a difference in the field of image processing). But these elements alone were not enough to effectively shape its new intended algorithm. In November 2013, the Group also needed an empirical basis that could serve as a fundamental substratum; it needed to *ground* a material coherence that could establish the veridiction of their future model. This was the whole benefit of the new ground truth—which should rather be called ground*ed* truth—as it was now possible to found and bring into existence a set of phenomena (here, saliency differentials) operating as an analytical referential. Once this scriptural fixation was achieved in March 2014, the world the Group inhabited was no longer the same: it was enriched and oriented by a set of relations materialized in a database. And the algorithm that finally came out from this database organized, reproduced, and in a sense, *consecrated* the relations embedded in it. From a static and particular ground truth emerged an operative algorithm potentially capable of reproducing and promoting the organizational rules of the ground truth in different configurations. By rooting the yet-to-be-constructed algorithm, the ground truth as assembled by the Group oriented the design of its algorithm in a particular direction. In that sense, the new ground truth was the contingent yet necessary *bias* of the group's algorithm.[19]

This propensity of computational models to be bound to and fundamentally biased by manually gathered and processed data is not limited to the field of digital image processing. For example, as Edwards (2013) showed for the case of climatology, the tedious collection, standardization, and compilation of weather data to produce accurate ground truths of the Earth's climate is crucial for both the parametrization and evaluation of General Circulation Models (GCMs).[20] Of course, just as in the field of image processing, the construction of ground truths by climatologists does not guarantee the definition of accurate and effective GCMs: crucial insights in fluid dynamics, statistics, and (parallel) computer programming are also required. Yet, without ground truths providing parameters and evaluations, no efficient and trustworthy GCM could come into existence. For the case of machine learning algorithms for handwriting recognition or spam filtering, Burrell (2016, 5–6) noted the importance of "test data" in setting the learning parameters of these algorithms as well as in evaluating their performances. Here as well, ground truths appear central, defining what is statistically learned by algorithms and allowing the evaluation of their learning performances.[21] The same seems also to be true of many algorithms for high-frequency trading: as MacKenzie (2014, 17–31) suggested, detailed analysis of former financial transactions as well as the authoritative literature of financial economics work as empirical bases for the shaping and evaluation of "execution" and "proprietary trading" algorithms.

Yet, despite growing empirical evidences, algorithms' tendency to be existentially linked to ground-truth databases that cannot, obviously, be reduced to mere *sets of data* remains little discussed in the abundant computer science literature on algorithms. The issue is generally omitted: mathematical analysis and programming techniques, sometimes highly complex, are discussed *after*, or *as if*, a ground truth has been constructed, accepted, distributed, and made accessible. The theoretical exploration of what I called in chapter 1 the standard conception of algorithms tends to take for granted the existence of stable and shared referential repositories. This omission may even be what makes such a vision of algorithms possible: considering algorithms as tools ensuring the computerized transition from problems to solutions might imply to suppose already defined problems and already assessable solutions.

Some sociologists—most of them STS-inspired—do consider the topic head on, though. In their critique of predictive algorithmic systems, Barocas and Selbst (2016) warned against the potentially harmful consequences of problem definition and training sets' collection. In a similar way, Lehr and Ohm (2017) emphasized on the handcrafted aspect of "playing with the data" for the design of statistical learning algorithms. More recently, Bechmann and Bowker (2019) built on these arguments to propose the notion of *value-accountability-by-design*: a call for systemic efforts to make arbitrary choices involved in algorithm-related data collection, preparation, and classification more explicit. In the wake of Ananny and Crawford (2018), they thus suggest that, to better appreciate algorithmic behavior, ex ante focus on ground-truthing processes might be more conclusive than ex post audits or source code scrutinization (as it is, for example, proposed in Bostrom [2017] and Sandvig et al. [2016]). In a similar way, Grosman and Reigeluth (2019) investigated the design of an algorithmic security system for the detection of threatening behaviors. They show that the definition of the problem that the algorithm will have to solve—and, therefore, the "true positives" it will have to detect—derive from collective problematization processes that include discussions and compromises among sponsors, competing interpretations of legal documents, and on-site simulations of threatening and inoffensive behaviors conducted by the project's engineers. They conclude that the normativity proper to algorithmic systems must also be considered in the light of the tensions that contributed to making this normativity expressible. In sum, all the above-mentioned authors have uncovered processes that resemble the one the Group had just gone through. Their investigations also show that what is called an "algorithm" often derives from collective processes expressed materially in contingent, but necessary, referential repositories.

At this early stage of the present inquiry, it would be unwise to define a general property common to all algorithms. Yet based on the preliminary insights of this chapter and the growing body of studies that touched on similar issues, one can make the reasonable hypothesis that behind many of these entities we like to call "algorithms" lie ground-truth databases that have made designers able to extract relevant numerical features and evaluate the accuracy of the automated transformations of inputs-data into output-targets. Consequently, as soon as such algorithms—once "in the wild," outside of their production sites—automatically process new data, their respective initial ground truths—along with the habits, desires, and values that participated in their shaping—are also invoked and, to a certain extent, promoted. As I will further develop at the end of this chapter, studying the performative effects of such algorithms in the light of the collective processes that constituted the output-targets these algorithms try to retrieve appears a stimulating, yet still underexplored, research topic when compared with the growing influence algorithms have on our lives.

Almost Accepted (Yet Rejected)

June 19, 2014: The reviewers rejected the Group's paper. The Group was greatly disappointed to see several months of meticulous work unrewarded by a publication that could have launched new research lines and generated many citations. But the feeling was also one of incomprehension and surprise in view of the reasons provided by the three reviewers.

Along with doubts about the usefulness of incorporating face information within saliency detection, the reviewers agreed on one seemingly key deficiency of the Group's paper: the performance comparisons of the computational model were only made with respect to the Group's new ground truth:

Assigned Reviewer 1
The paper does not show that the proposed method also performs better than other state-of-the-art methods on public benchmark ground truths. ... The experiment evaluation in this paper is conducted only on the self-collected face images. More evaluation datasets will be more convincing. ... More experiment needs to be done to demonstrate the proposed method.

Assigned Reviewer 2
The experiments are tested only on the ground truth created by the authors. ... It would be more insightful if experiments on other ground truths were carried out, and results on face images and non-face images were reported, respectively. This way one can more thoroughly evaluate the usefulness of a face-importance map.

Assigned Reviewer 3
The discussion is still too subjective and not sufficient to support its scientific insights. Evaluation on existing datasets would be important in this sense.

The reviewers found the technical aspects of the paper to be sound. But they questioned whether the new best saliency-detection model—as the Group presented it in the paper—could be confronted only with the ground truth used to create it. Indeed, why not confront this new model with the already available ground truths for saliency detection? If the model were really "more efficient" than the already published ones, it should also be more efficient on the ground truths used to shape and evaluate the performances of the previously published saliency-detection models. In other words, since the

Group presented its model as *commensurable* with former models, the Group should have—according to the reviewers—more thoroughly compared its performances. But why did the Group stop halfway through its evaluation efforts and compare its model only with respect to the new ground truth?

Discussion with BJ on the terrace of the CSF's cafeteria, June 19, 2014

FJ: The committee didn't like that we created our own ground truth? [22]

BJ: No. I mean, it's just that we tested on this one but we did not test on the other ones.

FJ: They wanted you to test on already existing ground truths?

BJ: Yes.

FJ: But why didn't you do that?

BJ: Well, that's the problem: Why did we not test it on the others? We have a reason. Our model is about face segmentation and multiple features. But in the other datasets, most of them do not have more than ten face images. ... In the saliency area, most people do not work on face detection and multiple features. They work on images where there is a car or a bird in the center. You always have a bird or something like this. So it just makes no sense to test our model on these datasets. They just don't cover what our model does. ... That's the thing: if you do classical improvement, you are ensured that you will present something at big conferences. But if you propose new things, then somehow people just misunderstand the concept.

It would not have been technically difficult for the Group to confront its model with the previous ground truths; they were freely available on the web, and such performance evaluations required roughly the same Matlab scripts as those used to produce the results shown in figure 2.9. The main reason the Group did not do such comparisons was that the previous models deriving from the previous ground truths would certainly have obtained better performance results. Since the Group's model was *not* designed to solve the saliency problem as defined by the previous ground truths, it would certainly have been outperformed by these ground truths' "native" models.

Due to a lack of empirical elements, I will not try to interpret the reasons why the Group felt obliged to frame the line of argument of its paper around issues of quantifiable performances. [23] Yet, in line with the argument of this chapter, I assume that this rejection episode shows again how image-processing algorithms can be bound to their ground truths. An algorithm deriving from a ground truth made of images whose targets are centered, contrastive objects will somehow manage to retrieve these targets. But when tested on a ground truth made of images whose targets are multiple decentered objects and faces, the same algorithm may well produce statistically poor results. Similarly, another algorithm deriving from a ground truth made of images whose targets are multiple decentered objects and faces will somehow manage to retrieve these targets. But when tested on a ground truth made of images whose targets are centered contrastive objects, it may well produce statistically poor results. Both such algorithms operate in different categories; their limits lie in the ground truths used to define their range of actions. As BJ suggested in a dramatic way, to a certain extent, *we get the algorithms of our ground truths.* Algorithms can be presented as statistically more efficient than others when they derive from the same—or very similar—ground truths. As soon as two algorithms derive from two ground truths with different targets, they can only be presented as different. Qualitative evaluations of the different ground truths in terms of methodology, data selection, statistical rigor, or industrial potentials can be conducted, but the two computational models themselves are irreducibly different and not commensurable. From the point of view of this case study—which may differ from the point of view of the reviewers—the Group's fatal mistake might have been to mix up quantitative improvement of performances with qualitative refinement of ground truths.

Interestingly, one year after this rejection episode, the Group submitted another paper, this time to a smaller conference in image processing. The objects of this paper were rigorously the same as those of the paper that was previously rejected: the same ground truth and the same computational model. Yet instead of highlighting the statistical performances of its model, the Group emphasized its ground truth and the fact that it allowed the inclusion of face segmentation within saliency detection. In this second paper that won the "Best Short Paper Award" of the conference, the computational model was presented as one example of the application potential of the new ground truth.

Problem Oriented and/or Axiomatic

This first case study accounted for a small part of a four-month-long project in saliency detection run by a group of young computer scientists in the Lab. Is it possible to draw on the observations of this exploratory case study? Could we use some of the accounted elements to make broader propositions and sketch analytical directions for the present book as well as for other potential future inquiries into the constitution of algorithms? More than just concerning a group of young computer scientists and a small prototype for saliency detection, I think indeed that this case study fleshes out important insights that deserve to be explored more thoroughly. For the remaining part of this chapter then, I will draw on this empirical case to tentatively propose two complementary research directions for the sociological study of algorithms.

I assume that this case study implicitly suggests a new way of seeing algorithms that still accepts their standard definition while expanding it dramatically. Indeed, we may now still consider an algorithm as being, at some point, a set of instructions designed to computationally solve a given problem. Though as explained at the end of chapter 1, I intentionally did not take this standard definition of algorithms as a starting point; at the end of the Group's project, once the numerical features were extracted from the training set and translated into machine-readable language, several Matlab files with thousands of lines of instructions constituted just such a set. From that point of view, the study of these sets of instructions at a theoretical level—as proposed, for example, by Knuth (1997a, 1997b, 1998, 2011); Sedgewick and Wayne (2011); Dasgupta, Papadimitriou, and Vazirani (2006); and many others—is wholly relevant to the problem at hand. How to use mathematics and machine-readable languages in order to propose a solution to a given problem in the most efficient way is indeed a fascinating question and field of study.

At the same time, however, we saw that the problem an algorithm is designed to solve does not preexist: it has to be produced during what one may call a "problematization process"—a succession of collective practices that aim to empirically define the *terms* of a problem to be solved. In our case study, the Group first drew on recent claims published in authoritative journals of cognitive biology to reframe the saliency problem as being face-related and continuous. As we saw, this first step of the Group's problematization process implied mundane and problematic practices such as the critique of previous research results (what did our opponents miss?) and the inclusion of some of the Lab's recent projects (how to pursue our recent developments?). The second step of the Group's problematization process implied the constitution of a ground truth that could operationalize the reframed problem of saliency. This second step also implied mundane and problematic practices such as the collection of a dataset on Flickr (what images do we choose?), the organization of a database (how do we organize our data?), the design of a crowdsourcing task (what question do we ask to the workers?), and the processing of the results (how do we get contours of features from rectangles?). Only at the very end of this process—once the laboriously constructed targets have been associated to the laboriously constructed dataset in order to form the final ground-truth database—was the Group able to formulate, program, and evaluate the set of Matlab instructions capable of transforming inputs into outputs by means of numerical computing techniques. In short, to design a computerized method of calculation that could solve the new saliency problem, the Group first had to define the boundaries of this new problem.

From these empirical elements, two complementary perspectives on the Group's algorithm seem to emerge. A first perspective might consider the Group's algorithm as a set of instructions designed to computationally solve a new problem in the best possible way. This first traditional view on the Group's algorithm would, in turn, put the emphasis on the mathematical choices, formulating practices, and programming procedures the Group used to transform the input-data of the new ground truth into their corresponding output-targets. How did the Group manipulate its training set to extract relevant numerical features for such a task? How did the Group translate mathematical operations into lines of code? And did it lead to *the most efficient result*? In short, this take on the Group's algorithm would analyze it in the light of its computational properties. Yet symmetrically, a second view on the Group's algorithm might consider it as a set of instructions designed to computationally retrieve, in the best possible way, output-targets that were designed during a specific problematization process. This second take on the Group's algorithm would, in turn, put the emphasis on the specific situations and practices that led to the definition of the terms of the problem the algorithm was designed to solve. How was the problem defined? How was the dataset collected? How was the crowdsourcing task conducted? In short, this second perspective—which this chapter endorsed—would analyze the Group's algorithm vis-à-vis the construction process of the ground truth it originally derived from (and by which it was biased).

If we tentatively expand the above propositions, we end up with two ways of considering algorithms that both pivot about these material objects called ground truths. What we may call an *axiomatic* perspective on algorithms would consider algorithms as sets of instructions designed to computationally *solve* in the best possible way a problem defined by a given ground truth. A second, and complementary, *problem-oriented* perspective on algorithms would consider algorithms as sets of instructions designed to computationally *retrieve* what has been defined as output-targets during specific problematization processes.

While I do think that both axiomatic and problem-oriented perspectives on algorithms are complementary and should thus be intimately articulated—specific numerical features being suggested by ground truths (and vice versa)—I also believe that they lead to different analytical efforts. By considering the terms of the problem at hand as given, the axiomatic way of considering algorithms facilitates the study of the actual mathematical and programming procedures that effectively end up transforming input sets of values into output sets of values in the best possible ways. This may sound like an obvious statement, but defining a calculating method requires minimal agreement on the initial terms and prospected results of the method (Ritter 1995). It is by assuming that the transformation of the input-data into the output-targets is desirable, relevant, and attestable that a step-by-step schema describing this transformation might be proposed. In the case of computer science, different areas of mathematics with many different certified rules and theorems can be explored, adapted, and enrolled to automate at best the passage from selected input-data to specified output-targets; linear algebra in the case of image processing (Klein 2013), probability theory in the case of data compression (Pu 2005), graph theory in the case of data structure (Tarjan 1983), number theory in the case of cryptography (Koblitz 2012), or statistics (and probabilities) in the case of the ever-popular machine-learning procedures supposedly adaptable to all fields of activity (Alpaydin 2016). As we will see in chapters 5 and 6, the exploration and teaching of these different certified mathematical bodies of knowledge must therefore be respected for what they are: powerful operators allowing the reliable transformative computation of ground-truth's input-data into their corresponding output-targets.

If the problem-oriented perspective on algorithms may not directly focus on the formation and computational effectiveness of algorithms, it may contribute to better documenting the processes that configure the terms of the problems these algorithms try to solve. Considering algorithms as retrieving entities may put the emphasis on the referential databases that define what algorithms try to retrieve and reproduce; the biases they build on in order to express their veracity. What ground truth defined the terms of the problem *this* algorithm tries to solve? How was *this* ground-truth database constituted? And when? And by whom? By pointing at moments and locations where outputs to be retrieved were, or are, being constituted within ground-truth databases, this analytical look at algorithms—that Bechmann and Bowker (2019) and Grosman and Reigeluth (2019) contributed to igniting—may suggest new ways of interacting with algorithms and those who design them. This avenue of research, which is still in its infancy, could moreover link its results to those of the more explicitly critical positions I mentioned in the introduction. If the investigations by Noble (2018) on the racist stereotypes promoted by the search engine Google or by O'Neil (2016) on how proxies used by proprietary scoring algorithms tend to punish the poorest have effectively acted as warning signs, practical ways to change the current situation still need to be elaborated. This is where the notion of *composition*, the keystone of this inquiry, comes again into play: at the time of (legitimate) indignation, the time of constructive confrontation must follow, which itself implies being able to present oneself *realistically*. As long as the practical work subtending the constitution of algorithms remains abstract and indefinite, modifying the ecology of this work will remain extremely difficult. Changing the biases that root algorithms in order to make them promote different values may, in that sense, be achieved by making the work practices that underlie algorithms' veracities more visible. If more studies could inquire into the *ground-truthing practices* algorithms derive from, then actual composition potentials may slowly be suggested.

*   *   *

Part I is now coming to an end. Let me then quickly recap the elements presented so far. In chapter 1, I presented the main setting of this inquiry: an academic laboratory I decided to call the "Lab" whose members spend a fair amount of time and energy assembling and publishing new image-processing algorithms, thus participating—at their own level—in the heterogeneous network of computer science industry. I also considered methodological issues and critically discussed the notion of algorithm as it is generally presented in the specialized literature.

In chapter 2, we dived into the daily work of the Lab and followed a group of young computer scientists trying to design a new algorithm for an important conference in image processing. Our initial encounter with the Group at the Lab's cafeteria was at first confusing, but after a quick detour via the image-processing literature on saliency detection, we were able to understand why the Group's project implied the shaping of a new referential database that could define the terms of the problem its desired algorithm should later try to solve. As we were accounting for these mundane yet crucial *ground-truthing practices*, we realized something very banal for practitioners of computer science but surprising to many others: it turns out that, to a certain extent, *we get the algorithms of our ground truths*. As the construction of image-processing algorithms implies the formation of *training sets* for formulating the relationships between input-images and output-targets as well as the formation of *evaluation sets* for measuring and comparing the performances of these formulated relationships, image-processing algorithms—and potentially many others—must rely, in one way or another, on manually constructed ground truths that precisely provide both sets. This half-discovery further suggested a research agenda that two complementary analytical perspectives on algorithms could irrigate. First, and in the wake of this chapter 2, a "problem-oriented perspective" could explore the collective processes leading to the formation and circulation of ground truths. This unconventional glance on algorithms may contribute to equipping broader topics related to data justice and algorithmic fairness. Yet to avoid reducing algorithms to the ground truths from which they derive, such studies of algorithms should be intimately articulated with an "axiomatic perspective" on algorithms that could further explore the formulation and evaluation of computational models from already constituted ground truths.

II Programming

It is sometimes difficult to say things that are quite simple.
—Hutchins (1995, 356)

If part I led, I hope, to interesting insights, it was nonetheless mundane-biased. Although I kept on insisting on the ordinary aspect of ground-truthing—criticizing previous papers, selecting data, defining targets, and so on—I remained very vague about less common practices that those who are not computer scientists generally expect to see in computer science laboratories. For example, where is the mathematics? If the Group managed to define relationships between input-data and output-targets, it certainly formulated them with the help of mathematical knowledge and inscriptions. And where are the cryptic lines of computer code? If the Group managed to first design a web application and later test its computational model on the evaluation set, it must have successfully written machine-readable lists of instructions. If I really want to propose a partial yet realistic constitution of algorithms, do I not need to account for these a priori exotic activities as well? The practices leading to the definition of mathematical models of computation will be the topic of part III. For now, I need to consider *computer programming*, this crucial activity that never stops being part of computer scientists' daily work.

Let us warm up with some basic assertions. Is it not a platitude to say that computer programming is a central activity? Every digital device that takes part in our courses of action required indeed the expert hands of "programmers" or "developers" who translated desires, plans, and intuitions into machine-readable lists of instructions. Banks, scientific laboratories, high-tech companies, museums, spare part manufacturers, novelists, eth-nographers: all indirectly rely on people capable of interacting with com-puters to assemble files whose content can be executed by processors at electronic speed. If by a mysterious black-magic blow all programmers who make computers compute in desired ways were removed from the collec-tive world, the remaining people would very soon end up yapping around powerless relics like, as Malraux says, *crowds of monkeys in Angkor temples*. The current importance of fast and reliable automated processing for most sectors of activity positions computer programming as an obligatory pas-sage point that cannot be underestimated.

Yet if the courses of action of computer programming are terribly impor-tant—without them, there would be no digital tools—their study does not always appear relevant. Most of the individuals of the collective world rightly have other things to do than spending time studying what animates the digital devices with which they interact. Moreover, those who study these individuals—for example, sociologists and social scientists—can also take programming practices for granted as political, social, or economic processes often appear *after* innumerable programming ventures have been successfully conducted. For many interesting activities and research topics, then, it makes perfectly sense *not* to look at how computer programs are empirically assembled.

In other situations, though, the activity of computer programming is more difficult to ignore. Computer scientists and engineers cannot, for example, take this activity for granted as it would imply ignoring an impor-tant and often problematic aspect of their work.[1] Unfortunately, as we shall see later, the methods they use to better understand their own practices tend to privilege the evaluation of the results of computer programming tasks rather than the practices involved in the production of these results. Programmers' insights resulting from the analysis of programming tasks thus remain distant from the actions of programming, for which they often remain unaccountable.

But programming practices are also difficult to ignore for *cognitive scien-tists* who work in artificial intelligence departments: as human cognition is—according to many of them—a matter of computing, understanding how computers become able to compute via the design of programs seems indeed to be a fruitful topic. But just like computer scientists and engineers, cognitive scientists have difficulties with properly accessing and inquiring into computer programming courses of action. For entangled reasons which I will cover in the following chapter, when cognitivists inquire into what makes programs exist, they cannot go beyond the form "program" that precisely needs to be accounted for. In a surprisingly vicious circle that has to do with the so-called computational metaphor of the mind, cognitivists end up proposing numerous (mental) programs to explain the development of (computer) programs.

Programming practices therefore appear quite tricky: terribly important but at the same time very difficult to effectively study. What makes these courses of action so elusive? Is it even possible to account for them? And if it is, what are their associative properties? And what do these properties suggest? The goal of this part II is to tackle some of these questions. The journey will be long, never straightforward, and sometimes, not developed enough. But let the reader forgive me: as you will hopefully realize, a full historical and sociological understanding of computer programming is a life project of its own. So many things have been said without much being shown! The reasons for dizziness are legitimate, the chances of success infinitesimal; yet, if we really care about these entities we tend to call algorithms, an exploratory attempt to better understand the practices required to make them effectively participate in our courses of action might not be, I hope, completely senseless.

Part II is organized as follows. In chapter 3, I start by retracing how the activity of programming was progressively made invisible before proposing conceptual means to help restore its practicality. I first focus on an important document written by John von Neumann in 1945 that presented computers as input-output devices capable of operating without the help of humans. This initial setting aside of programming practices from electronic computing systems further seemed to depict them as self-sufficient "electronic brains." In the second section of the chapter I present academic attempts to make sense of the incapacity of "electronic brains" to operate meaningfully. As we shall see, for intricate reasons related to the computational metaphor of the mind, I assume that researchers conducting these studies did not manage to properly approach computer programming practices, thus further contributing to their invisibilization. In the last section of the chapter where I progressively try to detach myself from almost everything that has been said about the practice of computer programming, I draw on contemporary work in the philosophy of perception to propose a definition of cognition as *enacted*. This enactive conception of cognition will further help us fully consider *actions* instead of *minds*. In chapter 4, I build on this unconventional conception of cognition as well as several other concepts taken from *Science and Technology Studies* to closely analyze a programming episode collected within the Lab. The study of these empirical materials makes me tentatively partition programming episodes into three intimately related sets of practices: *scientific* with the alignment of inscriptions, *technical* with the work-arounds of impasses, and *affective* with the shaping of scenarios. The need for constant shifting among these three modes of practices might be a reason why computer programming is a difficult yet fascinating experience. The last section of chapter 4 will be a brief summary.

3   Von Neumann's Draft, Electronic Brains, and Cognition

Many things have been written regarding computer programming—often, I believe, in problematic ways. To avoid getting lost in this abundant literature, it is important to start this chapter with an operational definition of computer programming on which I could work and eventually refine later. I shall then temporally define computer programming as the situated activity of inscribing numbered lists of instructions that can be executed by computer processors to organize the movement of bits and to modify given data in desired ways. This operational definition of computer programming puts aside other practices one may sometimes describe as "programming," such as "programming one's wedding" or "programming the clock of one's microwave."

If I place emphasis on the practical and situated aspect of computer programming in my operational definition, it is because important historical events have progressively set it aside. In this first section that draws on historical works on early electronic computing projects, we will see that once computer systems started to be presented as input-output instruments controlled by a central unit—following the successful dissemination of the so-called von Neumann architecture—the entangled sociotechnical relationships required to make these objects operate in meaningful ways had begun to be placed in the background. If electronic computing systems were, in practice, intricate and highly problematic sociotechnical processes, von Neumann's modelization made them appear as functional devices transforming inputs into outputs. The noninclusion of practices—hence their *invisibilization*—in the accounts of electronic computers further led to serious issues that suggested the first academic studies of computer programming in the 1950s.

A Report and Its Consequences

One cornerstone of what will progressively be called "von Neumann architecture" is the *First Draft of a Report on the EDVAC* that John von Neumann wrote in a hurry in 1945 to summarize the advancement of an audacious electronic computing system initiated during World War II at the Moore School of Electrical Engineering at the University of Pennsylvania. As I believe this report has had an important influence on the setting aside of the practical instantiations of computer systems, we first need to look at the history and dissemination of this document as well as the world it participated in enacting.

World War II: An Increasing Need for the Resolution of Differential Equations

An arbitrary point of departure could be President Franklin D. Roosevelt's radio broadcast on December 29, 1940, that publicly presented the United States as the main military supplier to the Allied war effort, therefore implying a significant increase in US military production spending.[1] Under the jurisdiction of the Army Ordnance Department (AOD), the design and industrial production of long-distance weapons were obvious topics for this war-oriented endeavor. Yet for every newly developed long-distance weapon, a complete and reliable *firing table* listing the appropriate elevations and azimuths for the reaching of any distant targets had to be calculated, printed, and distributed. Indeed, to have a chance to effectively reach targets with a minimum of rounds, every long-distance weapon had to be equipped with a booklet containing data for several thousand kinds of curved trajectories.[2] More battles, more weapons, and more distant shots: along with the mass production of weapons and the enrollment of soldiers capable of handling them, the US's entry into another world war in 1942 further implied an increasing need for the resolution of differential equations.

These practical mathematical operations—which can take the form of long iterative equations that require only addition, subtraction, multiplication, and division—were mainly conducted in the premises of the Ballistic Research Laboratory (BRL) at Aberdeen, Maryland, and at the Moore School of Electrical Engineering in Philadelphia. Hundreds of "human computers" (Grier 2005), mainly women (Light 1999), along with mechanical desk calculators and two costly refined versions of Vannevar Buch's differential analyzer (Owens 1986)—an analogue machine that could compute mathematical equations[3]—worked intensely to print out ballistic missile firing tables. Assembling all of the assignable factors that affect the trajectories of a projectile shot from the barrel of a gun (gravity; the elevations of the gun; the shell's weight, diameter, and shape; the densities and temperatures of the air; the wind velocities, etc.)[4] and aligning them to define and solve messy differential equations[5] was a tedious process that involved intense training and military chains of command (Polachek 1997). But even this unprecedented ballistic calculating endeavor could not satisfy the computing needs of this wartime. Too much time was required to produce a complete table, and the backlog of work rapidly grew as the war intensified. As Campbell-Kelly et al. (2013, 68) put it:

> The lack of an effective calculating technology was thus a major bottleneck to the effective deployment of the multitude of newly developed weapons.

In 1942, drawing on the differential analyzer and on the pioneering work of John Vincent Atanasoff and Clifford Berry on electronic computing (Akera 2008, 82–102; Burks and Burks 1989) as well as on his own research on delay-line storage systems,[6] John Mauchly—an assistant professor at the Moore School—submitted a memorandum to the AOD that presented the construction of an electronic computer as a potential resource for faster and more reliable computation of ballistic equations (Mauchly [1942] 1982).[7] The memorandum first went unnoticed. But one year later, thanks to the lobbying of Herman Goldstine—a mathematician and influential member of the BRL—a meeting regarding the potential funding of an eighteen-thousand-vacuum-tube electronic computer was organized with the BRL's director. And despite the skepticism of influent members of the National Defense Research Committee (NDRC),[8] a $400,000 research contract was signed on April 9, 1943.[9] At this point, the construction of a computing system that could potentially solve large iterative equations at electronic speed and therefore accelerate the printing out of the firing tables required for long-distance weapons could begin. This project, initially called "Project PX," took the name of ENIAC for *Electronic Numerical Integrator and Computer*.

The need to quickly demonstrate technical feasibility forced Mauchly and John Presper Eckert—the chief engineer of the project—to make irreversible design decisions that soon appeared problematic (Campbell-Kelly et al. 2013, 65–87). The biggest shortcoming was related to the new computing capabilities of the system: If delay-line storage could potentially make the system add, subtract, multiply, and divide electric translations of numbers at electronic speed, such storage prevented the system from being instructed via punched cards or paper tape. This common way of both temporally storing data and describing the logico-arithmetic operations that would compute them was well adapted for electromechanical devices, such as the Harvard Mark I that proceeded at three operations per second.[10] But an electronic machine such as the ENIAC that was supposed to perform five thousand operations per second could not possibly handle this kind of paper material. The solution that Eckert and Mauchly proposed was then to set up both data and instructions manually on the device by means of wires, mechanical switches, and dials. This choice led to two related impasses. First, it constrained the writable electronic storage of the device; more storage would have indeed required even bigger machinery, entangled wires, and unreliable vacuum tubes. Second, the work required to set up all the circuitry and controllers and start an iterative ballistic equation was extremely tedious; once the data and the instructions were laboriously defined and checked, the whole operating team needed to be briefed and synchronized to set up the messy circuitry (Campbell-Kelly et al. 2013, 73). Moreover, the passage from diagrams provided by the top engineers to the actual setup of the system by lower-ranked employees was by no means a smooth process—the diagrams were tedious to produce, hard to read, and error-prone, and the number of switches, wires, and resistors was quite confusing.[11]

Two important events made an alternative appear. The first is Eckert's work on mercury delay-line storage, which built upon his previous work on radar technology. By 1944, he became convinced that these items could be adapted to provide more compact, faster, and cheaper computing storage (Haigh, Priestley, and Rope 2016, 130–132). The second event is one of the most popular anecdotes of the history of computing: the visit of John von Neumann at the BRL in the summer of 1944. Contrary to Eckert, Mauchly, and even Goldstine, von Neumann was already an important scientific figure in 1944. Since the 1930s, he was at the forefront of mathematical logic, the branch of mathematics that focuses on formal systems and their abilities to evaluate the consistencies of statements. He was well aware of the works on computability by Alonzo Church and Alan Turing, with whom he collaborated at Princeton.[12] As such, he was one of the few mathematicians who had a formal understanding of computation. Moreover, by 1944, he had already established the foundations of quantum mechanics as well as game theory. Compared with him and despite their breathtaking insights on electronic computing, Eckert and Mauchly were still provincial engineers. Von Neumann was part of another category: he was a scientific superstar of physics, logics, and mathematics, and he worked as a consultant on many classified scientific projects, with the more notable one certainly being the Manhattan Project.

Von Neumann's visit was part of a routine consulting trip to the BRL and therefore was not specifically related to the ENIAC project. In fact, as many members of the NDRC expressed defiance toward the ENIAC, von Neumann was not even aware of its existence. But when Goldstine mentioned the ENIAC project, von Neumann quickly showed interest:

> It is the summer of 1944. Herman Goldstine, standing on the platform of the railroad station at Aberdeen, recognizes John von Neumann. Goldstine approaches the great man and soon mentions the computer project that is underway in Philadelphia. Von Neumann, who is at this point deeply immersed in the Manhattan Project and is only too well aware of the urgent need of many wartime projects of rapid computations, makes a quick transition from polite chat to intense interest. Goldstine soon brings his new friend to see the project. (Haigh, Priestley, and Rope 2016, 132)

By the summer of 1944, it was accepted among Manhattan Project's scientific managers that a uniform contraction of two plutonium hemispheres could make the material volume reach critical mass and create, in turn, a nuclear explosion. Yet if von Neumann and his colleagues knew that the mathematics of this implosion would involve huge systems of partial differential equations, they were still struggling to find a way of defining them. And for several months, von Neumann had been seriously considering electronic computing for this specific prospect (Aspray 1990, 28–34; Goldstine [1972] 1980, 170–182).

After his first visit to the ENIAC, von Neumann quickly realized that even though the ENIAC was by far the most promising computing system he had seen so far, its limited storage capacity could by no means help define and solve the very complex partial differential equations related to the Manhattan Project.[13] Convinced that a new machine could overcome this impasse—notably by using Eckert's insights about mercury delay-line storage—von Neumann helped design a new proposal for the construction of a post-ENIAC system. He moreover attended a crucial BRL board meeting where the new project was evaluated. His presence definitely helped with attaining the final approval of the project and its new funding of $105,000 by August 1944. The new hypothetical machine—whose design and construction would fall under the management of Eckert and Mauchly—was initially called "Project PY" before being renamed EDVAC for *Electronic Discrete Variable Automatic Computer.*

Different Layers of Involvement

The period between September 1944 and June 1945 is crucial for my adventurous story of the setting aside of computer programming practices. It was indeed during this short period of time that von Neumann proposed considering computer programs as input lists of instructions, hence surreptitiously invisibilizing the practices required to shape these lists. As this formal conception of electronic computing systems was not unanimously shared among the participants of both ENIAC and EDVAC projects, it is important at this point to understand the different layers of involvements in these two projects that were intimately overlapping. One could schematically divide them into three layers: the engineering staff, the operating team, and von Neumann himself.

The first layer of involvement included the engineering staff—headed by Mauchly, Eckert, Goldstine, and Arthur W. Burks—that was responsible for the logical, electronic, and electromechanical architectures and implementations of both the ENIAC and the EDVAC. The split of the ENIAC into different units, the functioning of its accumulators—crucial parts for making the system compute electric pulses—and the development and testing of mercury delay-line storage for the future EDVAC were part of the prerogatives of the engineering staff. It is difficult to see now the blurriness of this endeavor that was swimming in the unprecedented. But besides the systems' abilities to compute more or less complex differential equations, one crucial element the engineering staff had to conceive and make happen was a way to instruct these messy systems. In parallel to the enormous scientific and engineering problems of the different parts of the systems, the shaping of readable documents that could describe the operations required to make these systems *do* something was a real challenge: How, in the end, could an equation be put into an incredibly messy electronic system? In the case of the ENIAC, the engineering staff—in fact, mostly Burks (Haigh, Priestley, and Rope 2016, 35–83)—progressively designed a workflow that could be summarized as such: assuming ballistic data and assignable factors had been adequately gathered and translated into a differential equation—which was already a problematic endeavor—the ENIAC's engineering staff would first have to transform this equation into a logical diagram; then into an electronic diagram that took into account the different unit as blocks; and then into another, bigger, diagram that took into account the inner constituents of each block. The end result of this tedious process—the final "panel diagram" drawn on large sheets of paper (Haigh, Priestley, and Rope 2016, 42)—was an incredible, yet necessary, mess.

This leads us to another layer that included the so-called operators—mainly women computers—who tried to make sense, correct, and eventually implement these diagrams into workable arrangements of switches, wires, and dials. Contrary to what the top engineers had initially thought, translating large panel diagrams into a workable configuration of switches and wires was not a trivial task. Errors in both the diagrams and the configurations of switches were frequent—without mentioning the fragility of the resistors—and this empirical "programming" process implied constant exchanges between high-level design in the office and low-level implementations in the hangar (Light 1999, 472; Haigh, Priestley, and Rope 2016, 74–83). Both engineers and operators were engaged in a laborious process to have ENIAC and, to a lesser extent, EDVAC produce meaningful results, and these computing systems were considered heterogeneous processes that indistinctly mixed problematic technical components, interpersonal relationships, mathematical modeling, and transformative practices.

Next to these two layers of involvement was von Neumann who certainly constituted a layer on his own. First, contrary to Mauchly, Eckert, Burks, and even Goldstine, he was well aware of recent works in mathematical logic and, in that sense, was prone to formalizing models of computation. Second, von Neumann was very interested in mathematical neurology and was well aware of the analogy between logical calculus and the brain as proposed by McCulloch and Pitts in 1943 (more on this later). This further made him consider computing systems as *electronic brains* that could more or less intelligently transform inputs into outputs (Haigh, Priestley, and Rope 2016, 141–142; von Neumann 2012). Third, if he was truly involved in the early design of the EDVAC, his point of view was that of a consultant, constantly on the move from one laboratory to another. He attended meetings—the famous "Meetings with von Neumann" (Stern 1981, 74)—and read reports and letters from the top managers of the ENIAC and EDVAC but was not part of the mundane tedious practices at the Moore School (Stern 1981, 70–80; Haigh, Priestley, and Rope 2016, 132–140). He was thus parallel to, but not wholly a part of, the everyday practices in the hangars of the Moore School. Finally, being deemed one of the greatest scientific figures of the time—which he certainly was—his visits were real trials that required preparation and cleaning efforts. If he visited the hangars of the Moore School several times, he mainly saw the *results* of messy setup processes, not the processes themselves. A lot was indeed at stake: at that time, the electronic computing projects of the Moore School were not considered serious endeavors among many important applied mathematicians at MIT, Harvard, or Bell Labs—notably Vannevar Buch, Howard Aiken, and George Stibitz (Stern 1981). Taking care of von Neumann's support was crucial as he gave legitimacy to the EDVAC project and even to the whole school.

All of these elements certainly contributed to shaping von Neumann's particular view on the EDVAC. In the spring of 1945, while the engineering and operating layers had to consider this post-ENIAC computing system as a set of problematic relations encompassing the definition of equations, the adequate design of fragile electromechanical units, and back-and-forth movements between hangars and offices, von Neumann could consider it as a more or less functional object whose inner relationships could be modeled.

Despite many feuds over the paternity of what has later been fallaciously called "the notion of stored program,"[14] it is clear now for historians of technology that the intricate relationships among these three layers of involvement in the EDVAC project collectively led to the design decision of storing both data and instructions as pulses in mercury delay lines (Campbell-Kelly et al. 2013, 72–87; Haigh, Priestley, and Rope 2016, 129–152). After several board meetings between September 1944 and March 1945, the top engineers and von Neumann agreed that, if organized correctly, the new storage capabilities of mercury delay lines could be used to temporally conserve not only numerical data but also the description of in-built arithmetical and logical operations that will later compute them. This initial characteristic of the future EDVAC further suggested, to varying degrees, the possibility of paper or magnetic-tape *documents* whose contents could be loaded, read, and processed at electronic speed by the device, without the intervention of a human being.

For the engineers and operators deeply involved in the ENIAC-EDVAC projects, the notion of lists of instructions that could automatically instruct the system was rather disconnected from their daily experiences of unreadable panel diagrams, electronic circuitry, and messy setup processes of switches and wires. To them, the differentiation between the computing system and its instructions hardly made sense: in practice, an electronic computing system was part of a broader sociotechnical process encompassing the definition of equations, the writing of diagrams, the adequate design of fragile electromechanical units, back-and-forth movements between hangars and offices, etc. To paraphrase Michel Callon (1999) when he talked about Air France, for these two layers of involvement, it was not an electronic calculator that could eventually compute an equation but a whole arrangement of engineers, operators, and artifacts in constant relationship.

The vision von Neumann had for both the ENIAC and EDVAC projects was very different: as he was constantly on the move, attending meetings and reading reports, he had a rather disembodied view of these systems. This process of disembodiment that often affects top managers was well described by Katherine Hayles (1999) when she compared the points of view of Warren McCulloch—the famous neurologist—and Miss Freed—his secretary—on the notion of "information":

> Thinking of her [Miss Freed], I am reminded of Dorothy Smith's suggestion that men of a certain class are prone to decontextualization and reification because they are in a position to command the labors of others. "Take a letter, Miss Freed," the man says. Miss Freed comes in. She gets a lovely smile. The man speaks, and she writes on her stenography pad (or perhaps on her stenography typewriter). The man leaves. He has a plane to catch, a meeting to attend. When he returns, the letter is on his desk, awaiting his signature. From his point of view, what has happened? He speaks, giving commands or dictating words, and things happen. A woman comes in, marks are inscribed onto paper, letters appear, conferences are arranged, books are published. Taken out of context, his words fly, by themselves, into books. The full burden of the labor that makes these things happen is for him only an abstraction, a resource diverted from other possible uses, because he is not the one performing the labor. (Hayles 1999, 82–83)

Hayles's powerful proposition is extendable to the case that interests us here: contrary to Eckert, Mauchly, Burks, and the operating team, von Neumann was not the one performing the labor. Whereas the engineering and operating teams were entangled in the headache of making the ENIAC and EDVAC do meaningful things, von Neumann was entangled in the different headache of providing relevant insights—notably in terms of formalization—to military projects located all around the United States. To a certain extent, this position, alongside his interest in contemporary neurology and his exceptional logical and mathematical insights, certainly helped von Neumann write a document about the implications of storing both data and instructions as pulses in mercury delay lines. Provided as a summary of the discussions among the EDVAC team between the summer of 1944 and the spring of 1945, he wrote the *First Draft of a Report on the EDVAC* ([1945] 1993) that, for the first time, modeled the logical architecture of a hypothetical machine that would store both the data and the instructions required to compute them. Unaware of, and not concerned with, its laborious instantiations within the Moore School, von Neumann presented the EDVAC as a system of interacting "organs" whose relationships could by themselves transform inputs into outputs. And despite the skepticism of Eckert and Mauchly about presenting their project with floating terms, such as "neurons," "memory," "inputs," and "outputs"—and eventually their fierce resentment to see that their names were never mentioned in the document[15]—thirty-one copies of the report were printed and distributed among the US computing-related war projects in June 1945.

Proofs of Concept and the Circulation of the Input-Output Model

The many lawsuits and patent-related issues around the *First Draft* are not important for my story. What matters at this point is the surreptitious shift that occurred and persistently stayed within the computing community: Whereas computing systems were, in practice, sociotechnical *processes* that could ultimately—perhaps—produce meaningful results, the formalism of the *First Draft* surreptitiously presented them as brain-like *objects* that could automatically transform inputs into outputs. And if these high-level insights were surely important to sum up the confidential work that had been undertaken at the Moore School during the war and share it with other laboratories, they also contributed to separating computing systems from the practices required to make them operate. The *First Draft* presented the architecture of a functioning computing machine and thus put aside the actions required to make this machine function. The translation operations from equations to logical diagrams, the specific configurations of electric circuitry and logic gates, the corrections of the diagrams from inaccurate electronic circulation of pulses; all of these sociotechnical operations were taken for granted in the *First Draft* to formalize the EDVAC at the logical level. Layers of involvement were relative layers of silence (Star and Strauss 1999); by expressing the point of view of the consultant who built on the results of intricate endeavors, the "list of the orders" (the programs) and the "device" (the computer) started to be considered two different entities instead of one entangled process.

But were the instructions really absent from the computing system as presented in the *First Draft*? Yes and no. The story is more intricate than that. In fact, the *First Draft* defined for the first time a quite complete set of instructions that, according to the formal definition of the system, could make the hypothetical machine compute every problem expressible in its formalism (von Neumann [1943] 1993, 39–43). But similarly to Turing's seminal paper on computable numbers (Turing 1937), von Neumann's set of instructions was integrally part of his formal system: the system constituted the set of all sets of instructions it could potentially compute. The benefits of this formalization were huge as it allowed the existence of all the infinite combinations of instructions. Yet, the surreptitious drawback was to consider these combinations as nonproblematic realizations of potentialities instead of costly actualizations of collective heterogeneous processes. While making a universal machine do something in particular was, and is, very different from formalizing such a universal machine, both practices were progressively considered equivalent.[16]

The diffusion of von Neumann's architecture as presented in the *First Draft* was not immediate. At the end of the war, several computing systems coexisted in an environment of mutual ignorance—most projects were classified during the war—and persistent suspicion—the Nazi threat was soon replaced with the communist (or capitalist) threat. During the conferences and workshops of the *Moore School Series* that took place in summer 1946, the logical design of the EDVAC was, for example, very little discussed as it was still classified. Nonetheless, several copies of the *First Draft* progressively started to circulate outside of the US defense services and laboratories, notably in Britain, where a small postwar research community could build on massive, yet extremely secret, code-breaking computing projects (Abbate 2012, 34–35; Campbell-Kelly et al. 2013, 83–84).

Contrary to Cold War–oriented American research projects, postwar British projects had no important funding as most of the UK government's money was being invested in the reconstruction of the devastated infrastructures. This forced British scientific managers to design rather small prototypes that could quickly show promising results. In June 1948, inspired by von Neumann's architecture as presented in the *First Draft*, Max Newman and Frederic Williams from the University of Manchester provided a first minimal proof of concept that the cathode-ray tube storage system could indeed be used to store instructions and data for computation at electronic speed in a desired, yet fastidious, way. One year later, Maurice Wilkes from the University of Cambridge—who also obtained a version of the *First Draft* and participated in the *Moore School Series* in 1946—successfully led the construction of an electronic digital computer with a mercury delay-line storage that he called the EDSAC for *Electronic Delay Storage Automatic Calculator*. Largely due to the programming efforts of Wilkes's PhD student David Wheeler (Richards 2005), the EDSAC could load data and instructions punched on a ribbon of paper and print the squares of the first one hundred positive integers. These two successful experiences participated in rendering electromechanical relays and differential analyzers obsolete in the emerging field of computer science research. But more importantly for the present story, these two successful experiments also participated in the diffusion of von Neumann's functional definition of electronic computing systems as input-output devices controlled by a central organ. As it ended up working, the model, and its encapsulated metaphors, were considered accurate.

At the beginning of 1950s, when IBM started to redefine computers as data-processing systems for businesses and administrations, von Neumann's definition of computing system further expanded. As cited in Haigh, Priestley, and Rope (2016, 240), an IBM paper written by Walker Thomas asserts, for example, that "all stored-program digital computers have four basic elements: the memory or storage element, the arithmetic element, the control element, and the terminal equipment or input-output element" (Thomas 1953, 1245). More generally, the broader inclusion of computing systems within *commercial arrangements* (Callon 2017) participated in the dissemination of their functional definition. It seems indeed that, to create new markets, intricate and very costly computing systems had better be presented as devices that automatically transform inputs into outputs rather than artefacts requiring a whole infrastructure to operate adequately. The noninclusion of the sociotechnical interactions and practices required to make computers compute seems, then, to have participated in their expansions in commercial, scientific, and military spheres (Campbell-Kelly et al. 2013, 97–117). But the putting aside of programming practices from the definition of computers further led to numerous issues related to the ad hoc labor required to make them function.

The Psychology of Programming (And Its Limits)

The problem with practice is that it is necessary to do things: essence is existence and existence is action (Deleuze 1995). And as soon as electronic computing systems started to be presented as input-output functional devices controlled by a central organ, the efforts required to make them function in desired ways quickly stood out: it was extremely tedious to make the devices do meaningful things. These intelligent electronic brains were, in practice, dull as dishwater. But rather than casting doubts on the input-output framework of the *First Draft* and considering it formally brilliant but empirically inaccurate, the blame was soon casted on the individuals responsible for the design of computer's inputs. In short, if one could not make electronic brains operate, it was because one did not manage to give them the inputs they deserved. What was soon called the "psychology of programming" tried, and tries, to understand *why* individuals interact so laboriously with electronic computers.

This emphasis on the individual first led to *aptitude tests* in the 1950s that aimed at selecting the appropriate candidates for programming jobs in a time of workforce scarcity. By the late 1970s, entangled dynamics that made Western software industry shift from scientific craft to gender-connoted engineering supported the launching of *behavioral studies* that typically consisted of programming tests whose relative results were attributed to controlled parameters. A decade later, the contested results of these behavioral tests as well as theoretical debates within the discipline of psychology led to *cognitive studies* of programming. Cognitive scientists put aside the notion of parameters as proposed by behaviorists to focus on the mental models that programmers should develop to construct efficient programs. As we shall see, these research endeavors framed programming in ways that prevented them from inquiring into what programmers do, thus perpetuating the invisibilization of their day-to-day work.

Personnel Selection and Aptitude Tests

By the end of the 1940s, simultaneous to the completion of the first electronic computing systems that the von Neumann architecture inspired, the problem of the actual handling of these systems arose: these automatons appeared to be highly heteronomous. This practical issue quickly arose in the universities hosting the first electronic computers. As Maurice Wilkes wrote in his memoirs about the EDSAC:

> By June 1949 people had begun to realize that it was not so easy to get programs right as at one time appeared. I well remember when this realization first came on me with full force. The EDSAC was on the top floor of the building and the tape-punching and editing equipment one floor below on a gallery that ran round the room in which the differential analyzer was installed. I was trying to get working my first non-trivial program, which was one for the numerical integration of Airy's differential equation. It was on one of my journeys between the EDSAC room and the punching equipment that "hesitating at the angles of stairs" the realization came over me with full force that a good part of the remainder of my life was going to be spent in finding errors in my own programs. (Wilkes 1985, 145)

Although the EDSAC theoretically included all possible programs, the actualization of these programs within specific situations was the main practical issue. And this became obvious to Wilke once he was directly involved in trying to make the functional device function.

In the industry, the heteronomous aspect of electronic computing systems also quickly stood up. A first example is the controversies surrounding the UNIVAC—an abbreviation for *Universal Automatic Computer*—an electronic computing system that Eckert and Mauchly developed after they left the Moore School in 1946 to launch their own company (which Remington Rand soon acquired). The potential of the UNIVAC gained a general audience when a whole programming team—which John Mauchly headed—made it run a statistical program that accurately predicted the results of 1952 American presidential election. This marketing move, whose costs were carefully unmentioned, further expanded the image of a functional electronic brain receiving inputs and producing clever outputs. But when General Electric acquired a UNIVAC computer in 1954, it quickly realized the gap between the presentation of the system and its actual enactment: it was simply impossible to make this functional system function. And it was only after two years and the hiring of a whole new programming team that a basic set of accounting applications could start producing some meaningful results (Campbell-Kelly 2003, 25–30). IBM faced similar problems with its computing system 701. The promises of smooth automation quickly faced the down-to-earth reality of practice: the first users of IBM 701—notably Boeing, General Motors, and the National Security Agency (Smith 1983)—had to hire whole teams specifically dedicated to making the system do useful things.[17]

US defense agencies were confronted with the same issue. After the explosion of the first Soviet atomic bomb in August 1949, the United States appeared dangerously vulnerable; the existing air defense system and its slow manual gathering and processing of radar data could by no means detect nuclear bombers early enough to organize counter operations of interceptor aircrafts. This threat—and many other entangled elements that are far beyond the scope of this chapter—led to the development of a prototype computer-based system capable of processing radar data in real time.[18] The promising results of the prototype further suggested in 1954 the realization of a nationwide defense system of high-speed data-processing systems—called *Semi-Automatic Ground Environment* (SAGE).[19] The US Air Force contacted many contractors to industrially develop this system of systems, with IBM being awarded the development of the 250 tons AN/FSQ-7 electronic computers.[20] But none of these renowned institutions—among them IBM, General Electric, Bell Labs, and MIT—accepted the development of the lists of instructions that would make such powerful computers usable. Almost by default, the $20 million contract was awarded to the RAND Corporation, a nonprofit (but nonphilanthropic) governmental organization created in 1948 that operated as a research division for the US Air Force. RAND had already been involved in the previous development of the SAGE project, but its team of twenty-five programmers was obviously far too small for the new programming task. So by 1956, RAND started an important recruiting campaign all around the country to find individuals who could successfully pursue the task of programming.

In this early Cold War period, the challenge for RAND was then to recruit a lot of programming staff in a short period of time. And to equip this massive personnel selection imperative, psychologists from RAND's *System Development Division* started to develop tests whose quantitative results could positively correlate with future programming aptitudes. Largely inspired by the Thurstone Primary Mental Abilities Test,[21] these aptitude tests—although criticized within RAND itself (Rowan 1956)—soon became

the main basis for the selection of new programmers as they allowed crucial time savings while being based on the statistically driven discipline of
psychometrics. The intensive use of aptitude tests helped RAND to rapidly
increase its pool of programmers, so much so that its *System Development
Division* was soon incorporated into a separate organization, the *System
Development Corporation* (SDC). As early as 1959, the SDC had "more than
700 programmers working on SAGE, and more than 1,400 people supporting them. ... This was reckoned to be half of the entire programming man-
power of the United States" (Campbell-Kelly 2003, 39). But besides enabling
RAND/SDC to engage more confidently in the SAGE project, aptitude tests
also had an important effect on the very conception of programming work.
Although the main goal of these tests was to support a quick and nationwide personnel selection, they also contributed to framing programming as
a set of abstract intellectual operations that can be measured using proxies.

The regime of aptitude testing as initiated by the SDC quickly spread
throughout the industry, notably prompting IBM to develop its own ques-
tionnaire in 1959 to support its similarly important recruitment needs. Well
in line with the computer-brain parallel inherited from the seminal period
of electronic computing, the IBM Programming Aptitude Test (PAT) typically asked job candidates to figure out analogies between forms, continue
lists of numbers, and solve arithmetic problems (see figure 3.1). Though
the correlation between candidates' scores to aptitude tests and their future
work performances was a matter of debate, aptitude tests quickly became
mainstream recruiting tools for companies and administrations that purchased electronic computers during the 1960s. As Ensmenger (2012, 64)
noted: "By 1962, an estimated 80 percent of all businesses used some form
of aptitude test when hiring programmers, and half of these used IBM PAT."
The massive distribution and use of these tests among the emerging computing industry further constricted the framing of programming practices
as measurable innate intellectual abilities.

Supposed Crisis and Behavioral Studies

By framing programming as an activity requiring personal intuitive qualities, aptitude tests have somewhat worked against gendered discriminations related to unequal access to university degrees. As Abbate (2012, 52)
noted: "A woman who had never had the chance to earn a college degree—
or who had been steered into a nontechnical major—could walk into a job

---

PART III (Cont'd)

13. During his first three years, a salesman sold 90%, 105%, and 120%, respectively, of his yearly sales quota which remained the same each year. If his sales totaled $252,000 for the three years, how much were his sales below quota during his first year?

   (a) $800   (b) $2,400   (c) $8,000
   (d) $12,000   (e) $16,000

14. In a large office, 2/3 of the staff can neither type nor take shorthand. However, 1/4 of the staff can type and 1/6 can take shorthand. What proportion of people in the office can do both?

   (a) 1/12   (b) 5/36   (c) 1/4
   (d) 5/12   (e) 7/12

15. A company invests $80,000 of its employee pension fund in 4% and 5% bonds and receives $3,360 in interest for the first year. What amount did the company have invested in 5% bonds?

   (a) $12,800   (b) $16,000   (c) $32,000
   (d) $64,000   (e) $67,200

16. A company made a net profit of 15% of sales. Total operating expense were $488,000. What was the total amount of sales?

   (a) $361,250   (b) $440,000   (c) $450,000
   (d) $488,750   (e) $500,000

17. An IBM Sorting Machine processes 1,000 cards per minute. However, 20% is deducted to allow for card handling time by the operator. A given job requires 5,000 cards to be put through the machine 5 times and 9,000 cards to be put through 7 times. How long will it take?

   (a) 1 hr. 10 min.   (b) 1 hr. 28 min.   (c) 1 hr. 45 min.
   (d) 1 hr. 50 min.   (e) 2 hrs. 10 min.

Figure 3.1
Sample of the 1959 IBM Programmer Aptitude Test. In this part of the test, the participant is asked to answer problems in arithmetic reasoning. *Source:* Reproduced by the author from a scanned 1959 IBM Programmer Aptitude Test by J. L. Hughes and W. J. McNamara. Courtesy of IBM.

interview, take a test, and instantly acquire credibility as a future programmer." From its inception, computer programming, unlike the vast majority of skilled technical professions in the United States, has involved women workers, some of whom had already taken part to computing projects during the war.

However, like most Western professional environments in the late 1950s, the nascent computing industry was fueled by pervasive stereotypes, often preventing women programmers from occupying upper managerial positions and encouraging them to do relational customer care work. These gender dynamics should not be overlooked as they help to understand the rapid, and often underappreciated, development of ingenious software equipment. Due to their unique position within the computer-related professional worlds—both expert practitioners and, often, representatives toward clients—women, given their rather small percentage within the industry, actively contributed to innovations aimed at making programming easier for experts and novices alike. The most notorious example is certainly Grace Murray Hopper, head of programming for UNIVAC, who developed the first compiler—a program that translates other programs into machine code[22]—in 1951 before designing the business programming language B-0 (renamed FLOW-MATIC) in 1955. But many other women actively took part to software innovations throughout the 1950s and 1960s, though often in the shadow of more visible male managers. Among these important figures are Adele Mildred Koss and Nora Moser who developed widely used code for data editing in the mid-1950s; Lois Haibt who was responsible for flow analysis of the FORTRAN high-level programming language; and Mary Hawes, Jean Sammet, and Gertrude Tierney who were at the forefront of the common business-oriented language (COBOL) project in the late 1950s (Abbate 2012, 79–81).

From the mid-1960s onward, refinements over compilers and high-level programming languages, which had often come from women, were added to the impressive tenfold increase in computing power (Mody 2017, 47–77). This combination of new promising software and hardware infrastructures prompted large iconic computer manufacturers to start building increasingly complex programs, such as operating systems and massive business applications. The resounding failures of some of these highly visible projects, like the IBM project System 360,[23] soon gave rise to a sense of uncertainty among commentators at the time, some of whom used the evocative expression of "software crisis" (Naur and Randell 1969, 70–73). Historians of computing have expressed doubts about the reality of this software crisis as precise inquiries have shown that, apart from some highly visible and nonstandard projects, software production in the late 1960s was generally on time and on budget (Campbell-Kelly 2003, 94). But the crisis rhetoric, which also fed on an exaggerated but popular discourse on software production costs,[24] nonetheless had tangible effects on the industry to the point of changing its overall direction and identity.

When compared with the related discipline of microelectronics, programming has long suffered from a lack of credibility and prestige. Despite significant advances throughout the 1950s and the 1960s, actors taking part to software production were often accorded a lower status within Western computing research and industry. This was true for women programmers since they were working in a technical environment. But it was also true for men programmers since they were working in a field that included women. Under this lens, the crisis rhetoric that took hold at the end of the 1960s—feeding on iconic failures that were not representative of the state of the industry—provided an opportunity to *reinvent* programming as something more valuable according to the criteria of the time (Ensmenger 2010, 195–222). This may be one of the reasons why the positively connoted term "engineering" started to spread and operate as a line of sight, notably via the efforts of the 1968 North Atlantic Treaty Organization (NATO) conferences entitled "Software Engineering" and the setting up of professional organizations and academic journals such as the Institute of Electrical and Electronics Engineers' *IEEE Transactions on Software Engineering* (1975) and the Association for Computing Machinery's *ACM Software Engineering Notes* (1976). Though contested by eminent figures who considered that software production was already rigorous and systematic, this complex process of disciplinary relabeling was supported by many programmers—women and men—who saw the title of engineer as an opportunity to improve their work conditions. However, as Abbate (2012, 104) pointed out: "An unintended consequence of this move may have been to make programming and computer science less inviting to women, helping to explain the historical puzzle of why women took a leading role in the first wave of software improvements but become much less visible in the software engineering era."

This stated desire to make software production take the path of engineering—considered the solution to a supposed crisis that itself built on a gendered undervaluation of programming work—has rubbed off on the academic analysis of programming. Parallel to this disciplinary reorientation, a line of positivist research claiming behaviorist tradition began to take an interest in programming work in the early 1970s. For these researchers, the analytical focus should shift: instead of defining the inherent *skills* required for programming and design aptitude tests, scholars should rather try to extract the *parameters that induce the best programming performances* and propose ways to improve software production. The introduction and dissemination of high-level programming languages as well as the multiplication of academic curricula in computer science highly participated in establishing this new line of inquiry. With programming languages such as FORTRAN or COBOL that did not depend on the specificities and brands of computers, behavioral psychologists along with computer scientists became able to design programming *tests* in controlled environments. Moreover, the multiplication of academic curricula in computer science provided relatively diverse *populations* (e.g., undergrads, graduates, faculty members) that could pass these programming tests. These two elements made possible the design of experiments that ranked different sets of parameters (age, experience, design aids) according to the results they assumedly produced (see figure 3.2).

This framework led to numerous tests on debugging performances (e.g., Bloom 1980; Denelesky and McKee 1974; Sackman, Erikson, and Grant 1968; Weinberg 1971, 122–189; Wolfe 1971), design aid performances (e.g., Blaiwes 1974; Brooke and Duncan, 1980a, 1980b; Kammann 1975; Mayer 1976; Shneiderman et al. 1977; Weinberg 1971, 205–281; Wright and Reid 1973), and logical statement performances[25] (e.g., Dunsmore and Gannon 1979; Gannon 1976; Green 1977; Lucas and Kaplan 1976; Sime, Green, and Guest 1973; Sime, Arblaster, and Green 1977; Sime, Green, and Guest 1977; Sheppard et al. 1979; Weissman 1974). But despite their systematic aspect, these studies suffered from the obviousness of their results, for as explained by Curtis (1988), without formally being engaged in behavioral experiments, software contractors were already aware that, for example, experienced programmers produced better results than inexperienced ones did, or that design aids such as flowcharts or documentation were helpful tools for the practice of programming. These general and redundant facts did not help programmers to better design lists of instructions. By the 1980s, the increasingly powerful computing systems remained terribly

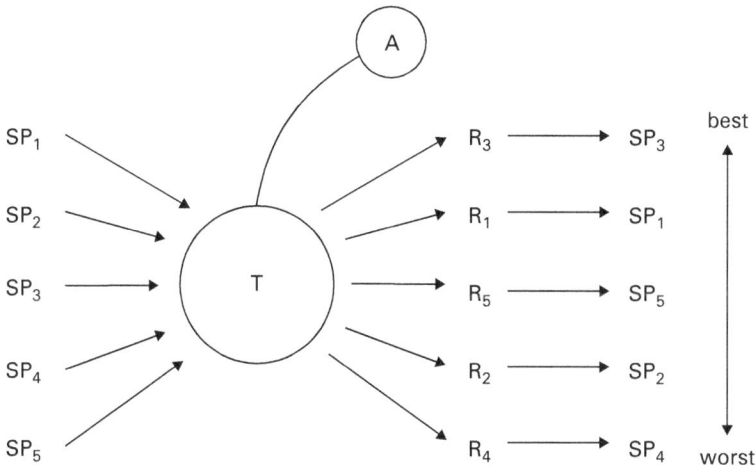

Figure 3.2
Schematic of behavioral studies of computer programming. Let us assume a programming test T, the test's best answers A, and five sets of parameters $SP_{1,...,5}$. SP_1 could, for example, gather the parameters "unexperimented, male, with flowcharts"; SP_2 could, for example, gather the parameters "experienced, female, without flowcharts," and so on. Once all SPs have passed T, the results Rs of each SP allow the ranking of all SPs from best to worst. In this example, R_3 (the results of SP_3) made SP_3 be considered the best set of parameters. Inversely, R_4 (the results of SP_4) made SP_4 be considered the worst set of parameters.

difficult to operate, be they instructed by software engineers working in more and more malely connoted environments.

The Cognitive Turn
By the end of the 1970s, the behavioral standpoint began to be criticized from inside the psychological field. To more and more *cognitive psychologists*, sometimes working in artificial intelligence departments, it seemed that the obviousness of behavioral studies' results was function of a methodological flaw, with many of the ranked sets of parameters gathering important individual variations of results. According to several cognitive researchers, the unit of analysis of behavioral studies was erroneous; since many results' disparities existed within the same sets of parameters, the ranking of these sets was simply senseless (Brooks 1977, 1980; Curtis 1981; Curtis et al. 1989; Moher and Schneider 1981). The solution that these cognitivists proposed to account for what they called "individual differences" was then to dive

inside the individuals' head to better understand the *cognitive processes* and *mental models* underlying the formation of computer programs.

The strong relationships between the notions of "program" and "cognition" also participated in making the study of computer programming attractive to cognitive scientists. As Ormerod (1990, 63–64) put it:

> The fields of cognition and programming are related in three main ways. First, cognitive psychology is based on a "computational metaphor," in which the human mind is seen as a kind of information processor similar to a computer. Secondly, cognitive psychology offers methods for examining the processes underlying performance in computing tasks. Thirdly, programming is a well-defined task, and there are an increasing number of programmers, which makes it an ideal task in which to study cognitive process in a real-world domain.

These three elements—the assumed-fundamental similarity between cognition and computer programs, the growing population of programmers, and the available methods that could be used to study this population—greatly contributed to making cognitive scientists consider computer programming as a fruitful topic of inquiry. Moreover, investing in a topic that behaviorists failed to understand was also seen as an opportunity to demonstrate the superiority of cognitivist approaches. To a certain extent, the aim was also to show that behaviors were a function of mental processes:

> [Behaviorists] attempt to establish the validity of various parameters for describing programming behavior, rather than attempting to specify underlining processes which determine these parameters. (Brooks 1977, 740)

The ambition was then to describe the mental processes that lead to *good* programming performances and eventually use these mental processes to train or select *better* programmers. The methodology of cognitive studies was, most of the time, not radically different from that of behavioral studies on programming, though. Specific programming tests were proposed to different individuals, often computer science students or faculty members. The responses, comments (oral or written), and metadata (number of key strokes, time spent on the problem, etc.) of the individuals were then analyzed according to the rights answers of the test as well as based on general cognitive models of human understanding that the computational metaphor of the mind has inspired (especially the models of Newell and Simon [1972] and, later, Anderson [1983]). From this confrontation among results, comments, and general models of cognition, different mental models specific

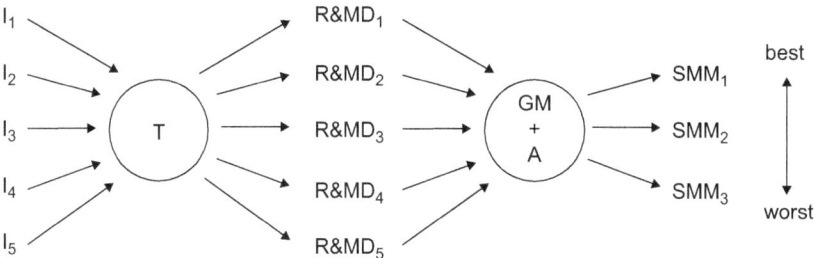

Figure 3.3
Schematic of cognitive studies of computer programming. Let us assume a programming test T, the test's best answers A, five individuals $I_{1,...,5}$, and a general model of cognition GM. Once all Is have passed T, the corresponding results Rs and metadata MD (for example, comments from I on T) are gathered together to form five R&MDs. All R&MDs are then evaluated and compared according to A and GM. At the end of this confrontation, specific mental models (SMMs) are proposed and ranked from best to worst according to their assumed ability to produce the best programming results.

to the task of computer programming were inferred, classified, and ranked according to their performances (see figure 3.3).

This research pattern on computer programming led to numerous studies proposing mental models for solving abstract problems (e.g., Adelson 1981; Brooks 1977; Carroll, Thomas, and Malhotra 1980; Jeffries et al. 1981; Pennington 1987; Shneiderman and Mayer 1979) and developing programming competencies (e.g., Barfield 1986; Coombs, Gibson, and Alty 1982; McKeithen et al. 1981; Soloway 1986; Vessey 1989; Wiedenbeck 1985). Due, in part, to their mitigated results—as admitted by Draper (1992), the numerous mental models proposed by cognitivists did not significantly contribute to better programming performances—cognitive studies have later reintegrated behaviorist considerations (e.g., controlled sets of parameters) to acquire the hybrid and management-centered form they have today (Capretz 2014; Ahmed, Capretz, and Campbell 2012; Ahmed et al. 2012; Cruz, da Silva, and Capretz 2015).

Limits

From the 1950s up to today, computer scientists, engineers, and psychologists have deployed important efforts in the study of computer programming. From aptitude tests to cognitive studies, these scholars have spent a fair amount of time and energy trying to understand what is going on when someone is programming. They certainly did their best, as we all do. Yet I think one can nonetheless express some critiques of, or at least reservations about, some of their methods and conceptual habits regarding the study of programming activity.

Aptitude tests certainly constituted useful recruiting tools in the confusing days of early electronic computing. In this sense, they surely helped counterbalance the unkeepable promises of electronic brains, themselves deriving—I suggest—from the dissemination of von Neumann's functional depiction of electronic computers and its setting aside of programming practices. Moreover, the weight of aptitude tests' results has also constituted resources for women wishing to pursue careers in programming, and some of these women have devised crucial software innovations. Yet as central as they might have been for the development of computing, aptitude tests suffer from a flaw that prevents them from properly analyzing the actions taking part in computer programming: they test candidates on what electronic computers should supposedly do (e.g., sorting numbers, solving equations) but not on the skills required to make computers do these things. They mix up premises and consequences: if the results of computer programming can potentially be evaluated in terms of computing and sorting capabilities, the way in which these results are achieved may require other units of analysis.

Behavioral studies suffer from a similar flaw that keeps them away from computer programming actions. By analyzing the relationships between sets of parameters and programming performances, behaviorist studies put the practices of programming into a black box. In these studies, the practices of programmers do not matter: only the practices' *conditions* (reduced to contextual parameters) and *consequences* (reduced to quantities of errors) are considered. One may object that this nonconsideration of practices is precisely what defines behaviorism as a scientific paradigm, its goal being to predict consequences (behaviors) from initial conditions (Watson 1930), an aim that well echoed the engineerization of software production in the 1970s. It is true that this way of looking at things can be very powerful, especially for the study of complex processes that include many entities, such as traffic flows (Daganzo 1995, 2002), migrations (Jennions and Møller 2003), or cells' behaviors (Collins et al. 2005). But inscribing numbered lists of symbols is a process that does not need any drastic reduction: a programming situation involves only one, two, perhaps three individuals whose actions can be accounted for without any insurmountable difficulties. For the study of such a process that engages few entities whose actions are slow enough to be accounted for, no need a priori exists to ignore what is happening in situation.

For cognitive studies, the story is more intricate. They are certainly right to criticize behavioral studies for putting into black boxes what precisely needs to be accounted for. Yet the solution cognitivists propose to better understand computer programming leads to an impasse we now need to consider.

As Ormerod (1990, 63) put it, "cognitive psychology is based on a 'computational metaphor' in which the human mind is seen as a kind of information processor similar to a computer." From this theoretical standpoint, cognition refers to the reasoning and planning models the mind uses to transform emotional and perceptual input information into outputs that take the form of thoughts or bodily movements. Similarly to a computer— or rather, similarly to *one specific and problematic image of computers*—the human mind "runs" mental models on inputs to produce outputs. The systematic study of the complex mental models that the mind uses to transform inputs into outputs is the very purpose of cognitive studies. Scientific methods of investigation, such as the one presented in figure 3.3, can be used for this specific prospect.

When cognitive science deals with topics such as literature (Zunshine 2015), religion (Barrett 2007), or even chimpanzees' preferences for cooked foods (Warneken and Rosati 2015), its foundations usually hold on: complex mental models describing how the mind processes input information in terms of logical and arithmetic statements to produce physical or mental behaviors can be proposed and compared without obvious contradictions. But as soon as cognitive science deals with computer programming, a short circuit appears that challenges the whole edifice: the cognitive explanation of the constitution of computer programs is tautological as the very notion of cognition already requires constituted computer programs.

To better understand this tricky problem, let us consider once again the computational metaphor of the mind. According to this metaphor, the mind "runs" models—or programs—on inputs to produce outputs. In that sense, the mind looks like a computer as described by von Neumann in the *First Draft*: input data are stored in memory where lists of logical and arithmetic instructions transform them into output. But as we saw in the previous sections, von Neumann's presentation of computers was functional in the sense that it did not take into consideration the elements required to make a computer function. In this image of the computer that reflects von Neumann's very specific position and status, the elements required to assemble the actual transformative lists of instructions—or programs— that command the functioning of an electronic computer's circuitry have already been gathered.

From here, an important flaw of cognitive studies on computer programming starts to appear: as the studies rely on an image of the computer that already includes constituted computer programs, these cognitive studies are not in a position to inquire into what constitutes computer programs. In fact, the cognitive studies are in a situation where they can mainly propose circular explanations of programming: if there are (computer) programs, it is because there are (mental) programs. Programs explain programs: a perfect tautology.

As long as cognitive science stays away from the study of computer programming, its foundations hold on: mental programs can serve as explicative tools for observed behaviors. But as soon as cognitive science considers computer programming, its limits appear: cognition and programs are of the same kind. Thunder in the night! Cognition, as inspired by the computational metaphor of the mind, works as a stumbling stone to the analysis of computer programming practices as its fundamental units of analysis are assembled programs. In such a constricted *epistemic culture* (Knorr-Cetina 1999), the in situ analysis of courses of action cannot but be omitted, despite their active participation in the constitution of the collective computerized world. This is an unfortunate situation that even the bravest propositions in human-computer interaction (HCI) have not been able to modify substantially (e.g., Flor and Hutchins 1991; Hollan, Hutchins, and Kirsh 2000). Is there a way to conceptually dis-constrict the empirical study of computer programming?

Putting Cognition Back to Its Place

Most academic attempts to better understand computer programming seem to have annoying flaws: aptitude tests mix up premises and consequences, behavioral studies put actions into black boxes, and cognitive studies are stuck in tautological explanations. If we want to consider computer programming as accountable practices, it seems that we need to distance ourselves from these brave but problematic endeavors.

Yet, provided that our critics are relevant, we are at this point still unable to propose any alternative. Do the actions of programmers not have a *cognitive* aspect? Do programmers not use their *minds* to computationally solve complex *problems*? The confusion between cognition and computer programs may well derive from a misleading history of computers—as I tried to suggest—its capacity to establish itself as a generalized habit commands respect. How can we not present empirical studies of computer programming practices as silly reductions? How can we justify the desire to account for, and thus make visible, the *courses of action* of computer programming, these practices that are obligatory passage points of any computerization project?

Fortunately, contemporary work in philosophy has managed to fill in the gap that has separated cognition from practices, intelligent minds from dull actions. It is thanks to these inspiring studies that we will become able to consider programming as a practice without totally turning our back on the notion of cognition. To do so, I will first need to quickly reconsider the idea that computers were designed in the image of the human brain and mind. As we already saw—though partially—this idea is relevant only in retrospect: what has concretely happened is far more intricate. I will then reconsider the philosophical frame that encloses cognition as a computational process. Finally, following contemporary works in the philosophy of perception, I will examine a definition of cognition that preserves important aspects of how we make sense of the things that surround us while reconnecting it to practices and actions. By positing the centrality of agency in cognitive processes, this *enactive conception of cognition* will further help us empirically consider what is happening during computer programming episodes.

A Reduction Process

The computational metaphor of the mind forces cognitivists to use programs to explain the formation of programs. The results of programming processes—programs—are thus used to explain programming processes. It is not easy to find another example of such an explicative error: it is like explaining rain with water, chicken poultry with the chicken dance ... But how did things end up this way? How did programs end up constituting the fundamental base of cognition, thus participating in the invisibilization of computer programming practices?

The main argument that justifies the computational metaphor of the mind is that "computers were designed in the image of the human" (Simon and Kaplan 1989, quoted in Hutchins 1995, 356). According to this view that spread in the 1960s in reaction to the behavioral paradigm (Fodor 1975, 1987; Putnam [1961] 1980), how the human brain works inspired the design of computers, and this can, in turn, provide a clearer view on how we think. Turing is generally considered the father of this argument, with the Universal Machine he imagined in his 1937 paper "On Computable Numbers" being able to simulate any mechanism describable in its formalism. According to this line of thought, it was Turing's self-conscious introspection that allowed him to define a device capable of any computation as he was looking "at what a mathematician does in the course of solving mathematical problems and distilling this process to its essentials" (Pylyshyn 1989, 54). Turing's demonstration would then lead to the first electronic computers, such as the ENIAC and the EDVAC, whose depiction as giant brains appears legitimate as how we think inspired these computers in the first place.

In line with the recent work of Simon Penny (2017), I assume that this conception of the origins of computers is incorrect. As soon as one considers simultaneously the process by which Turing's thought experiment was reduced to an image of the brain *and* the process by which the EDVAC was reduced to an input/output device controlled by a central organ, one realizes that the relationship between computers and the human brain points to the other direction: the human brain was designed in a very specific image of the computer that already included all possible programs.

Let us start with Turing as he is often considered the father of the computational metaphor of the mind. It is true that Turing compared "a man in the process of computing a real number" with a "machine which is only capable of a finite number of conditions" (Turing 1937, 231). Yet his image of human computation was not limited to what is happening inside the head: it also included hands, eyes, paper, notes, and sets of rules defined by others in different times and locations. As Hutchins put it: "The mathematician or logician was [for Turing] materially interacting with a material world" (Hutchins 1995, 361). By modeling the properties of this sociomaterial arrangement into an abstract machine, Turing could distinguish between computable and noncomputable numbers, hence showing that Hilbert's *Entscheidungsproblem* was not solvable. His results had an immense impact on the mathematics of his time as they suggested a class of numbers calculable by finite means. But the theoretical machine he invented to define this class of numbers was by no means designed *only* in the image of the human brain; it was a theoretical device that expressed the sociomaterial process enabling the computation of real numbers.

What participated in reducing Turing's theoretical device to an expression of a mental process was the work of McCulloch and Pitts on neurons. In their 1943 paper entitled "A logical Calculus of the Ideas Immanent in Nervous Activity," McCulloch and Pitts built upon Carnap's (1937) propositional logic and a simplified conception of neurons as all-or-none firing entities to propose a formal model of mind and brain. In their paper, neurons are considered units that process input signals sent from sensory organs or from other neurons. In turn, the outputs of this neural processing feed other neurons or are sent back to sensory organs. The novelty of McCulloch and Pitts's approach is that, thanks to their simplified conception of neurons, the input signals that are processed by neurons can be represented as propositions or, as Gödel (1931) previously demonstrated, as numbers.[26] From that point, their model could consider configurations of neural networks as logical operators processing input signals from sensory organs and outputting *different* signals back to sensory organs. This way to consider the brain as a huge network of neural networks able to express the laws of propositional calculus on binary signals allowed McCulloch and Pitts to hypothetically consider the brain as a Turing machine capable of computing numerical propositions (McCulloch and Pitts [1943] 1990, 113). Even though they did not mathematically prove their claim and recognized that their model was computationally less powerful than Turing's model, they nonetheless infused the conception of mind as the result of the brain's computational processes (Piccinini 2004).

At first, McCulloch and Pitts's paper remained unnoticed (Lettvin 1989, 17). It was only when von Neumann used some of their propositions in his 1945 *First Draft* (von Neumann [1945] 1993, 5–11) that the equivalence between computers and the human mind started to take off. As we saw earlier, von Neumann had a very specific view on the EDVAC: his position as a famous consultant who mainly sees the clean results of laborious material processes allowed him to reduce the EDVAC as an input-output device. Once separated from its instantiation within the hangars of the Moore School of Electric Engineering, the EDVAC, and especially the

ENIAC, effectively *looked like* a brain as conceived by McCulloch and Pitts. From that point, the reduction process could go on: von Neumann could use McCulloch and Pitts' reductions of neurons and of the Turing machine to present his own reductive view on the EDVAC. However, it is important to remember that von Neumann's goal was by no means to present the EDVAC in a realistic way: the main goal of the *First Draft* was to formalize a model for an electronic computing system that could inspire other laboratories without revealing too many classified elements about the EDVAC project. All of these intricate reasons (von Neumann's position, wartime, von Neumann's interest in mathematical biology) made the EDVAC appear in the *First Draft* as an input-output device controlled by a central organ whose configuration of networks of neurons could express the laws of propositional calculus.

As we saw earlier, after World War II, the *First Draft* and the modelization of electronic computers it encapsulates began to circulate in academic spheres. In parallel, this conception of computers as giant electronic brains fitted well with their broader inclusion in commercial arrangements: these very costly systems had better be presented as functional brains automatically transforming inputs into outputs rather than intricate artifacts requiring great care, maintenance, and an entire dedicated infrastructure. Hence there were issues related to their operationalization as the buyers of the first electronic computers—the Air Force, Boeing, General Motors (Smith 1983)—had to select, hire, and train and eventually fire, reselect, rehire, and retrain whole operating teams. But despite these initial failures, the conception of computers as electronic brains held on, well supported, to be fair, by Turing's (1950) paper "Computing Machinery and Intelligence," the 1953 inaugural conferences on artificial intelligence at Dartmouth College (Crevier 1993), Ashby's book on the neural origin of behavior (Ashby 1952), and von Neumann's posthumous book *The Computer and the Brain* ([1958] 2012). Instead of crumbling, the conception of computers as electronic brains started to concretize to the point that it even supported a radical critique of behaviorism in the field of psychology. Progressively, the mind became the product of the brain's computation of nervous inputs. The argument appeared indeed indubitable: as human behaviors are the results of (computational) cognitive processes, psychology should rather describe the composition of these cognitive processes—a real tour de force whose consequences we still experience today.

But this colossus of the computational metaphor of the mind has feet of clay. As soon as one inquires sociohistorically into the process by which brains and computers have been put into equivalence, one sees that the foundations of the argument are shaky; a cascade of reductions, as well as their distribution, surreptitiously ended up presenting the computer as an image of the brain. Historically, it was first the reduction of the Turing machine as an expression of mental processes, then the reduction of neurons as on/off entities, then the reduction of the EDVAC as an input-output device controlled by a central organ, then the distribution of this view through academic networks and commercial arrangements that *allowed* computers to be considered as deriving from the brain. It is the collusion of all of these *translations* (Latour 2005), along with many others, that made computers appear as the consequences of the brain's structure.

Important authors have finely documented how computer-brain equivalences contributed, for better or worse, to structuring Western subjectivities throughout the Cold War period (e.g., Dupuy 1994; Edwards 1996; Mirowski 2002). For what interests me here, the main problem of the conception of computers as an image of the brain is that its correlated conception of cognition as computation contributed to further invisibilizing the courses of actions taking part in computer programming. According to the computational metaphor of the mind, the brain *is* the set of all the combinations of neural networks—or logic circuits[27]—that allow the computation of signals. The brain may choose one specific combination of neural networks for the computation of each signal, but the combination itself is already assembled. As a consequence, the study of how combinations of neural networks are assembled and put together to compute specific signals—as it is the case when someone is programming—cannot occur as it would imply to go beyond what constitutes the brain. Cognitive studies may involve inquiring about *which* program the brain uses for the computation of a specific input, but the way this program was assembled remains out of reach: it was already there, ready to be applied to the task at hand. In short, similarly to von Neumann's view on the EDVAC but with far less engineering applications, the brain as conceived by the computational metaphor of the mind *selects the appropriate mental program from the infinite library of all possible programs*. But as this library is precisely what constitutes the brain, it soon becomes senseless to inquire into how each program was concretely assembled.

The cognitivist view on computers as designed in the image of the brain seems then to be the product of at least three reductions: (1) neurons as on/ off firing entities, (2) the Turing machine as an expression of mental events, and (3) the EDVAC as an input/output device controlled by a central organ. The further distribution of this view on computers through academic, commercial, and cultural networks further legitimatized the conception of cognition as computation. But this cognitive computation was a holistic one that implied the possibility of all specific computations: the brain progressively appeared as the set of all potential instruction sets, hence preventing inquiries into the constitution of actual instruction sets. The tautological impasse of cognitive science when it deals with computer programming seems, then, to be deriving from a delusive history of the computer. The ones who inherit from a nonempirical history of electronic computers might consider cognition as computation and programming as a mental process. Yet the ones who inherit from an empirical history of the constitution of electronic computing systems and who pay attention to translation processes and distributive networks have no other choice but to consider cognition differently. But how?

The Classical Sandwich and Its Consequences

We now have a clearer—yet still sketchy—idea of the formation of the computational metaphor of the mind. An oriented "double-click" history (Latour 2013, 93) of electronic computers that did not pay attention to the small translations that occurred at the beginning of the electronic computing area enabled cognitive scientists—among others—to retroactively consider computers as deriving from the very structure of the brain. But historically, what has happened is far more intricate: McCulloch and Pitts's work on neurons and von Neumann's view on the EDVAC echoed each other to progressively form a powerful yet problematic depiction of computers as giant electronic brains. This depiction further legitimized the computational metaphor of the mind—also coined *computationalism*—that yet paralyzed the analysis of the constitution of actual computer programs since the set of all potential programs constituted the brain's fundamental structure. At this point of the chapter, then, to definitively turn our back on computationalism and propose an alternative definition of cognition that could enable us to consider the task of computer programming as a practical activity, we need to look more precisely at the metaphysics of this computational standpoint.

If computationalism in cognitive science derives from a quite recent nonempirical history of computers, its metaphysics surely belongs to a philosophical lineage that goes back at least to Aristotle (Dreyfus 1992). Susan Hurley (2002) usefully coined the term "classical sandwich" to summarize the metaphysics of this lineage—also referred to as "cognitivism"— that considers perception, cognition, and agency as distinct capacities. For the supporters of the classical sandwich, human perception first grasps an input from the "real" world and translates it to the mind (or brain). In the case of computationalism, this perceptual input takes the shape of nervous pulses that can be expressed as numerical values. Cognition, then, "works with this perceptual input, uses it to form a representation of how things are in the subject's environment and, through reasoning and planning that is appropriately informed by the subject's projects and desires, arrives at a specification of what the subject should do with or in her current environment" (Ward and Stapleton 2012, 94). In the case of computationalism, the cognitive step implies the selection and application of a mental model—or mental program—that outputs a different numerical value to the nervous system. Finally, agency is considered the output of both perception and cognition processes and takes the form of bodily movements instructed by nervous pulses.

This conception of cognition as "stuck" in between perception and action as meat in a sandwich has many consequences. It first establishes a sharp distinction between the mind and the world. Two realms are then created: the realm of "extended things" that are said to be material and the realm of "thinking things" that are said to be abstract and immaterial.[28] If matter thrones in the realm of "extended things" by allowing substance and quantities, mind thrones in the realm of "thinking things" by allowing thoughts and knowledge.

Despite the ontological abyss between them, the realms of "thinking things" and "extended things" need to interact: after all, we, as individuals, are part of the world and need to deal with it. But a sheet of paper cannot go through the mind, a mountain is too big to be thought, a spoken sentence has no matter: some transformation has to occur to make these things possible for the mind to process. How, then, can we connect both "extended" and "thinking" realms? The notions of *representation* (without hyphen) and *symbols* have progressively been introduced to keep the model viable. For the mind to keep in touch with the world of "real things," it needs to work with *representations* of real things. Because these representations happen in the head and refer to extended things, they are usually called *mental representations of things*.

Mental representations of things need to have at least two properties. They first need a *form* on which the mind could operate. This form may vary according to different theories among cognitivism. For the computational metaphor of the mind, this form takes, for example, the shape of electric nervous pulses that the senses acquire and that are then routed to the brain. The second property that mental representations of things require is *meaning*; that is, the distinctive trace of what representations refer to in the real world. Both properties depend on each other: a form has a meaning, and a meaning needs a form. The notion of *symbol* is often used to gather both the half-material and semantic aspects of the mental representations of things. In this respect, cognition, as considered by the proponents of the classical sandwich, processes symbolic representations of things that the senses offer in their interactions with the real world. The result of this processing is, then, another representation of things—a statement *about* things—that further instructs bodily movements and behaviors.

The processing of symbolic representations of things does not always lead to accurate statements about things. Some malfunctions can happen either at the level of the senses that badly translate real things or at the level of the mind that fails to interpret the symbols. In both cases, the whole process would lead to an inaccurate, or *wrong*, statement about things. These errors are not desirable as they would instruct inadequate behaviors at the end of the cognitive process. It is therefore extremely important for cognition to make *true* statements. If cognition does not manage to establish adequate correspondences between our minds and the world, our behaviors will be badly instructed. Conversely, by properly acquiring *knowledge* about the real world, cognition can make us behave adequately.

I assume that the symbolic representational thesis that derives from cognition as considered by the classical sandwich leads to two related issues. The first issue deals with the amalgam between *knowledge* and *reality* it creates, hence refusing giving any ontological weight to entities whose trajectories are different from scientific facts. The second issue deals with the thesis's incapacity to consider practices *in the wild*, with most of the models that take symbolic representational thesis to the letter failing the test of ecological validation.

Let us start with the first issue, certainly the most difficult. We saw that, according to cognitivism, the *adaequatio rei et intellectus* serves as the measure of valid statements and behaviors. For example, if I say "the sun is rising," I make an invalid statement and thus behave wrongly because what I say does not refer adequately to the real event. Within my cognitive process, something went wrong: in this case, my senses that made me believe that the sun was moving in the sky probably deceived me. In reality, thanks to other mental processes that are better than mine, we know as a matter of fact that it is the earth that rotates around the sun; some "scientific minds"—in this case, Copernicus and Galileo, among others—managed indeed to adequately process symbolic representations to provide a true statement about the relations between the sun and the earth, a relation that the laws of Reason can demonstrate. My statement and behavior can still be considered a joke or some form of sloppy habit: what I say/do is not *true* and therefore does not really *count*.

The problem of this line of thought that only gives credit to scientific facts is that it is grounded on a very unempirical conception of science. Indeed, as STS authors have demonstrated for almost fifty years, many material networks are required to construct scientific facts (Knorr-Cetina 1981; Lynch 1985; Latour and Woolgar 1986; Collins 1992). Laboratories, experiments, equipment, colleagues, funding, skills, academic papers: all of these elements are necessary to laboriously construct the "chains of reference" that give access to remote entities (Latour 1999b). In order to know, we need equipment and collaboration. Moreover, as soon as one inquires into science in the making instead of ready-made science, one sees that both the knowing mind and the known thing start to exist only at the very end of practical scientific processes. When everything is in place, when the chains of reference are strong enough, when there are no more controversies, I am becoming able to look at the majestic Californian sunrise and meditate about the power of habits that makes me go against the most rigorous fact: the earth is rotating. Thanks to numerous scientific networks that were put in place during the sixteenth and seventeenth centuries, I *gain* access to such—poor—meditation. Symmetrically, when everything is in place, when the chains of reference are strong enough, the sun *gains* its status of known thing as one part of its existence—its relative immobility—is indeed being captured through scientific work and the maintenance of chains of reference. In short, what others have done and made durable enables me to think directly about the objective qualities of the sun. As soon as I can follow solidified scientific networks that gather observations, instruments, experiments, academic papers, conferences, and educational books, I *become* a knowing mind, and the sun *becomes* a known object. Cognitivism started at the wrong end: the possibility of scientific knowledge starts with practices and ends with known objects and knowing minds. As Latour (2013, 80) summarized it:

> A knowing mind and a known thing are not at all what would be linked through a mysterious viaduct by the activity of knowledge; they are the progressive result of the extension of chains of reference.

One result of this relocalization of scientific truth within the networks allowing its production, diffusion, and maintenance is that reality is not the sole province of scientific knowledge anymore: other entities that go through different paths to come into existence can also be considered *real*. Legal decisions (McGee 2015), technical artifacts (Simondon 2017), fictional characters (Greimas 1983), emotions (Nathan and Zajde 2012), or religious icons (Cobb 2006): even though these entities do not require the same type of networks as scientific facts in order to emerge, they can also be considered *real* since the world is no longer reduced to sole facts. As soon as the dichotomy between knowledge and mind is considered one consequence of chains of reference, as soon as *what is happening* is distinguished from *what is known*, there is space for many varieties of existents. By disamalgamating reality and knowledge, the universe of the real world can be replaced with the multiverse of performative beings (James 1909)—an ontological feast, a breath of fresh air.

Besides its problematic propensity to posit correspondence between things and minds as the supreme judge of what counts as real, another problem of cognitivism—or *computationalism*, or *computational metaphor of the mind*; at this point, all of these terms are equivalent—is its mitigated results when it comes to support so-called expert systems (Star 1989; Forsythe 2002).

A first example concerns what Haugeland (1989) called "Good Old Fashioned Artificial Intelligence" (GOFAI), an important research paradigm in artificial intelligence that endeavored to design intelligent digital systems from the mid-1950s to the late 1980s. Although the complex algorithms implied in GOFAI's computational conception of the mind soon appeared very effective for the design of computer programs capable of complex tasks, such as playing chess or checkers, these algorithms symmetrically appeared very problematic for tasks as simple as finding a way outside a room without running into its wall (Malafouris 2004). The extreme difficulty for expert systems to reproduce very basic human tasks started to cast doubts on computationalism, especially since cybernetics—an cousin view on intelligence that emphasizes "negative feedback" (Bowker 1993; Pickering 2011)—effectively managed to reproduce such tasks without any reference to symbolic representation. As Malafouris (2004, 54–55) put it:

> When the first such autonomous devices (*machina speculatrix*) were constructed by Grey Walter, they had nothing to do with complex algorithms and representational inputs. Their kinship was with W. Ross Ashby' Homeostat and Norbert Wiener's cybernetic feedback … On the basis of a very simple electromechanical circuitry, the so-called 'turtles' were capable of producing emergent properties and behavior patterns that could not be determined by any of their system components, effecting in practice a cybernetic transgression of the mind-body divide.

Another practical limit of computationalism when applied to computer systems is the so-called frame problem (Dennet 1984; Pylyshyn 1987). The frame problem is "the problem of generating behaviour that is appropriately and selectively geared to the most contextually relevant aspects of their situation, and ignoring the multitude of irrelevant information that might be counterproductively transduced, processed and factored into the planning and guidance of behaviour" (Ward and Stapleton 2012, 95). How could a brain—or a computer—adequately select the inputs relevant for the situation at hand, process them, and then instruct adequate behaviors? Sports is, in this respect, an illuminating example: within the mess of a cricket stadium, how could a batter process the right input in a very short amount of time and behave adequately (Sutton 2007)? By what magic is a tennis player's brain capable of selecting the conspicuous input, processing it, and—eventually—instructing adequate behaviors on the fly (Iacoboni 2001)? To date, the only satisfactory computational answer to the frame problem, at least with regard to perceptual search tasks, is to consider it NP-complete, thus recognizing it should be addressed by using heuristics and approximations (Tsotsos 1988, 1990).[29]

Finally, the entire field of HCI can be considered an expression of the limits of computationalism as it is precisely because human cognition is not equivalent to computers' cognition that innovative interfaces need to be imagined and designed (Card, Moran, and Newell 1986). One famous example came from Suchman (1987) when she inquired into how users interacted with Xerox 8200 copier: as the design of Xerox's artifact included an equivalence between computers' cognition and human cognition, interacting with the artifact was a highly counterintuitive experience, even for those who designed it. Computationalism made Xerox designers forget about important features of human cognition, such as the importance of action and "situatedness" for many sense-making endeavors (Suchman 2006, 15). Besides refusing giving any ontological weight to nonscientific entities, computationalism thus also appears to restrain the development of intelligent computational systems intended to interact with humans.

Enactive Cognition

Despite its impressive stranglehold on Western thought, cognitivism has been fiercely criticized for quite a long time.[30] For the sake of this part II—whose main goal is, remember, to document the practices of computer programming because they are nowadays central to the constitution of algorithms—I will deal only with one line of criticisms recently labeled "enactive conception of cognition" (Ward and Stapleton 2012). This reframing of human cognition as a local attempt to engage *with* the world is here crucial as it will—finally!—enable us to consider programming in the light of situated experiences.

Broadly speaking, proponents of enactive cognition consider that agency drives cognition (Varela, Thompson, and Rosch 1991). Whereas cognitivism considers action as the output of the internal processing of symbolic representations about the "real world," enactivism considers action as a relational co-constituent of the world (Thompson 2005). The shift in perspective is thus total: it is as if one were speaking two different languages. Whereas cognitivism deals with an ideal world that is being accessed indirectly via representations that, in turn, instruct agency, enactivism deals with a becoming environment of transformative actions (Di Paolo 2005). Whereas cognitivism considers cognition as computation, enactivism considers cognition as adaptive interactions with the environment whose properties are offered to and modified through the actions of the cognizer. For enactivism, the features of the environment with which we try to couple are then not fixed nor independent: they are continuously provided as well as specified based on our ability to attune with the environment.

With enactivism, the cognitivist separations among perception, cognition, and agency are blurred. Perception is no longer separated from cognition because cognizing is precisely about perceiving the takes that the environment provides: "The *affordances* of the environment are what it *offers* the animal, what it *provides* or *furnishes*, for either good or ill" (Gibson 1986, cited in Ward and Stapleton 2012, 93). Moreover, cognition does not need to be stuck in between perception and agency, processing inputs on representations to instructively define actions: for enactivism, the cognizer's effective actions both participate in, and are functions of, the takes that the sensible situation provides (Noë 2004; Ward, Roberts, and Clark 2011). Finally, agency cannot be considered the final product of a well or badly informed cognition process because direct perception itself is also part of agency: the way we perceive grips also depends on our capacities to grasp them. But the environment does not structure our capacity to perceive either; actions also modify the environment's properties and affordances, thus allowing a new and always surprising "dance of agency" (Pickering 1995). Perceptions suggest actions that, in turn, suggest new perceptions. From take to take, as far as we can perceive: this is what enactive cognition is all about.

This very minimal view on cognition that considers it "simply" as our capability to grasp the affordances of local environments has many consequences. First, enactivism implies that cognition (and therefore, to a certain extent, perception) is *embodied* in the sense that "the categories about the kind and structure of perception and cognition are constrained and shaped by facts about the kind of bodily agents we are" (Ward and Stapleton 2012, 98). Notions such as "up," "down," "left," and "right" are not anymore necessarily features of a "real" extended world: they are contingent effects of our bodily features that suggest a spatially arrayed environment. We experience the world through a body system that supports our perceptual apparatus (Clark 1998; Gallagher 2005; Haugeland 2000). Cognition is therefore multiple: to a certain extent, each body cognizes in its own way by engaging itself differently with its environment.

Second, enactivism implies that cognition is *affective* in the sense that "the form of openness to the world characteristic of cognition essentially depends on a grasp of the affordances and impediments the environment

offers to the cognizer with respect to the cognizer's goal, interest and projects" (Ward and Stapleton 2012, 99). Evaluation and desires thus appear
crucial for a cognitive process to occur: no affects, no intelligence (Ratcliffe
2009, 2010). "Care" is something we take; what "shows up" concerns us.
Again, it does not mean that our inner desires structure what we may perceive and grasp; our cognitive efforts also suggest desires to grasp the takes
our environment suggests.

Third, enactivism considers that cognition can sometimes be extended:
nonbiological elements, if properly embodied, can surely modify the boundaries of affective perceptions (Clark and Chalmers 1998). It does not mean
that every nonbiological item would increase our capability to grasp affordances: some artifacts are, of course, constraining ongoing desires (hence
suggesting new ones). But at any rate, the combinations of human and nonhuman apparatus, the *association* of biological and nonbiological substrates
fully participate in the cognitive process and should therefore also be taken
into account.

The fourth consequence of enactivism is the sudden disappearance of the
frame problem. Indeed, although this problem constitutes a serious drawback for cognitivism by preventing it from understanding—and thus from
implementing—the initial selection of the relevant input for the task at
hand, enactive cognition avoids it by positing *framing* as part of cognition.
Inputs are not thrown at cognizers anymore; their embodied, affective, and,
eventually, extended perception tries to grasp the takes that the situations
at hand propose. Cricket batters are trained, equipped, and concerned with
the ball they want to hit; tennis players inhabit the ball they are about to
smash. In short, whereas cognitivism deals with procedural *classifications*,
enactivism deals with bodily and affective *intuitions* (Dreyfus 1998).

The fifth consequence is the capacity to consider a wide variety of existents. This consequence is as subtle as it is important. We saw that one del-
eterious propensity of cognitivism was to amalgamate truth (or knowledge)
and reality: what counts as real for cognitivism is a behavior that derives
from a true statement about the real world. Cognition is, then, considered
the process by which we know the world and—hopefully—act accordingly. The picture is very different for enactivism. As enactive cognition is
about interacting with the surrounding environment, grasping the takes it
offers and therefore participating in its reconfiguration, knowledge can be
considered as an eventual, very specific, and very delightful by-product of

cognitive processes. Cognition surely helps scientists to align inscriptions and construct chains of reference according to the veridiction mode of the scientific institution; however, cognition also helps writers to create fictional characters, lawyers to define legal means, or devout followers to be altered via renewed yet faithful messages. In short, by distinguishing knowledge and cognition—cognizers do not *know* the world but *interact* with it, hence participating in its reconfiguration—enactivism places the emphasis on our local attempts to couple with what surrounds us and reconfigure it, hence sometimes creating new existing entities.

Finally, enactivism makes the notions of symbols and representations useless for cognitive activities. Indeed, since the world is now a local environment whose properties are constantly modified by our attempts to couple with it, no need exists to posit an extra step of mental representations supported by symbols. For enactivism, there may be symbols—in the sense that a take offered by the environment may create a connection with many takes situated elsewhere or co-constructed at another time—but agency is always first. When I see the hammer and sickle on a red flag on a street of Vientiane, Laos, I surely grasp a symbol but only by virtue of the connections this take is making with many other takes I was able to grasp in past situations: TV documentaries about the Soviet revolution, school manuals, movies, and so on. In that sense, a symbol becomes a network of many solidified takes. Similarly, some takes may re-present other takes, but these re-presentations are always takes in the first place. For example, I may grasp a romantic re-presentation of a landscape at the second floor of Zürich's *Kunsthaus*, but this re-presentation is a take that the museum environment has suggested in the first place. This take may derive from another take—a pastoral view from some country hill in the late eighteenth century—but, at least at the cognitive level, it is a take I am grasping at the museum in the first place.

To sum up, enactive cognition starts with agency; affective and embodied *actions* are considered our way of engaging with the surrounding environment. This environment is not considered a preexisting realm; it is a collection of situations offering takes we may grasp to configure other take-offering situations. From this minimal standpoint, cognition infiltrates every situation without constituting the only ingredient of what exists. Scientists surely need to cognize to conduct experiments in their laboratories; lawyers for sure need to cognize to define legal means in their offices;

programmers surely need to cognize to produce numbered lists of instructions capable of making computers compute in desired ways; yet facts, legal decision, or programs cannot be reduced to cognitive activities as they end up constituting existents that populate the world. With enactive cognition, the emphasis is made on the interactions among local situations, bodies, and capabilities that, in turn, participate in the formation of what is *existing*, computer programs included. Cognition, then, appears crucial as it provides grips but also remains very limited as it is constantly overflowed: there is always something more than cognition. May computer programming be considered as part of this *more*. This could make it finally appear in all its subtleties.

4   A Second Case Study

The journey was convoluted, but we are now finally in a position to consider computer programming as a practical, situated activity. In chapter 3, I first questioned von Neumann's architecture; for fundamental yet contingent reasons, its definition of computers as functional devices took for granted the situated practices required to make them function. If this unempirical presentation of electronic systems was certainly useful at the beginning of the computer area by sharing classified work and proposing a research agenda, it nonetheless misled the understanding of what makes computers actually compute. I then distanced myself from the different academic answers to the nonfunctional aspects of electronic computers as functionally defined by von Neumann. Aptitude tests for the selection of programmers started at the wrong end as they tried to select people without inquiring into the requirements for such tasks. Behavioral studies aiming to isolate the right parameters for efficient programming implied looking at the results of actions and not at the actions themselves. Finally, I tried to show how the cognitivist response to behavioral studies had, and has, problematic limitations: as mainstream cognitivism relies on the computational metaphor of the mind that itself needs already assembled programs, many cognitivists cannot go beyond the form "program" that ends up explaining itself. A process is being explained by its own result; programs need programs, a perfect tautology. Yet in the last section of chapter 3, I suggested that the very notion of cognition, once freed from the throes of *computationalism*, could still be a useful concept for rediscovering experience. Once cognition is considered an *enactive* process of grasping the affordances of local environments, the emphasis is placed on specific situations, places, bodies, desires, and capabilities.

From this point, we are ready to grasp programming in all of its materiality without being obtruded by the notions of "representations" (without hyphen), "mental models," or "computation." All of these things—and more generally von Neumann's functional presentation of computers—are the *results* of the situations we want to account for. To a certain extent, we are back in 1943 at the Moore School of Electrical Engineering: no mental models, no internal cognition, no von Neumann architecture, no programs; only actions, desires, and artifacts that interactively try to make meaningful electronic computations occur. Even though the following case study is based on data collected in the Lab between 2015 and 2016, I will try to study them as if the unempirical conceptions of electronic computing did not occur.

Presentation of the Empirical Materials

The development of an image-processing algorithm intended for academic publication is a process that involves many different activities and situations. But along the gathering of relevant data; the construction of ground truths; the formulation of transformative relationships between input-data and output-targets; and the numerous Group meetings, informal discussions, seminars, and coffee breaks that help all these things to happen, there are more or less long *computer programming episodes* when numbered lists of instructions have to be written in order to make an electronic device adequately compute digital data. It is these courses of action that have a beginning and an end that I will try to account for in this case study.

The problem that quickly stood out during my ethnographic endeavor within the Lab was how to document these courses of action. First, as the code being written during programming episodes was very cryptic, it was in the beginning difficult to have a grip on what was going on. Second, the configurations of these cryptic signs on the screens were constantly changing; new characters were added, other erased, other corrected, and so on. Third, these situations appeared quite engaging for the people involved, which prevented me from asking them questions about what they were doing. During these moments that looked particularly intense, I was clearly out of place.

To palliate these methodological issues, I designed my own image-processing project with the help of the Lab's members. After several Lab meetings, we collectively decided that I should try to design a preprocessing model that could sort images whose pixel configurations would fit further specific segmentation processes that were under development within the Lab. This modest project was explicitly designed to force me learn the basics of several computer programming languages and become more familiar with image processing in general. Importantly, the project also included a "helping clause" that allowed me to ask members of the Lab for help when I was stuck in a programming impasse. This somewhat unusual method turned out immensely fruitful. It first made me become more comfortable with several programming languages;[1] little by little, all these cryptic signs started to make more sense. It also made the members of the Lab more comfortable during the programming episodes I tried to document and account for. As the project had been designed collectively and could potentially be used for future projects, the members of the Lab found it somewhat relevant. And as the so-called helping sessions did not directly concern their own projects, they also felt more at ease with me taking notes and asking questions while they were programming. Finally—and perhaps more importantly—this method allowed me to better equip and document programming episodes: along with notes describing the movements and gestures of the one who was programming next to me, I could video record my monitors and audio record the discussions. For the eight helping sessions I needed for this project, I then ended up with descriptions, screen recordings, and audio recordings I could thoroughly analyze.

Though insightful in many respects, the materials collected during these helping sessions nonetheless had limitations. As the small programs resulting from these sessions were primarily intended for my own specific use, they were not directly designed to circulate within a professional community of programmers as it is typically the case in corporate software settings. In this sense, important topics such as program reading for the in situ shaping of intelligibility, as considered by Button and Sharrock (1995) in their paper on computer programming practices, could not be specifically investigated. Nevertheless, as we will see later in the chapter, some of my analytical propositions may well be related to Button and Sharrock's conclusions.

The following materials are taken from one helping session during which DF—a PhD student of the Lab—wrote a small program that I will from now on call PROG that dealt with data I had previously collected via a crowdsourcing task. The crowdsourcing task was divided into ten rounds. For each round, twenty to thirty unknown workers were shown fifty "natural pictures" of landscapes, faces, birds, buildings, and so on. The content of these pictures was extremely varied. For each image, each worker was asked to draw one or several rectangles around the parts of the image that first attracted their attention. Before switching to the next image, each worker also had to grade from one to seven how straightforward it had been for them to choose what specific parts of the image to label. After the ten rounds of this crowdsourcing task, 254 different workers each labeled fifty images for a total of five hundred images. The data collected from the activity of the workers (the IDs of the images they processed, the coordinates of the rectangles they drew, and the grades they gave for each labeling task) via a web application were gathered in .txt files organized as in figure 4.1. The content of these .txt files along with the natural images used for the crowd-sourcing task were the data on which PROG had to work.

If this small project was explicitly designed to better document program-ming practices, it also had an image-processing goal. This secondary goal was to find correspondences between the contents of the natural images—in terms of arrangement of numerical pixel-values—and both the rectangles and grades provided by the workers. In short, the assumption was that for

---

16714267603_cd60601b7f_b.jpg 1 startX: 25px startY: 32px width: 450px height: 361px
16705290404_d8de298f0e_b.jpg 5 startX: 430px startY: 76px width: 260px height: 414px
                             startX: 234px startY: 227px width: 189px height: 216px

---

Figure 4.1
Excerpt of a .txt file named "worker_05Waldave56jm9815.txt" as provided by the web application at the end of each session of the crowdsourcing task. The name of the file ("worker_05Waldave56jm9815.txt") corresponds to the ID given to the worker by the web application. Only two rows of the file are presented here. The first element of each row is a string of text that ends with ".jpg"; it corresponds to the ID of the image that had been processed by the worker. The second element of each row corresponds to the numeral grade given to the labeling task by the worker. The subsequent elements of each row correspond to the coordinates of the rectangle(s) drawn by the worker. Every rectangle is defined by four values part of the coordi-nate space of the image that was being processed. The first value of each rectangle ("startX: $n$px") corresponds to the horizontal coordinate of the picture. The second value ("startY: $n$px") corresponds to the vertical coordinate of the picture. The third value ("width: $n$px") corresponds to the pixel width of the drawn rectangle. The fourth value ("height: $n$px") corresponds to the pixel height of the drawn rectangle. Altogether, these four values allow to reconstruct—later—the rectangle(s) drawn by the user. Moreover, as indicated by the second row of the excerpt, the workers could draw several rectangles.

images with high grades and very dispersed rectangles, it may not make sense to divide their content into smaller parts. Symmetrically, for images with low grades and very compact rectangles, it may eventually make sense to divide their content into smaller parts (see figure 4.2). Being able to automatically sort pictures whose contents may or may not be divided into smaller parts could be useful for further lossy compression schema based on segmentation processes. In that sense, the computational method I tried to define could eventually serve as a preprocessing step for further, more complex, segmentation/compression methods that members of the Lab were developing at that time. But at any rate, to propose such a preprocessing method, many intermediary programs—including PROG—had to be assembled.

The design of the web application that enabled the crowdsourcing task and the gathering of data as shown in figure 4.2 required the completion of many different programs. First, a Python web-scrapping program had to be designed in order to browse and download heterogeneous, high-definition, and Creative-Commons-licenced images made available by the API of the Flickr website. The design of this small yet not-so-trivial program first required a "helping session" with a member of the Lab. Second, several programs using html, JavaScript, and PHP computer programming languages

Figure 4.2
Two views on the data collected during the crowdsourcing task. Both views were made possible by a Matlab program that parsed the data of the .txt files and related them to the corresponding .jpg images. On the left, workers roughly labeled the same part of the image and gave a very low grade to this labeling task (average of 1.16). One may then assume that it would make sense to divide the content of this image into smaller parts (in this case, the bird and the rest). On the right, the opposite situation: the workers labeled the image almost randomly and gave a high grade to this labeling task (average 5.25). One may them assume that it would make little sense to divide the content of this image into smaller parts.

had to be designed to allow workers to interact with a specific number of images and store their IDs, labels, and grades within .txt files. The design of this web application required two "helping sessions" with members of the Lab. Third, a first Matlab program was required in order to read the textual and numerical contents of all the .txt files and reorganize them within Matlab software environment. Because of its agility to design problems of linear algebra—all integers being considered scalars—Matlab is widely used for research and industrial purposes in computer science, electrical engineering, and economics. Yet if Matlab programming language is known for being well adapted for the computation of matrices and arrays, it is also known for being badly adapted for the reorganization of .txt data into matrices and arrays. This reorganization of data into matrices and arrays was generally called "parsing" by the members of the Lab. Again, a fourth helping session was required to help me assemble parsing programs that further enabled views such as those presented in figure 4.2.

The program whose formation we are about to follow—PROG—dealt with the analysis of the data as reorganized by previous parsing programs. The shaping of PROG required a fifth "helping session" with DF. The specifications of PROG can be summarized as such: for reasons we will cover at length in the next sections, PROG should be able to transform each labeled digital image as presented in figure 4.2 into another less complex digital image as presented in figure 4.3. The value of the pixels of each

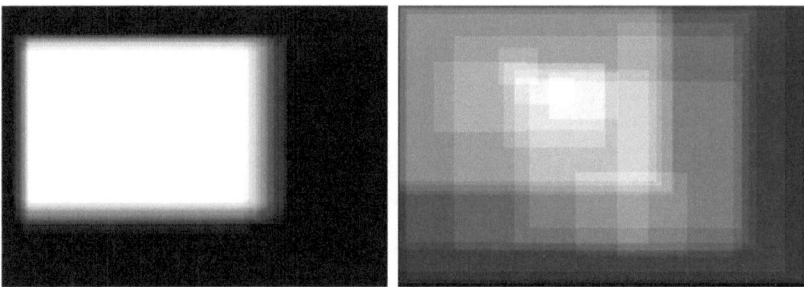

Figure 4.3
Two views on the results of PROG. Both simplified matrices are translations of the labeled images of figure 4.2. PROG was intended to select one part of the parsed data in order to transform the labeled images of figure 4.2 into much less complex matrices. These matrices allowed further analysis, notably in terms of histograms and frequencies.

less complex image should correspond to the number of rectangles each pixel is part of. For example, if a given pixel is part of zero rectangle, PROG should attribute the value zero to this pixel. But if another given pixel is part of, say, six rectangles, PROG should attribute the value six to this pixel. PROG was thus intended to gather together different values (dimensions of the natural image, dimensions of each rectangle drawn by the participants of the crowdsourcing task, incrementing values of each pixel) in order to create new images or, as usually coined in image processing, new *matrices*.

At this point, it is not necessary to fully understand the goals and specifications of PROG as we will closely consider them in the next sections. What is more important for now is to understand that PROG was designed in the Matlab software environment. Like other popular high-level programming languages, such as Python or C, Matlab is generally used in conjunction with an *integrated development environment* (IDE) that includes visualization and file organization functionalities (see figure 4.4). But unlike Python, C, and some of their compatible IDEs (e.g., PyCharm, Eclipse), Matlab—as a programming language in its own right *and* as an IDE—is owned and maintained by MathWorks Inc. and is distributed on a license basis. At the time of this inquiry, Matlab's proprietary feature was criticized by a growing number of Lab members who tended to prefer Python, which is open-source and supported by an active community of developers. However, notably because of its internal organization natively designed for matrix processing, Matlab was and still is frequently used. For reasons of readability, my follow-up of the practical formation of PROG will only focus on the Editor and the Command Window of the Matlab IDE. In the next sections, the content of figure 4.4 will then be presented as in figure 4.5.

Even if PROG was by far the smallest program of the project, I will not be able to account for its entire formation process. Instead of accounting for the whole programming episode that established PROG, I will only focus on specific *sequences* that are particularly instructive. My follow-up of the programming sequences is chronological, starting at Time 0 (T0) and ending at Time *n*. Yet the sampling of each T does not follow a fixed period of time but rather the modifications of both the Editor and the Command Window. Let us assume, for example, that figure 4.5 is the first expression of PROG during the programming sequence we are following (T0). As soon as

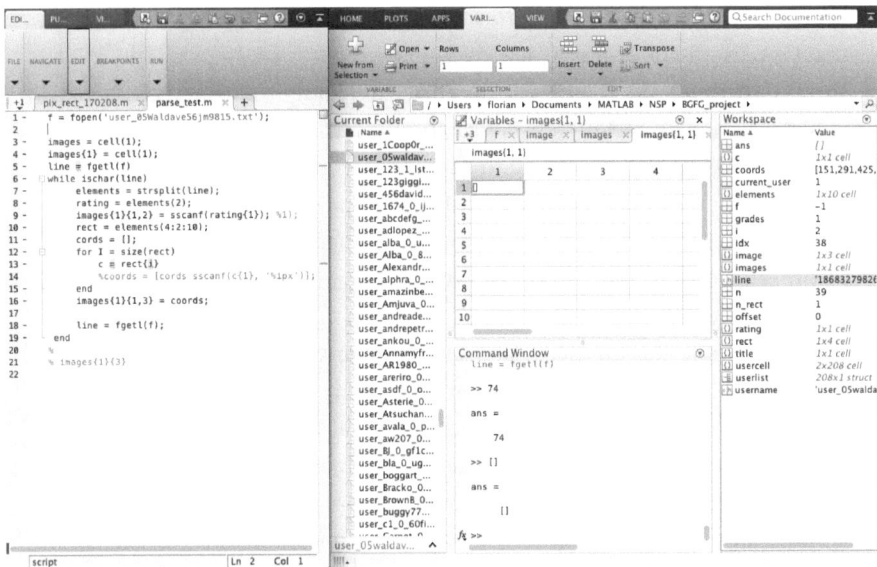

Figure 4.4

Screenshot of the Matlab IDE. The far-right window is called the Workspace. It gathers all the variables the programmer creates during their session. To the left of the Workspace, the Variables Window allows the programmer to visualize in spreadsheets the variables she created. In this screenshot, the variable "images[1,1]" is being visualized. Below it, to the left of the Workspace, there is the Command Window that shows the results of the operations conducted by the programmer. In this screenshot, the Command Window shows the answer "[ ]". The long window in the middle of the screenshot is the Current Folder Window that shows the content of the folder currently accessed by the software. On the left, the Editor is the window that allows the programmer to write Matlab programs—also called scripts—that is, numbered lists of instructions written in the Matlab programming language. When the programmer clicks on the Run icon (on the top middle of the Editor) or uses an equivalent personalizable shortcut key, the results of the script are printed in the Command Window. In this screenshot, the running of the script made "[ ]" appear in the Command Window. The spatial arrangements of these different windows can be modified according to the programmer's preferences.

```
1.   f = fopen('user_05Waldave56jm9815.txt');
2.
3.   images = cell(1);
4.   images{1} = cell(1);
5.   line = fgetl(f)
6.   while ischar(line)
7.      elements = strsplit(line);
8.      rating = elements(2);
9.      images{1}{1,2} = sscanf(rating{1}, '%1');
10.     rect = elements(4:2:10)
11.     cords = [];
12.     for I = size(rect)
13.         c = rect{i}
14.         %coords = [cords sscanf(c{1}, '%ipx')];
15.     end
16.     images{1}{1,3} = coords;
17.
18.     line = fgetl(f);
19. end                                              ans =
20. %
21. %images{1}{3}                                       []
```

Figure 4.5
Simplified Matlab IDE as it will be presented for the remainder of the analysis. To make the follow-up of programming sequences more readable, only the content of the Editor and the Command Window will be displayed. Here, the figure expresses (part of) the content of figure 4.4.

the programmer makes changes in both the Editor and the Command Window, these changes will be documented and highlighted as in figure 4.6.

In between the different Ts, the sayings and actions of the programmer (DF) and me (FJ) will be transcribed. To keep things readable, I may omit some small actions, such as quick mistypes or hesitation disfluencies. Following T1 (figure 4.6), the programming sequence would, for example, go on like this:

DF: "Hum, it doesn't work anymore."

FJ: "Apparently. ..."

DF: "Tssssss."

[at line 14, DF deletes "{1}"]
[DF runs the script]
[figure 4.7—T2]

DF: "OK. But why are there only two of them? I don't get it. Difficult today!"

[laughs]

```
1.  f = fopen('user_05Waldave56jm9815.txt');
2.
3.  images = cell(1);
4.  images{1} = cell(1);
5.  line = fgetl(f)
6.  while ischar(line)
7.     elements = strsplit(line);
8.     rating = elements(2);
9.     images{1}{1,2} = sscanf(rating{1}, '%1');
10.    rect = elements(4:2:10)
11.    cords = [];
12.    for I = size(rect)
13.        c = rect{i}
14.        coords = [cords sscanf(c{1}, '%ipx')];
15.    end
16.    images{1}{1,3} = coords;
17.
18.    line = fgetl(f);
19. end
20. %
21. %images{1}{3}
```
```
>> parse

Cell contents

reference from a non-

cell array object

Error in parse(line

14)

coords = [coords

sscanf(c{1}, '%ipx')]
```
~~14. %~~

Figure 4.6
The Editor and the Command Window at T1, when modified by the programmer. In the caption's title, the term "T1" indicates that it is the first change of the programming sequence being followed. The instructions that have been removed or added in the Editor are highlighted in gray. The content of the Command Window is updated. Finally, the instructions that have been deleted are indicated as strikeout text in the bottom cell. The line numbers of the deleted instructions are those of Tn-1 (here T0).

```
1.  f = fopen('user_05Waldave56jm9815.txt');
2.
3.  images = cell(1);
4.  images{1} = cell(1);
5.  line = fgetl(f)
6.  while ischar(line)
7.     elements = strsplit(line);
8.     rating = elements(2);
9.     images{1}{1,2} = sscanf(rating{1}, '%1');
10.    rect = elements(4:2:10)
11.    cords = [];
12.    for I = size(rect)
13.        c = rect{i}
14.        coords = [cords sscanf(c    , '%ipx')];
15.    end
16.    images{1}{1,3} = coords;
17.
18.    line = fgetl(f);
19. end
20. %
21. %images{1}{3}
```
```
ans =

83   74
```
~~14. {1}~~

Figure 4.7
Editor and Command Window at T2.

Here and then, I will also intervene to clarify things and analyze what is happening. Before we start with the first sequence, it is important to keep in mind that one does not need to understand everything that is said in the transcriptions nor all the elements within each T. What is important in this close analysis of computer programming practices is what is happening *in between each T*. It is by focusing on the relative differences between each T that we will manage to understand some of the issues at stake during these unconventional courses of actions.

I need to mention one last thing before we dive into the practices of computer programming. One may easily object that the following case study and its subsequent tentative propositions are not *representative* of programming practices in general. To this, I answer that representativeness is simply not at stake here. Representativeness is indeed a powerful and important concept but only when the boundaries of a population are clearly defined. Inhabitants *of a town*, cells *of a tissue*, words *of a book*: all can be related to a very costly and equipped *set*—the administrative and geographical limits of a towns, the physical limits of a sample, the hardcover of a book—that subsequently defines a territory and a population. In these specific—but very rare and often controversial—cases, the concept of representativeness can be used to extract statistically meaningful results. But when there is no territory, no set, the very notion of representativeness loses its raison d'être. What is programming? Who are programmers when they program? We do not know as there were very few studies of computer programming practices. This is typically where ethnography can be useful: the exploration of nondefined—or problematically defined—territories may provide takes for the design of subsequent boundaries to be explored statistically. And while I do think that the young street artist in Leipzig who is writing a small JavaScript program to animate the menus of her personal website, the engineer of Boeing who is working on the last Ada's update for cabin pressurization modules, or the computer scientist who tries to parse .txt files with the Matlab IDE *differ* in many ways—they have different problems, affects, environments, equipment—I also think that (almost) *none* of these situations have yet been accounted for ethnographically. We still have to start somewhere. The following case study is then one of the very first steps into, I hope, more systematic studies of programming courses of action; hence the exploratory aspect of its propositions.

Aligning Inscriptions

Let us focus on PROG. Building on what I presented in the last section, I will document a very short programming sequence that took less than five minutes in real time. I will stay as close as possible to the formatted-yet-empirical material, using the presentation method I introduced above as well as several concepts developed in STS in the course of the analysis. My hope is to show that one set of practices that are terribly important for programmers deal with the proliferation and alignment of *inscriptions* in order to pave out an access to a remote entity and, simultaneously, *identify a*

```
1.   I = imread(images{1});
2.   R = zeros(size(I));
3.   users = images {1,2};
4.   for i = 1:size(users), 1)
5.      j = 0;
6.      while 1
7.          rect = users{i,j+3};
8.          if size(rect,2) == 0
9.              break
10.         end
11.         j = j+1;
12.         x = rect(1):rect(1)+rect(3);
13.         y = rect(2):rect(2)+rect(4);
14.         R(y,x) = R(y,x) + 1;
15.     end
16. end
```

>>

Figure 4.8
Editor and Command Window at T0.

```
1.   I = imread(images{1});
2.   R = zeros(size(I));
3.   users = images {1,2};
4.   for i = 1:size(users), 1)
5.      j = 0;
6.      while 1
7.          rect = users{i,j+3};
8.          if size(rect,2) == 0
9.              break
10.         end
11.         j = j+1;
12.         x = rect(1):rect(1)+rect(3);
13.         y = rect(2):rect(2)+rect(4);
14.         R(y,x) = R(y,x) + 1;
15.     end
16. end
```

Index exceeds matrix dimensions

Figure 4.9
Editor and Command Window at T1.

location. Hopefully, this odd proposition will become clearer as the chapter goes on. For the moment, let us start *in medias res* with figures 4.8 and 4.9:

[figure 4.8—T0]
[DF runs the script]
[figure 4.9—T1]

DF: "OK. So it tells me it doesn't work."

FJ: "Apparently."

What is happening between T0 and T1? After DF runs the script, a red (here, gray) inscription appears in the Command Window, indicating that "Index exceeds matrix dimensions." Where does this text come from? Who wrote it? To better understand the origin of this cryptic notification, I have to introduce an important participant to the sequence: the interpreter (INT). For the sixteen lines of code in the Editor to generate electric pulses that would further allow the hardware of the computer to effectively compute the data of the .txt files, many steps have to be taken. Fortunately, for the case that interests us here, only the very first step is important. This first step consists in translating every line of code into something else—in this case, subroutines compiled into machine code—that would, in turn, generate electric pulses and the effective computation of the data. One of the entities responsible for this complex translation is INT. Every time DF runs the script, INT is surreptitiously triggered to translate the content of the Editor, byte by byte. We do not need to know exactly what INT does during its translating processes: even for DF, the very functioning of INT remains obscure. In fact, we just need to understand four characteristics of INT:

1. INT has its own *trajectory* that is fully understood by almost nobody: highly specialized teams employed by the company MathWorks, editors of Matlab, were required to shape it and are still currently maintaining it. In that sense—at least from the point of view of DF—INT can be considered *a being that takes the risk of existence* (James [1912] 2003; Latour 2013), just as a cat or an elephant seal.

2. INT translates one line of the Editor after the other.[2]

3. As soon as INT successfully translates a line, if this line instructs the printing of an inscription, INT prints this inscription in the Command Window.

4. As soon as INT cannot translate one line, it stops and prints a red (here, gray) inscription in the Command Window.

This leads us to the important notion of *inscription* that we have already encountered in the introduction where I emphasized the world-generative capabilities of these durable, mobile, and re-presentable entities. There are, of course, many different types of inscriptions: books, WhatsApp messages, shopping lists, or even tattooed bodies can be considered inscriptions, some being more durable, mobile, and re-presentable than others (Gitelman 2014). But in any case, inscriptions are translated manifestations of more or less attributable events and thus constitute, at least potentially, *takes* offered by the environment in specific situations. These inscriptions are not representations (without hyphen) of "real things" that feed mental computations. They are formatted re-presentations of events that may be grasped and, in turn, configure other world-generative takes. This is why I needed to tediously introduce enactive cognition at the end of chapter 3: as we are now aware that agency *precedes* cognition, documents and inscriptions can be considered no more but also no less than takes that may suggest other actions—from take to take, as far as we can perceive and make sense (Penny 2017).

Inscriptions-takes are sometimes grasped by cognizing individuals; other times, they are not. In our case, the inscription "Index exceeds matrix dimensions" is indeed grasped by DF. In fact, as DF ran the script, he expected an inscription to appear in the Command Window. Moreover, as DF is well aware—just as we are now—that any red inscription in the Command Window manifests that INT could not translate all the lines of the script, DF knows that the inscription "Index exceeds matrix dimensions" is the trace of an event related to INT.

From this point, we are able to better understand what the first inscription does to DF. At T1, the inscription "Index exceeds matrix dimensions" is a take grasped by DF that manifests that something—but what?—is affecting the trajectory of INT: *it tells me it doesn't work.*

Let us continue:

DF: "It doesn't go through. I'll just check the size of the image."

[DF creates a new line at 2 in the Editor; types "size(I)"]

INT has a problem: it doesn't go through the script. But what part of PROG is affecting INT? At this point, it is difficult to know exactly. In fact, understanding what is happening to INT is, from now on, necessary to the realization of PROG.

```
1.  I = imread(images{1});
2.  size(I)
3.  R = zeros(size(I));
4.  users = images {1,2};
5.  for i = 1:size(users), 1)
6.     j = 0;
7.     while 1
8.        rect = users{i,j+3};
9.        if size(rect,2) == 0
10.           break
11.       end
12.       j = j+1;
13.       x = rect(1):rect(1)+rect(3);
14.       y = rect(2):rect(2)+rect(4);
15.       R(y,x) = R(y,x) + 1;
16.    end
17. end
```

```
ans =
  Columns 1 through 2
  1024    712

  Column 3
  3

Index exceeds matrix
dimensions
```

Figure 4.10
Editor and Command Window at T2.

For DF, the initial red inscription indicates—though quite vaguely—that INT is affected by the size of something. The terms "exceeds" and "dimensions" of the red inscription attest for such a size-related problem. In order to have a better grip on what size-related problem is affecting the trajectory of INT, DF starts by examining the size of the image. To do this, DF adds the small line of code "size(I)" at the second line of the script and then runs it, thus triggering INT (figure 4.10—T2).

By adding the line of code "size(I)" at line 2 and then triggering INT, DF makes a new inscription appear in the Command Window:

```
ans =
Columns I through 2
1024 712
Column 3
3
```

This new inscription printed by INT in the Command Window is not red and can therefore be considered an actual translation of the code. This is taken for granted: decades of engineering developments allow DF to be certain that this new inscription is an unproblematic expression of INT. But still, is this inscription expressing the dimension of the right image? If not, the whole script should be reconsidered. To verify that INT is indeed failing to process the right image, DF uses the second non-red inscription to create a third one, this time emanating from me:

DF: "OK, so the size is 1024×712. Does that sound right to you?"

FJ: "Yes, it is correct for this image."

DF: "Ok. So it's happening *after*."

The oral statement "Yes, it is correct for this image"—itself deriving from inscriptions I had previously produced and encountered during a former unsuccessful programming attempt—allows DF to consider that the non-red inscription refers adequately to the image INT is failing to process. The certitude emanating from the articulation of the non-red inscription and the inscription-derived oral statement further allows DF to infer that "it's happening *after*." The "after" is here crucial. Indeed, since the second inscription is not red and appears above the red inscription in the Command Window, DF can conclude that whatever is affecting the trajectory of INT, it lies somewhere after the instruction "size(I)" he has just added at line 2. By adding and articulating two new inscriptions—the non-red inscription and the inscription relayed by my confirmatory oral statement—DF already gets a clearer view on INT: what is affecting its trajectory lies after the second line of the script.

Let us continue:

[DF examines the Command Window of figure 4.10—T2]

DF: "Ah, but it indicates also the colors! Typical Matlab."

[DF puts the cursor on "Column 3" in T2 Command Window]

DF: "See? [to FJ] We should take only the first two values for "R." Otherwise, it blocks."

FJ: "Because now 'R' has three values?"

DF: "I guess so."

[DF deletes line 2; at the end of "new" line 2, he types ".1), size(I,2"]

By pursuing his inspection of the non-red inscription in the Command Window at T2, DF notices that the size of the image INT fails to process is expressed by three values: "1024," "712," *and* "3." Where does this "3" come from? Difficult to say. It may come from Matlab systematic consideration of the data that structure a digital color image. Indeed, these specific matrices are bound to a width, a height, *and* three layers of RGB values. Most high-level programming languages do not take into consideration this third value as it generally does not express useful information about the actual dimensions of an image. But Matlab—in its fussy fashion—apparently expresses it, and this may be, according to DF, the source of the problem affecting INT.

At this point, DF believes that the documentation he gathered about INT's trajectory through the piling up and alignment of three inscriptions—the red inscription, the non-red inscription, and the auditory statement (itself being a translation of written inscriptions considered in the past)—is accurate enough to complete the script; according to DF, based on the evidences he produced, collected, and aligned, INT does not support the third value of "size(I)." This information about INT that points toward line 3 may, in turn, allow the modification of the script and smooth the trajectory of INT. DF also deletes "size(I)" at line 2 that mainly served for him as an instrument for the probing of INT. Then, in line with his insight about the provenance of the problematic phenomenon that affects the trajectory of INT, he types ",1),size(I,2" in the Editor in order to define "R" according to only two values: "1024" and "712," for the case of the first image of the ground truth. He then runs the script:

[DF runs the script]

[figure 4.11—T3]

DF: "Ah no. It's not here, apparently."

Unfortunately for DF, these modifications do not change the state of INT. As we can see in the Command Window at T3 (figure 4.11), DF's new

```
1.   I = imread(images{1});
2.   R = zeros(size(I,1), size(I,2));
3.   users = images {1,2};
4.   for i = 1:size(users), 1)
5.       j = 0;
6.       while 1
7.           rect = users{i,j+3};
8.           if size(rect,2) == 0
9.               break
10.          end
11.          j = j+1;
12.          x = rect(1):rect(1)+rect(3);
13.          y = rect(2):rect(2)+rect(4);
14.          R(y,x) = R(y,x) + 1;
15.      end
16.  end

2.   size(I)
```

```
Index exceeds matrix
dimensions
```

Figure 4.11

Editor and Command Window at T3.

triggering of INT does not lead to the disappearance of the red inscription: something is still affecting INT, and it was not the image size defined by three values instead of only two.[3] Using a scientific expression, we can say that "INT-being-affected-by-the-third-value-of-size(I)" was an *artifact*: it does not participate in the phenomenon that affects INT's trajectory. In turn, the problematic location is not line 2; it is *somewhere else*. More experiments are therefore needed; more inscriptions have to be produced, compared, and aligned.

The artifact "INT-being-affected-by-the-third-value-of-size(I)" was not totally worthless for DF, though. Thanks to it, DF is now certain that INT is being affected by a size-related problem that occurs *after* line 2. But this certainty about INT is for the moment too thin; it does not allow DF to precisely identify what is affecting INT and therefore modify the code accordingly.

Let us continue:

DF: "OK. Well, we'll print the rectangle then. And just compare."

[DF deletes ";" at the end of line 8; he creates a new line at 3 in the Editor; he types "size(R)" at line 3]

PROG deals with natural images on which rectangles have been previously drawn by workers during a crowdsourcing task. As we saw in the previous section that presented the empirical materials of this chapter, the drawn rectangles are not strictly speaking *on* the images: they are stored as coordinates within .txt files. The script we are now examining is intended to use the width and height values of each natural image as well as its rectangles in order to create a new image that is less complex and easier to analyse. These new simplified images—that I will from now on call matrices—should only express the number and the position of the rectangles that the workers drew on the initial color images. In this respect, the workflow of the script is quite straightforward: first, an empty matrix is created using the width and height values of the initial natural image, then a rectangle is created using the workers' data in the .txt file related to this image, then the rectangle is added to the empty matrix. Progressively, as more and more rectangles are added to the matrix, the matrix acquires more values. In the field of image processing, we say that the matrix is *incremented*. Figure 4.3 provides two examples of PROG's final outputs; that is, matrices that have been incremented according to the coordinates of the rectangles related to their IDs in .txt files. But we are not there yet; at this point of the programming episode, INT—this lively entity on which it is difficult to have a grip, at least for biped mammals—is affected by something that prevents it from translating the code adequately.

What is affecting INT is not clear. But the previous inscriptions DF managed to handle and align have made him see that INT's problem has to do with some size and dimension. Moreover, DF is also aware of the general workflow of the script since he mostly designed it (more on this later). In this respect, what if the first rectangle that is added to the first matrix exceeds the boundaries of the matrix? It would be very problematic as it would signify that some .txt data are corrupted. But as the rectangle is indexed to .txt data, this would satisfy the red inscription "Index exceeds matrix dimension." But how could DF be certain of that? Just as before, by producing more inscriptions and compare them.

To print the size of the first rectangle, DF deletes ";" at the end of line 8.[4] In order to print the dimension of the first image of the dataset, he writes "size(R)" on line 3. He then runs the script:

[DF runs the script]

[figure 4.12—T4]

[DF examines the Command Window of figure 4.12—T4]

DF: "So, 197 and 323. Makes less than 1024, obviously. And same for height. Alright. It's strange because it doesn't exceed."

Two new non-red, and thus a priori nonproblematic, inscriptions appear in the Command Window at T4 (figure 4.12). The first one "ans = 1024 712"

```
1.   I = imread(images{1});
2.   R = zeros(size(I,1), size(I,2));
3.   size(R)
4.   users = images {1,2};
5.   for i = 1:size(users), 1)
6.      j = 0;
7.      while 1
8.         rect = users{i,j+3}
9.         if size(rect,2) == 0
10.           break
11.        end
12.        j = j+1;
13.        x = rect(1):rect(1)+rect(3);
14.        y = rect(2):rect(2)+rect(4);
15.        R(y,x) = R(y,x) + 1;
16.     end
17. end
```

```
ans =
     1024    712

rect=
     197   91   323   371

Index exceeds matrix
dimensions
```

Figure 4.12

Editor and Command Window at T4.

describes the dimension of the first image of the collection. The second one "rect = 197 91 323 371" describes the dimensions of the first rectangle drawn by the first worker as well as the location of this rectangle within the first image. The first value of rect, "197," refers to its horizontal coordinate within the image, and the second value, "91," refers to its vertical coordinate. These two numbers therefore indicate that the rectangle starts at pixel [197:91] of the image. The third value of rect, "323," expresses the width of the rectangle and the fourth value, "371," expresses its height. These two last numbers therefore indicate that the width of the rectangle is 323 pixels and that its height is 371 pixels.

At T4, DF is already aware of what all these values refer to; before this programming episode, I explained to him the conventions I used to structure the data of the .txt files. But once these values are printed and compared with the width and height of the image, basic yet terribly important arithmetic evaluations can be undertaken: "$197 + 323 < 1024$" and "$91 + 371 < 712$." These are crucial clues as they do not corroborate the red inscription of the Command Window; the rectangle doesn't exceed the dimensions of the image. The size and position of the rectangle is not what is affecting INT. Something else is disrupting INT in its relation with PROG. But what? And where is it? More inscriptions are required to better document what affects INT and modify the script accordingly.

What we see at T4 is a perfect example of the process I'm here trying to highlight: by printing the size of the image and the coordinates of the rectangle, DF acquires a better grip on the process at hand. He can articulate these two new inscriptions and align them to the previous ones. In that sense, he is enactively paving out some access to INT and its red inscription. Even though this production and alignment of inscriptions do not work as DF hoped—the dimensions of the rectangle do not exceed the dimensions of the image—this gives him another clue about the phenomenon under scrutiny: what is affecting INT lies *somewhere else*. This practice of grasping, producing, and aligning inscriptions in order to identify the origin of a problematic phenomenon is, I believe, central to programming. As we will see, it is not the only type of practices that are deployed during computer programming sequences. But in some specific situations, when an important entity is blocked in its trajectory, thus preventing the computation of data by means of electric pulses, the handling and aligning of inscriptions remains crucial. In these situations when a problematic *location* has to be found, the design of experiments and the articulation of their results appear necessary to pave a very specific path, itself providing very specific *information* about some small, scattered, and very swift entities we may call "interpreters," "compilers," or even "processors" in the case of microcode. I will come back to this proposition at the end of this programming sequence. But already at this point, it is important to note that the mundane addition and alignment of inscriptions DF is currently making might be central to the very activity of computer programming.

With these preliminary elements in mind, let us continue:

DF: "I'll just try something else. We'll see if the rectangle corresponds."

[DF creates a new line at 13 in the Editor; on this new line, he types "imshow(I(y,x,:))"]

DF needs a new inscription: if the relationship between the rectangle and the image is not problematic for INT, something else must be. But what? As is often during programming episodes, the situation starts to be confusing. To be sure that the rectangle expressed in the Command Window at T4 is the right one and not some sort of not-yet-identified artifact, DF needs to see this first rectangle when superimposed over the first image. To do so, he creates a new line in the Editor and types the small instruction "imshow(I(y,x,:))." He then runs the script:

[DF runs the script]
[figure 4.13—T5]
[figure 4.14]
[DF examines figure 4.14]

```
1.   I = imread(images{1});
2.   R = zeros(size(I,1), size(I,2));
3.   size(R)
4.   users = images {1,2};
5.   for i = 1:size(users), 1)
6.       j = 0;
7.       while 1
8.           rect = users{i,j+3}
9.           if size(rect,2) == 0
10.              break
11.          end
12.          j = j+1;
13.          imshow(I(y,x,:))
14.          x = rect(1):rect(1)+rect(3);
15.          y = rect(2):rect(2)+rect(4);
16.          R(y,x) = R(y,x) + 1;
17.      end
18. end
```

```
ans =
   1024    712

rect=
   197    91    323    371

Index exceeds matrix
dimensions
```

Figure 4.13
Editor and Command Window at T5.

Figure 4.14
Output of PROG at T5.

> **DF:** "OK. So theoretically, this should be the first rectangle labeled by
> the first worker."

The new inscription triggered by DF at T5 (figure 4.14) is this time a little
different. Instead of text, it is a part of an image. More precisely, it is the
expression of the first rectangle the first worker drew on the first image.
And just like between T2 and T3, this new inscription allows DF to create
another inscription, this is time emanating from me:

> **DF:** "Does it correspond?"
>
> **FJ:** "Yes, yes, it does."
>
> **DF:** "OK good. So it definitely blocks somewhere else. Maybe it can't
> define the second rectangle."

Having worked on the data of the ground truth for a couple days, I am a
trustworthy reference: at least for the first image, I know quite well the
position of the different rectangles. Once again, the articulation and alignment of two inscriptions—the first rectangle over the first image and my
own verification (informed by inscriptions I had previously encountered)—
allow DF to pursue his inquiry into the problematic phenomenon engaging
INT. If the first rectangle and the part of the code responsible for defining
it are not what is affecting INT, the problem should lie somewhere else.
Perhaps in the second rectangle and, more generally, the part of the code
responsible for defining it? Once again, new inscriptions are required:

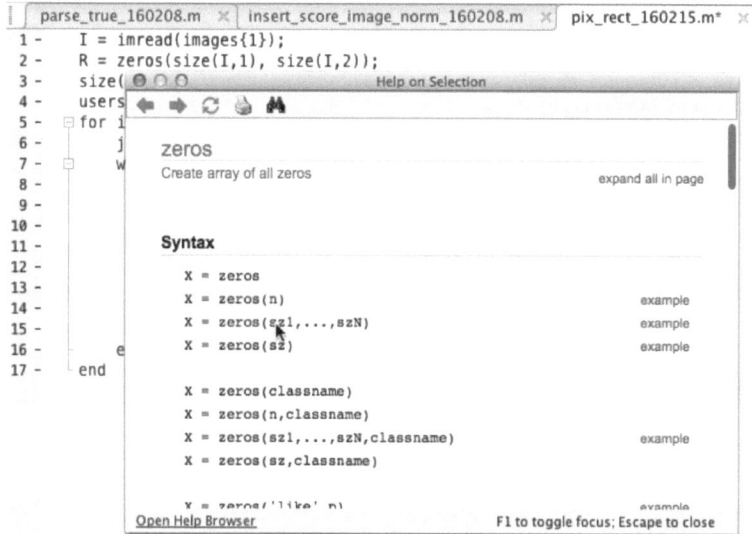

Figure 4.15
Screenshot of "help on selection" as triggered by DF at T5.

DF: "It might be when we define the empty matrix."

[DF deletes "`imshow(I(y,x,:))`" on line 13; on line 2, he selects the function "`zeros`," right clicks on it, and selects "help on selection"] [figure 4.15]

The new inscription (figure 4.15) is again a little different from those appearing in the Command Window. It turns out indeed that the Matlab IDE provides access to a "Help on Selection" database that, if connected to the internet, displays the correct syntax for each selected function. This pop-up window being aligned with the suspect function at line 2, DF can use the mouse cursor to compare the correct syntax of the help menu with what is written in the Editor:

DF: "No, no, we did it right. It is somewhere else."

[DF closes the "help on selection" window]

The comparison between the help menu and the script allows DF to be certain that INT is not affected by this line of code; the syntax is right, so INT is able to understand it. The problem lies somewhere else:

[DF runs the script]
[figure 4.16—T6]

1.	`I = imread(images{1});`		0	0	0	0	
2.	`R = zeros(size(I,1), size(I,2));`		0	0	0	0	
3.	`size(R)`		0	0	0	0	
4.	`users = images {1,2};`		0	0	0	0	
5.	`for i = 1:size(users), 1)`		0	0	0	0	
6.	`  j = 0;`		0	0	0	0	
7.	`  while 1`		0	0	0	0	
8.	`      rect = users{i,j+3};`		0	0	0	0	
9.	`      if size(rect,2) == 0`		0	0	0	0	
10.	`          break`		0	0	0	0	
11.	`      end`		0	0	0	0	
12.	`      j = j+1;`		0	0	0	0	
			0	0	0	0	
			0	0	0	0	
13.	`      x = rect(1):rect(1)+rect(3);`		0	0	0	0	
14.	`      y = rect(2):rect(2)+rect(4);`						
15.	`      R(y,x)`						
16.	`  end`		Index exceeds matrix				
17.	`end`		dimensions				

```
13. imshow(I(x,y,:))
16. R(y,x) + 1;
```

Figure 4.16
Editor and Command Window at T6.

DF: "Huh, I don't get it … There's only the empty matrix."

At T6 (figure 4.16), DF is getting a little lost. The new inscription he has just produced is difficult to grasp; how does it relate to the previous ones? The zeros only refer to the empty matrix "R" that, by definition, cannot become too big. This inscription is "not eligible" as one says in law; no relationship between this inscription and the previous ones can be established. Something else has to be tried:

DF: "It's so stupid. Sorry, I'm a bit rusty … I'll just try another way."

[at the end of line 15, DF types "`= R(y,x) + ones(numel(y), numel(x));`"]

DF: "So basically [to FJ], I do a 1×1 matrix that contains one and then I repeat it according to the size of the region. It's very stupid, but at least I'm sure it will work. We'll see if it changes anything."

[DF runs the script]
[figure 4.17—T7]

DF: "Well, at least it doesn't change anything. It doesn't block here either."

The experiment of DF is conclusive. At T6 (figure 4.16), he was not totally convinced by the instruction at line 15. At T7 (figure 4.17), he tries another equivalent "stupid" way to express it. We do not need to dig too far into this

```
 1.  I = imread(images{1});
 2.  R = zeros(size(I,1), size(I,2));
 3.  size(R)
 4.  users = images {1,2};
 5.  for i = 1:size(users), 1)
 6.      j = 0;
 7.      while 1
 8.          rect = users{i,j+3};
 9.          if size(rect,2) == 0
10.              break
11.          end
12.          j = j+1;
13.          x = rect(1):rect(1)+rect(3);
14.          y = rect(2):rect(2)+rect(4);
15.          R(y,x) = R(y,x) + ones(numel(y),
                     numel(x));
16.      end
17. end
```

```
ans =
   1024    712

rect=
   197   91   323   371

Index exceeds matrix
dimensions
```

Figure 4.17
Editor and Command Window at T7.

affective aspect of code since we are going to consider it later on in the chapter. At this point, what is more important is that DF used an instruction he was certain INT could translate. The solidity of this *fact*, certainly consolidated during his previous experiences with Matlab programming language, allows him to equip a new experiment. Once again, when articulated with the previous inscriptions, the two new inscriptions "ans = 1024 712" and "rect = 197 91 323 371" are instructive; as they are similar to the ones that appeared at T4, DF can conclude that the problematic phenomenon engaging INT does not derive from the line 15 of the script. It has to be somewhere else, again:

> **DF:** "OK, I'll do something very, very stupid but I just want to see if it's here."

[DF creates a new line at 7; types "1"; creates a new line at 10; types "2"]
[DF runs the script]
[figure 4.18—T8]
[DF examines the Command Window of figure 4.18—T8]

> **DF:** "OK. It's here [at line 9 of figure 4.18—T8]. See? [DF puts the cursor on line 9] It gives '1,' then 'rect,' then '2,' then '1,' then stops. It's this 'j+3' that becomes too big after the first rectangle. It takes the first rectangle, and if the second rectangle is bigger, it just can't increment."

At T8 (figure 4.18), the *stupid thing* pays off: the new inscription successfully identifies the source of the problematic phenomenon engaging INT. At

```
1.   I = imread(images{1});
2.   R = zeros(size(I,1), size(I,2));
3.   size(R)                                          ans =
4.   users = images {1,2};                             1024    712
5.   for i = 1:size(users), 1)
6.       j = 0;                                        ans =
7.       while 1                                        1
8.            1;
9.            rect = users{i,j+3};
10.           2                                        rect =
11.           if size(rect,2) == 0                      197   91   323   371
12.               break
13.           end                                      ans =
14.           j = j+1;                                  2
15.           x = rect(1):rect(1)+rect(3);
16.           y = rect(2):rect(2)+rect(4);            ans =
17.           R(y,x) = R(y,x) + ones(numel(y),         1
                 numel(x));
18.       end                                         Index exceeds matrix
19.  end                                              dimensions
```

Figure 4.18
Editor and Command Window at T8.

line 9, "j+3" *becomes too big after the first rectangle*, thus disrupting INT in its translation efforts. But how does DF make this inference? How does he confidently attribute to line 9 the responsibility of disrupting INT? If we look attentively at the Command Window of T8, just as DF does, we see that its first series of numbers—"1024" and "712"—expresses the size of "R" as line 3 of the script in the Editor instructs it. If we continue our examination, we see that the subsequent number "1" expresses the instruction "1" as line 8 instructs it. Then we see that the third series of numbers—"197," "91," "323," and "371"—expresses the size of the first rectangle as line 9 instructs it. Then the fourth number in the Command Window—"2"—expresses the instruction "2" as instructed at line 10. The fifth number—"1"—expresses, again, the instruction "1" on line 8. This element is crucial because it shows that, at this specific moment, INT is about to deal with the *second* rectangle. And as the last element of the Command Window indicates, as soon as INT tries to translate line 9 *for the second time*, it blocks and prints a red error. By sequentially examining the Command Window, what is affecting INT becomes for us—as for DF—identifiable: at the second round of the script, INT is not able to translate line 9. This last inscription allows DF to attribute the origin of the INT-related phenomenon to one specific location.

At this point, it is important to remember that this last inscription—even though crucial—did not allow by itself the constitution of a connection between INT's red inscription and line 9. It is the addition and the alignment of all the previous inscriptions that progressively led to the definition of this last inscription. The whole aligning process allowed DF to pinpoint the provenance of the phenomenon affecting INT: it cannot translate "j+3" at line 9 for the second time.

As some readers may have noticed, in order to account for this small programming sequence I used several notions that have been developed in the STS literature to describe an a priori very different process: experimental practices in scientific laboratories. I now need to discuss this connection between laboratory practices and computer programming practices I have surreptitiously drawn.

For the last fifty years, many studies of scientific work have underlined the centrality of textual documents (Latour and Woolgar 1986), diagrams (Netz 2003), graphs (Dennis 1989; Gooday 1990), and notes (Lynch 1985; Garfinkel 1981) that I gather here—following Latour (2013)—under the umbrella term "inscriptions." Other important studies also showed the centrality of the instruments and experiments required to produce, confront, and articulate these inscriptions (Hacking 1983; Knorr-Cetina and Mulkay 1983; Collins 1975; Dear 1987; Gooding, Pinch, and Schaffer 1989). And still other studies further emphasized the importance of the manipulation and circulation of these inscriptions (Latour 1987; Knorr-Cetina 1999) that, through comparison, confrontation, alignment—in short, *articulation*— sometimes end up forming what Latour (1999a) calls "chains of reference": more or less solidified paths that document, when everything is in place, the behavior of some remote entity (e.g., a planet, a virus, a particle). These important studies present certified knowledge as being produced and objective at the same time: thanks to scientific practices—and scientific *institutions* that support the expression of these practices—knowledge *is* objective.[5]

As this short programming sequence seems to indicate, programming practices may sometimes—not always—resemble some of the practices required for the construction of certified knowledge. Indeed, the production of inscriptions—via experiments and instruments—and their comparison and alignment in order to produce even *more* inscriptions echo well with what has been observed in scientific laboratories. Little by little, through the manipulations, comparisons, and alignments of inscriptions, some access is paved out that may allow the characterization of a phenomenon engaging a remote entity. In the case of computer programming, this remote entity may vary: it can be, for example, a Matlab interpreter, a C compiler, or an Intel microprocessor. At any rate, the common characteristic of these different entities is the incredible swiftness of their constitutive relationships. Indeed, how is it possible to have a grip on an interpreter, a compiler or—worst—a processor that executes billions of operations per second? Once assembled, these entities are very difficult to grasp; hence the relevance of the scientific mode of veridiction to better understand what is affecting them. Moreover, I assume that the adoption of laboratory practices during computer programming episodes is not a result of the miniaturization of electronic components that followed the development of planar process at the end of the 1950s (Lécuyer, Brock, and Last 2010). As shown by historical studies of early electronic computers made of two-meter-high accumulators and multipliers—themselves made of hundreds of resistors connected with wires and soldered joints—every short circuit, carry errors, or divider fault that occurred during computation episodes had to be identified and located through the tedious formation of error reports, inscriptions, and experiments (Haigh, Priestley, and Rope 2014; 2016, 60–83). In these early days of electronic computing, programmers also had to align inscriptions to pave out an access to the affected component of the system.

Another similarity between scientific practices and the practices of computer programming is a common tendency to forget about the instruments that enabled the characterization of the phenomenon under scrutiny. In both cases, when the source of a phenomenon has been identified thanks to a specific laboratory setting, the practices, instruments, and experiments that allowed the formation of the chain of reference are generally put aside (Latour and Woolgar 1986, 105–155). This characteristic of science can make its history difficult to conduct. As established facts are purified from the scaffoldings that allowed them to be assembled and solidified in the first place, great may be the temptation to start from established facts and extrapolate backward (Collins 1975). To empirically grasp the practice of science, it is therefore crucial to consider facts as *consequences* of specific processes rather than *causes* of prior events (Bloor 1981). To a lesser extent, the same is true for computer programming. When the phenomenon engaging the remote entity is characterized; when the problematic location in the script is identified, most of the instruments (small bits of code, questions to FJ, "stupid things") are put aside and soon forgotten. At the end of the programming episode, when the script is *functional* and performs as desired, most of these *intermediary objects* (Vinck

2011) are generally left behind. Consequently, if one takes completed scripts or programs as starting points for the study of programming, the greater is the risk to miss what has been necessary to complete these scripts or programs.[6]

For the case of computer programming, one may imagine different expressions of the alignment practices I have documented above. Even though I conjecture that these expressions still consist in forming chains of reference in order to access remote entities and point at specific locations within numbered lists of inscriptions, they may not necessary deploy themselves in a spatio-temporal landmark that is similar to the one of DF. If we consider for example "program testing"—an important industrial process that consists in detecting and documenting errors in order to modify lines of code—this work can be highly distributed in space and time (Parrington and Roper 1989; Myers, Sandler, and Badgett 2011).[7] The "bug reports" we often encounter when one of our software programs crash for mysterious reasons are other expressions of this necessity to align inscriptions because they consist precisely in documenting at what time and following what actions the program fatally affected the interpreter, compiler, or processor. These reports serve as first inscriptions that will, in turn, be articulated with another one, and then another one, until eventually it indicates one origin of the phenomenon within the source code of the program. Moreover, alignment practices can also be automated and integrated within the programming languages themselves. This is typically the case when an interpreter or compiler indicates by itself its *breakpoint*, the line of the script that negatively affects its trajectory. But if these error reports appear automatic to the programmer, it should not be forgotten that they are the product of *hetero*matic processes as the programming teams involved in the maintenance and enhancement of programming languages have to cope with alignment of inscriptions in order to establish what type of errors should be indexed in the first place.[8] While different in terms of extension and labor involved, these processes of program testing, bug reporting, and programming language design are also, possibly, about aligning inscriptions and producing chains of reference.

The practice of aligning inscriptions to identify locations within numbered lists of written symbols may also explain, at least in part, the obsession of professional programmers with program intelligibility.[9] This topic has been well documented by Button and Sharrock (1995) in their admirable, yet solitary, study of computer programming practices. As they showed, making a program intelligible to other programmers involves conventional *naming* of variables and functions to make its structure readable as an organized and referenced document. It also involves *formatting* and *laying out* the different functions and parameters of the code to make it easily browsable from its visual organization. This also typically includes *commenting* on the program by means of small explicative sentences whose initial symbols ("%" for the case of Matlab) allow them to be ignored by interpreters or compilers. If the programming sequence we have just been following does not directly deal with formatting, laying out, and commenting, it nonetheless specifies what these practices are striving toward. In view of the elements presented above, naming, formatting, and commenting all point to future moments when they can operate as landmarks directly enrollable in the constitution of chains of reference. These marks may thus form an additional referential infrastructure capable of accelerating alignment work in the event of a future negative affection of an interpreter or a compiler (which is likely to happen in corporate settings where complex programs have to be maintained and enhanced).

But are the alignment practices of computer programming equivalent to the laboratory practices in the sciences? Of course not, and it is now time to present an important difference between them. Whereas the alignment practices of programming lead to the identification of a location within a script, scientific laboratory practices generally lead to the definition of *new objects* whose properties and contours are later presented in academic papers and discussed among peers. We will come back to this crucial aspect of the formation of scientific knowledge when we will consider mathematics in chapters 5 and 6. For now, suffice it to say that whereas both impetuses and outcomes of alignment practices in computer programming mainly concern programmers who try to complete adequate scripts, alignment practices in scientific laboratories are turned toward the completion of persuasive written claims. Scientific laboratories are always *counter-laboratories* (Latour 1987, 79–100): they are also to be understood as a means to publish stronger claims than their competitors. The agonistic aspect of laboratory practices in the sciences that constantly try to establish what should count as natural must then be demarcated from the self-referential aspect of laboratory practices in computer programming: While scientists try to make a case for the objective reality of the phenomena they practically make appear, programmers try to follow a *scenario* they are attached to (more on this later). In short, the networks in which scientists and programmers participate are, I believe, quite dissimilar. Whereas alignment of inscriptions in the sciences support the publication of claims, alignment practices in computer programming support the completion of a technical artifact that yet needs to be intelligible in corporate settings.

The analogy between scientific and programming practices therefore has its limits. Yet I also believe that both practices share some crucial—and quite surprising—similarities, both allowing the formation of chains of reference and access to remote beings. And just like scientific work, computer programming cannot be reduced to this specific type of practice. Indeed, once the remote entity has been reached, once the problematic location has been localized, many operations still need to be conducted. In this respect, aligning inscriptions is only a small part of the activity of programming.

Technical Detours

We saw in the previous section that sometimes, during programming episodes, when a small, swift, and difficult-to-grasp entity (e.g., an interpreter, a compiler, a microprocessor) is affected in its trajectory to the point of not being able to trigger electric pulses for the computation of data anymore, programmers need to multiply inscriptions, align them, and pile them up until the inscriptions constitute some access to the entity—access that, in turn, indicates a location within the script. But what happens next?

In this section, we will focus on another set of practices deployed during programming episodes. While this set of practices surely goes along the alignment of inscriptions, it has different implications. Whereas the scientific aspect of programming involves the addition and alignment of inscriptions (experiments, confirmations, "stupid things") in order to reach a remote entity, what I shall call the *technical aspect of programming* involves the inclusion and substitution of entities to get around impasses. Once again, this odd sentence will hopefully become clearer as the chapter goes on. For now, we shall continue to follow PROG, starting exactly when the previous sequence ended:

[DF examines the Command Window of figure 4.18—T8]

DF: "It is this 'j+3' that becomes too big after the first rectangle. It takes the first rectangle and if the second rectangle is bigger, it just can't increment. So I'll just put in some order."

```
1.   I = imread(images{1});
2.   R = zeros(size(I,1), size(I,2));

3.   users = images {1,2};
4.   for i = 1:size(users), 1)
5.       j = 0;
6.       while 1

7.               rect = users{i,j+3}

8.               if size(rect,2) == 0
9.                   break
10.             end
11.             j = j+1;
12.             x = rect(1):rect(1)+rect(3);
13.             y = rect(2):rect(2)+rect(4);
14.             R(y,x) = R(y,x) + ones(numel(y),
                    numel(x));
15.     end
16.  end
```

Index exceeds matrix
dimensions

```
3.   size(R)
8.   1
10.  2
```

Figure 4.19
Editor and Command Window at T9.

[DF deletes lines 3, 8 and 10; deletes ";" at the end of line 9]

[DF runs the script]

[figure 4.19—T9]

DF: "OK, we just need to change a few things."

As we saw in the previous section, DF managed to localize the line of the script that is badly affecting INT. Several inscriptions had to be produced and aligned in order to establish this certified knowledge. But these inscriptions are now useless; they were only relevant as part of DF's quasi-scientific inquiry into INT. It is now time for DF to really *change a few things* in the script. To do so, he starts by putting in *some order* and deleting the instructions that were used to him as experimental instruments (figure 4.19).

At this point of the chapter, to account for what happens next, I need to introduce a complementary notation that will allow us to have a better grip on the *technical innovations* DF is about to conduct. Following results of historical and sociological studies of technical projects, the notation I will draw on has been proposed during the 1990s as an attempt to illustrate the evolution of technical projects without using the traditional and problematic distinction between nature and society (Latour, Mauguin, and Teil

1992). We do not need to understand all the subtleties of this mapping that, by the way, never really took off.[10] For what interests us here, we shall only cover the basic principles of these so-called sociotechnical graphs (STGs).

One of the results of the studies of sociotechnical projects was to show that the trajectories of such projects are a function of their capacity to enroll new *actants*—human or nonhuman entities—in order to overcome critical impasses (Akrich 1989; Callon 1986; Latour 1993a). Historical examples of such enrollments are legion: in order for American Bell to prevail over Western Union in the development of the telephone network in the United States, it had to enroll—after many lawsuits—crucial telephone patents within its sociotechnical network (Brooks 1976). By enabling the production of highly reliable and flexible switching transistors, the planar process allowed Fairchild Semiconductor to become a commercial partner of the US Air Force (Lécuyer, Brock, and Last 2010). By enrolling the time-sharing technology as developed at MIT at the beginning of the 1960s, John Kemeny and his team were able to pursue the development of the BASIC programming language at Dartmouth College (Montfort et al. 2013, 158–194). For each example, a specific actant—a set of telephone patents, the planar process, the time-sharing technology—is enrolled, and this, in turn, makes the project slightly shift. One important credit to the history and sociology of technologies is to have successively demonstrated how crucial the inclusion of new actants for the development of technical projects is—may they be huge as the electrification of the United States at the end of the nineteenth century (Hughes 1983; Nye 1992) or small as the installation of a road bump (Latour 2006).

Yet, this "latitudinal" dimension of technical projects enrolling new actants in order to develop and expand would be incomplete without an orthogonal "longitudinal" dimension expressing the transformations suggested by the newly enrolled actants. Another crucial result of the history and sociology of technical projects is indeed that the inclusion of new actants simultaneously modifies the relationships among the previous actants of the project, thus potentially creating new impasses. Using the examples of the previous paragraph, Bell's technical system was transformed by the inclusion of telephone patents: the previously tiny network became a potential monopoly over telephone communications in the United States, hence necessitating further reconfigurations so as not to be the target of antitrust lawsuits by the US Department of Justice (Gertner 2013). Fairchild Semiconductor was fundamentally transformed by the inclusion of the planar process: it became a powerful entity soon capable of industrial production of integrated circuits. These production capacities participated, in turn, in the development of intercontinental ballistic missiles, and this further created an explosion of the demand for integrated circuits and the progressive formation of serious competitors (most notably, Texas Instruments and Motorola; see Campbell-Kelly et al. 2013, 210–225). Similarly, the inclusion of time-sharing technology within Dartmouth's computer system greatly participated in the design of the BASIC programming language by considerably increasing its beta testing. But the inclusion of the actant "time sharing" also transformed Dartmouth's computing infrastructure, which, by allowing its extensive utilization by students, soon started to be used for original computer-game experiments (Montfort et al. 2013, 165–194). More than just enrolling (or losing) actants, technical projects are also modified by them. And just like the latitude—inclusive—axis of technical projects, this longitude—transformative—axis does not only concern large and highly complex technological systems: small mundane projects are also affected by it (Latour 1992).

Building on this dual aspect of technical projects as well as concepts borrowed from linguistics, the proponents of STGs proposed a way to map the development of technical projects according to two dimensions: a *syntagmatic* dimension and a *paradigmatic* dimension. The first dimension (syntagmatic) of STG is defined by specific assemblages of actants at a certain time T. This configuration of actants at a time T is specific to each technical project and should therefore be supported by a narrative that exposes the whys and wherefores of the project being considered. As this dimension expresses association among variables, it can be called the AND dimension. The configuration of actants in the AND dimension is separated into two branches: the "allies" whose configuration participates in the development of the project and the "opponents" whose configuration constitutes an obstacle to the completion of the project. Again, which actant counts as an ally or as an opponent to the development of the project depends on the narrative the STG is only summarizing (Latour, Mauguin, and Teil 1992, 39). The boundary that separates allies' configuration of actants and opponents' configuration of actants constitutes the "frontline" of the technical project at time T.

The second (paradigmatic; nothing to do with Thomas Kuhn's notion) dimension is defined by the substitutions that have occurred in both allies' and opponents' configurations at time T + 1. Since this dimension expresses substitution of variables, it can be called the OR dimension. Depending on the fluctuation of allies' and opponents' configurations at $T+1$, the front-line of the technical project may also fluctuate. Once again, which actant is substituted by another, thus potentially making the frontline fluctuate, depends on the narrative of the technical project.

Two other elements are necessary to translate the narrative of a technical project into an STG: a specified point of view and what I call a "scenario." First, the point of view of the actant whose view on the project is being summarized by the STG has to be specified. In that sense, for any given narrative about a technical project, if this narrative takes the point of view of many different actants, each point of view can (potentially) be mapped by one specific STG. Second, the desire of the actant whose point of view is being mapped also has to be specified. This topic is a tricky one and will be further developed in the next section of this chapter. For now, suffice it to say that what the actant wants to achieve, the future it wants to live in, the *scenario to which it is attached* should be specified in each STG.

Let us now try to adapt these theoretical elements to the project that interests us here: DF's project to complete PROG. If we consider T8 and the whole narrative that precedes it, we might be able to translate it into an STG summarizing DF's allies and opponents. The first element of the graph should indicate the point of view that it re-presents. Contrary to most narratives about large technical systems where many points of view are considered and confronted, our small narrative only accounts for the point of view of DF. The second element of the graph should be the *scenario* to which DF is attached. As already touched upon in the previous section, we know that DF's scenario for PROG can be summarized as such: "Creating a matrix whose pixel-values correspond to the numbers of rectangles drawn by workers on each pixel." Concerning the actants: every instruction of the script can be considered an actant as they all make INT do things. But other actants might also be included in the graph as long as they impact on the project as framed by its scenario. In that sense, the red inscriptions printed in the Command line and what these inscriptions refer to according to DF as well as the final actions the script is intended to accomplish on the data of the .txt file can also be included in the STG. Moreover, as the narrative of the script-project indicates that several instructions are now stabilized, we may consider these "stable packages" of instructions as one single actant. If we consider these elements altogether and adapt them for T8, we end up with a diagram that looks like figure 4.20.

Point of view of DF

Scenario:
Create a matrix whose pixel-values correspond to the numbers of rectangles
drawn by workers on each pixel

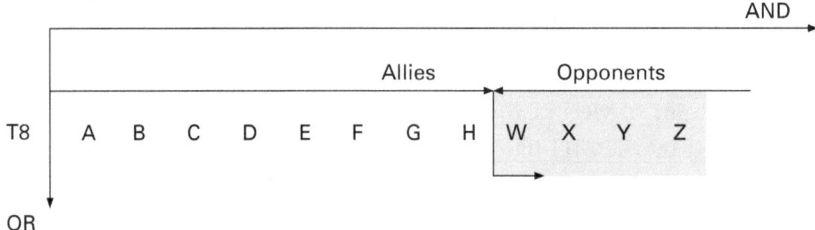

Figure 4.20
STG of T8. "A" refers to PROG lines 1, 2, and 4 (stabilized since T0); "B" refers to
line 3; "C" refers to lines 5, 6, and 7 (stabilized since T0); "D" refers to line 8; "E"
refers to line 9; "F" refers to line 10; "G" refers to lines 11, 12, 13 (stabilized since
T0); "H" refers to lines 14, 15, 16, 17, 18, 19 (stabilized since T6); "W" refers to the
inscription "Index exceeds matrix dimensions"; "X" refers to DF's assertions "the
second rectangle is too big for INT"; "Y" refers to DF's assertion "rectangles cannot
increment the values of the matrix"; and "Z" refers to the script's incapacity to follow
the desired scenario.

It is important to remember that the STG mapping of T8 is a simplification of T8 as initially presented in its Matlab view and enriched by DF's
sayings. As any simplification, it omits many elements. But as many sim-
plifications, it may also work as an instrument to identify key features of
messy processes (Star 1983).

From this point, based on the narrative presented above, we can include
T9 in the STG graph, thus slightly modifying the configurations of allies
and opponents (see figure 4.21).

For each remaining T of this programming sequence, I will first present
its complete narrative (simplified Matlab IDE and transcriptions of DF's
sayings), discuss it shortly, and then present its STG translation. As both
"point of view" and "scenario" will not change throughout the programming sequence, I will ignore them from now on. Moreover, in every new
STG, I shall highlight the newly enrolled actant in bold. At the very end
of the programming sequence, when DF will have completed PROG, the
succession of all the STGs should allow us to detect another set of practices
deployed by programmers that goes along with the alignment of inscriptions while being, I believe, fundamentally different.

Point of view of DF

Scenario:
Create a matrix whose pixel-values correspond to the numbers of rectangles drawn by workers on each pixel

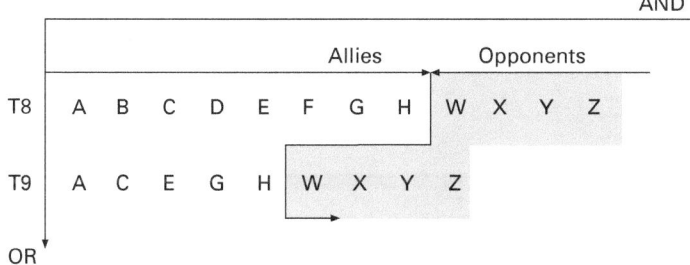

Figure 4.21
STG of T8 and T9.

Let us continue to follow DF as he tries to shape PROG:

DF: "We're gonna do it like this."

[DF creates a new line at 7]

DF: "If 'j+3' is larger"

[at line 7, types "if j+3 >"]

DF: "than the size of the cell of the user"

[at line 7, types "size(users{j})"]

DF: "then it goes over it"

[DF creates a new line at 8; types "break"]
[DF runs the script]
[figure 4.22—T10]
[DF examines Command Window of figure 4.22—T10]

DF: "Argh, of course. I shouldn't take 'j.' Can't define anything that way."

At T10 (figure 4.22), DF enrolls a new actant: the "if" statement that starts at line 7 and ends at line 9. Since, at this point, he knows for a fact that INT is blocked if the second rectangle is bigger than the first one, the addition of a conditional statement that could ask INT to *go over* this dimension problem makes complete sense. The addition of an "if" statement would thus allow INT to continue its interpretation of the script even though it encounters a rectangle bigger than the previous one. But as the red inscription and DF's

```
1.   I = imread(images{1});
2.   R = zeros(size(I,1), size(I,2));
3.   users = images {1,2};
4.   for i = 1:size(users), 1)
5.      j = 0;
6.      while 1
7.          if j+3 > size(users, {j})
8.                  break
9.          end
10.         rect = users{i,j+3};;
11.         if size(rect,2) == 0
12.                 break
13.         end
14.         j = j+1;
15.         x = rect(1):rect(1)+rect(3);
16.         y = rect(2):rect(2)+rect(4);
17.         R(y,x) = R(y,x) + ones(numel(y),
                numel(x));
18.     end
19.  end
```

Cell contents indices
must be greater than
0

Figure 4.22
Editor and Command Window at T10.

saying indicate, the statement was inappropriately expressed: DF should not have taken "j" as the size variable of "users" since it already equals to zero at line 5. The consequence of this attribution mistake is that INT *cannot define anything.* No rectangle can be defined, and the matrix cannot, in turn, be incremented.

If we map T10 as an STG in line with T8 and T9, we obtain figure 4.23. Looking at it, we can see that new actants have appeared and created differences in the project, slightly altering its frontline. In the allies' configuration, "I" has been added by DF in order to get around the configuration of "W," "X," "Y," and "Z." But if this new actant made "W" and "X" disappear—that is, the index does not exceed the matrix dimension *anymore,* and the second rectangle is not too big *anymore*—it is only by making two new opponents appear: "V" and "U." "Y" and "Z" are then still solidly opposing resistance to DF's project since, at this point, no rectangle can be defined.

Let us continue:

[at line 7, DF deletes "users, {j})"]

DF: "Actually, the size of the cell should just be 'users, 2'"

[at line 7, types "users,2)"]
[runs the script],
[figure 4.24—T11]

DF: "OK, it may work."

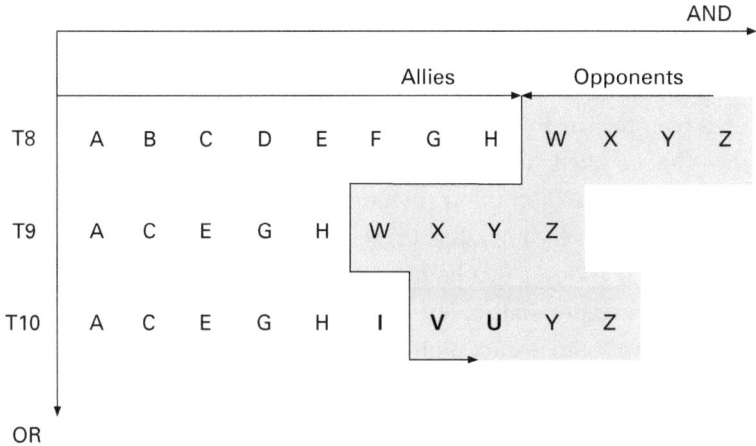

Figure 4.23
STG of T8, T9, and T10. At T10, "I" refers to lines 7 to 9; "V" refers to the inscription
"cell contents indices must be greater than 0"; and "U" refers to DF's assertion "nothing can be defined."

```
1.  I = imread(images{1});
2.  R = zeros(size(I,1), size(I,2));
3.  users = images {1,2};
4.  for i = 1:size(users), 1)
5.     j = 0;
6.     while 1
7.         if j+3 > size(users, 2)
8.             break
9.         end
10.        rect = users{i,j+3};;
11.        if size(rect,2) == 0
12.            break
13.        end
14.        j = j+1;
15.        x = rect(1):rect(1)+rect(3);
16.        y = rect(2):rect(2)+rect(4);
17.        R(y,x) = R(y,x) + ones(numel(y),
                  numel(x));
18.    end
19. end

7.  size(users, {j})
```

```
>>
```

Figure 4.24
Editor and Command Window at T11.

At T11 (figure 4.24), DF modifies the conditional instruction: instead of referring to "j," the size of the new rectangle now refers to the second value of the cell, "users." We do not need to understand precisely what this value and cell refer to. The important thing at T11 is that the inclusion of a new actant—the modified "if" statement—creates an important difference: INT does not print a red inscription anymore. This indicates that INT has managed to translate every line, thus triggering electronic computation on the data of the .txt files. At this point, then, *it may work*: the rectangles may increment the empty matrix. But it is not over yet since, symmetrically, *it may also not work*. Since the Command Window does not provide any indication about the incrementation of the empty matrix, something else may also have happened.

If we continue our STG re-presentation of this programming sequence by including T11, we obtain figure 4.25. Several changes affected the allies' configuration at T11. "I" disappeared: DF deleted it because it made opponents disappear only by making new ones appear. But two new actants are included: "J" that corresponds to the new conditional statement and "K" that corresponds to the absence of any error inscription within the Command Window (and, corollary, to DF's assertion that "it may work"). Did this new configuration of allies managed to get around the configuration of opponents? Only partially since the incertitude suggested by "K" has its corollary: as there is no indication in the Command Window, the script may also not work ("T"), that is, it may not increment the empty matrix properly. As a consequence, "Z"—"the script does not follow the scenario"—holds on. At this point, DF still needs to include something else; he still needs to pursue his project *by other means* in order to *get around* the impasse constituted by "T" and "Z."

Let us continue to follow DF:

DF: "But I just need to be sure."

[creates a line 20; types "imshow(R)"]
[runs the script]
[figure 4.26—T12 and figure 4.27]

FJ: "This is close!"

DF: "Yep. But it clips after the value 1."

FJ: "Clips?"

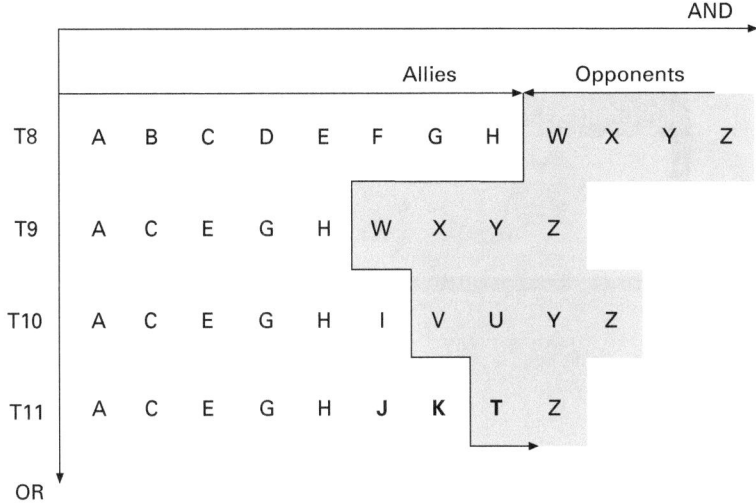

Figure 4.25
STG of T8, T9, T10, and T11. At T11, "J" refers to the new "if" statement at lines 7 to 9; "K" refers to DF's assertion that "it may work"; and "T" refers to DF's implicit assertion that, symmetrically, "it may not work."

```
1.   I = imread(images{1});
2.   R = zeros(size(I,1), size(I,2));
3.   users = images {1,2};
4.   for i = 1:size(users), 1)
5.       j = 0;
6.       while 1
7.           if j+3 > size(users, 2)
8.                   break
9.           end
10.          rect = users{i,j+3};;
11.          if size(rect,2) == 0
12.                  break
13.          end
14.          j = j+1;
15.          x = rect(1):rect(1)+rect(3);
16.          y = rect(2):rect(2)+rect(4);
17.          R(y,x) = R(y,x) + ones(numel(y),
                 numel(x));
18.      end
19.  end                                        >>
20.  imshow(R)
```

Figure 4.26
Editor and Command Window at T12.

Figure 4.27
Screenshot of the output of PROG at T12.

> **DF:** "Yes, it often does that. Basically, it doesn't consider anything above 1. I mean, the matrix may have values more than one, but it does not show it on the image."

At T12 (figure 4.26), DF adds a new instruction—"imshow(R)"—that asks INT to print an image of the incremented matrix (figure 4.27). The results are convincing as well as disappointing. The positive thing is that a matrix has effectively been incremented. The image printed by INT attests to this: it has differentiated values that together form a white shape. But the negative thing is that this image has only binary values: ones forming the white shape and zeros forming the black background. According to DF, INT is once again the cause of this problem: by *clipping* after the value 1 the printed image can only be binary. In these conditions, it is difficult to know what values constitute the incremented matrix. At this point, again, DF needs to include something else in the script in order to make it follow the desired scenario.

Let us have a look on the STG to get a condensed look on what has just happened (figure 4.28). The configuration of allies has again expanded: "L" and "M" allowed DF to be sure that the rectangles increment the matrix.

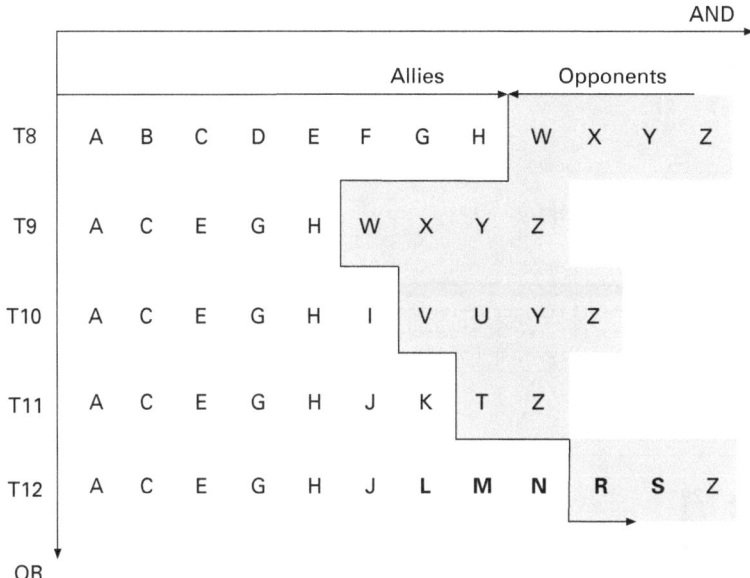

Figure 4.28
STG of T8, T9, T10, T11, and T12. At T12, "L" refers to the instruction "`imshow(R)`" at line 20; "M" refers to the binary image of the matrix output by PROG; "N" refers to DF's conclusion that rectangles do increment the matrix; "R" refers to DF's assertion that INT "clipps" after 1; and "S" refers to the DF's saying that the matrix should not have only binary values.

This, in turn, made "T" disappear so that no incertitude remains concerning this aspect of the project. But the binary characteristic of "M" made "R" appear in the configuration of opponents: for unknown reasons, INT *clips* after one. This, in turn, creates "S," the incertitude about the incrementing capability of the script that may stop after "1." In these conditions, Z remains, and the script is still not following the desired scenario. Once again, DF has no choice: he has to enroll something else to the configuration of allies; he has to *delegate* the work-around of "R," "S," and "Z" to a new actant.

With these elements in mind, let us continue:

DF: "So I'll just try to divide the value of 'R' by the maximal value of the matrix. If it has other values than one, it should show it."

[at line 20, types "`/max(R(:))`"]
[runs the script]
[figure 4.29—T13 and figure 4.30]

DF: "All right, this is the right image of the matrix. This is it."

```
1.   I = imread(images{1});
2.   R = zeros(size(I,1), size(I,2));
3.   users = images {1,2};
4.   for i = 1:size(users), 1)
5.      j = 0;
6.      while 1
7.          if j+3 > size(users, 2)
8.                      break
9.          end
10.         rect = users{i,j+3};;
11.         if size(rect,2) == 0
12.                break
13.         end
14.         j = j+1;
15.         x = rect(1):rect(1)+rect(3);
16.         y = rect(2):rect(2)+rect(4);
17.         R(y,x) = R(y,x) + ones(numel(y),
                numel(x));
18.     end
19.  end
20.  imshow(R / max(R(:)))
```

>>

Figure 4.29
Editor and Command Window at T13.

Figure 4.30
Screenshot of the output of PROG at T13.

By including this last small bit of code—"/max(R(:))"—DF manages to complete the script (figure 4.29). No incertitude remains: the matrix is correctly incremented as the new output image shows (figure 4.30). DF thus successfully managed to make INT design an empty matrix according to width and height values; define rectangles from width, height, and position values; and use these rectangles to successively increment the pixel-values of the matrix. Several *technical* operations had to be conducted but, in the end, the project fulfilled its initial ambitions. At this point, the script can be considered a technical artifact that *does* something definable.

If we take a look at the STG (figure 4.31), we see that the inclusion of "/max(R(:))" managed to get around the impasse previously formed by "R," "S," and "Z." At T13, the inclusion of "O" and its corollary "P" made "R," "S," and "Z" disappear. The addition of the instruction "/max(R(:))" made

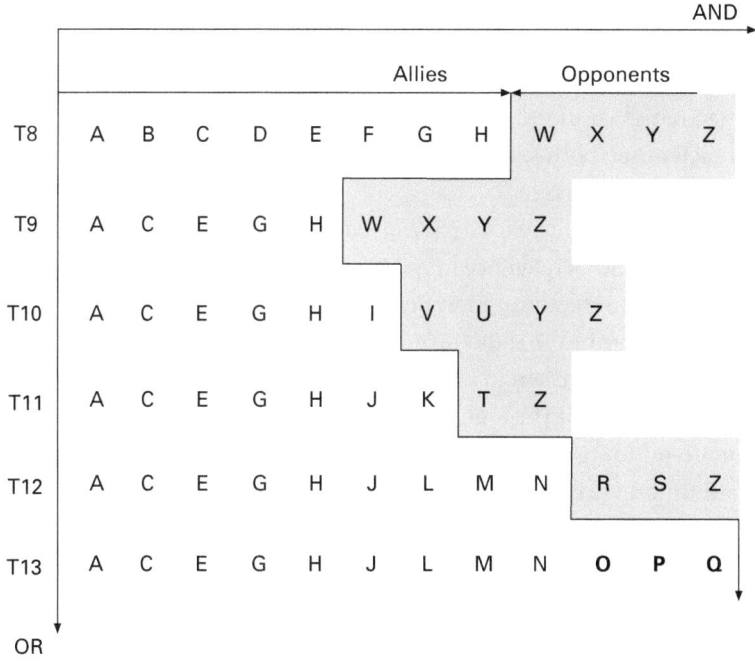

Figure 4.31
STG of T8, T9, T10, T11, T12, and T13. At T13, "O" refers to the instruction "/max(R(:))"; "P" refers the output image generated by PROG; and "Q" refers to the fulfillment of PROG's scenario: now, the pixel-values of the new matrix correspond to the number of rectangles drawn by workers on each pixel.

INT print a gray-scale image of the matrix, hence showing DF that its values do indeed variate between zero and the total number of rectangles drawn by the crowdworkers. All the opponents to the project have been replaced by allies; all dead-ends have been bypassed. The scenario is followed. As DF puts it, "this is it." The programming sequence is over.

What do these STGs add to our analysis of this programming sequence? What does this simplification allow us to see? While the previous section put the emphasis on the *scientific* moment of programming practices, I assume that the present section puts the emphasis on the *technical* moment of programming practices. Are *scientific* and *technical* practices different? In the middle of the action, they surely overlap to the point of appearing similar. But, following Latour (2013), I nonetheless assume that both express themselves quite differently.

We saw that the surprising similitude between the laboratory practices of science and the practices of programming lies in that they both multiply and align inscriptions in order to shape chains of reference, thereby allowing the assemblage of information about remote entities. Even though both activities cannot be considered equivalent, I believe they echo well with each other: both sometimes produce and align inscriptions in order to access remote beings.

Although the sequence we have just documented required the formation of a (small) chain of reference in order to be initiated, I assume the sequence also expressed something radically different. At T9, DF needed to *change things* in the script. What did he do? At each T, he *included* new actants and *delegated* actions to them in order to *get around* impasses that were obstructing the following of the scenario. The practices involved in this sequence did not tend toward gaining knowledge about these impasses; they tended toward finding ways to get around them. This is precisely why STGs were, in the end, instructive tools: by simplifying the narrative, they allowed to follow these successive shifts, this constant zigzag that expressed the enrollment of new entities, the delegation they implied, and the work-arounds they triggered. The script, once completed at T13, became a *technical artifact*. But it was only through technical practices, ingenious inclusions, delegations, and work-arounds that such an artifact could come to existence. Along with the finished script, thanks to the simplification provided by the STGs, we can glance at the lightning strike drawn by DF and its technical actions (figure 4.32).

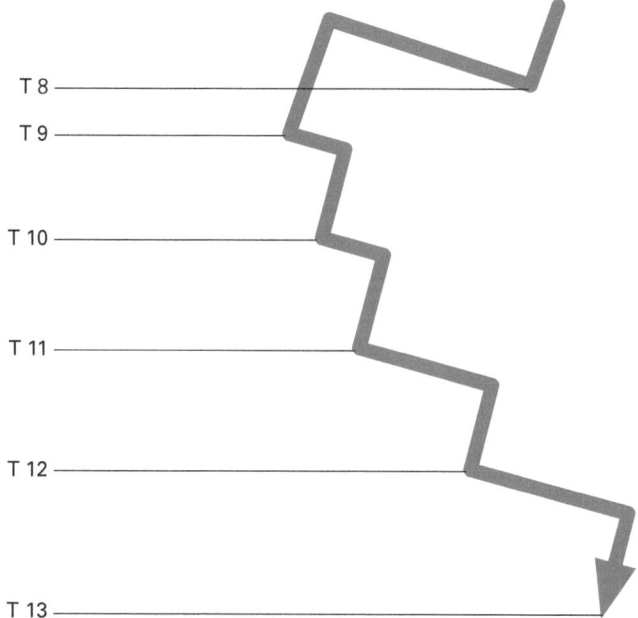

Figure 4.32
Technical zigzag of DF while assembling PROG.

The sequence was not linear; it was rhythmed by breaks of continuity that vanished at soon as the script was completed. Just as chains of reference are ignored as soon as they allowed the constitution of an information about a remote being, the constant shifts, inclusions, delegations, and work-arounds of technical practices are made invisible once they allowed the completion of the artifact. Here lies, I believe, a serious limitation of the studies of programming that only consider the results of programming tests (see chapter 3). By only considering the final technical object (the finished script), they cannot grasp the practices that were necessary to the technicality of this object. It is only by going backward from the artifact to the detours that have constantly modified its form, thus making it singular, that we may capture the technical aspect of computer programming. Any working script holds thanks to all the now-invisible allies that were added to each configuration in order to get around—one may even say, in order to *hack* (Nissenbaum 2004)—now also-invisible opponents. Just as the proliferation and alignment of inscriptions made DF become *knowledge-able*, the technical detours made him *in-genious*: by catching entities—*jinns*—and enrolling them *in* work-arounds, he was able to include allies and get around opponents, thus drawing a dazzling zigzag.

It is interesting to note that these types of technical moments, when programming is about the drawing of a zigzag, are often the most appreciated ones. While the construction of chains of reference can be very frustrating—the inscriptions keep piling up without forming any reliable chain of reference—the practices involved in the drawing of zigzags often appears more playful. Unfortunately, I cannot support this claim by any empirical materials; this would imply the presentation of many other programming figures that are already too numerous at this point in the chapter. But in one of her literary accounts of programming *affects*, Ellen Ullman nicely expressed this feeling programmers often experience when they are engaged into technical detours that are very difficult to catch once the artifact is completed:

> "Damn! The NULL case!"
> "And if not we're out of the text field and they hit space—"
> "—yeah, like for—"
> "—no parameter—"
> "Hell!"
> "So what if we space-pad?"
> "I don't know…. Wait a minute!"
> "Yeah, we could space-pad—"
> "—and do space as numeric."
> "Yes! We'll call SendKey(space) to—"
> "—the numeric object."
> …
> "No, no, no, no. What if the members of the set start with spaces. Oh, God."
> He is as near to naked despair as has ever been shown to me by anyone not in a film. Here, in *that place*, we have no shame. He has seen me sleeping on the floor, drooling. We have both seen Danny's puffy, white midsection—young as he is, it's a pity—when he stripped to his underwear in the heat of the machine room. I have seen Joel's dandruff, light coating of cat fur on his clothes, noticed things about his body I should not. And I'm sure he's seen my sticky hair, noticed how dull I look without make-up, caught sight of other details too intimate to mention. Still, none of this matters anymore. Our bodies were abandoned long ago, reduced to hunger and sleeplessness and the ravages of sitting for hours at a keyboard and a mouse. Our physical selves have been battered away. Now we know each other in one way and one way only: the code.
>
> Besides, I know I can now give him pleasure of an order which is rare in any life: I am about to save him from despair.

> "No problem," I say evenly. I put my hand on his shoulder, intending a gesture of reassurance. "The parameters never start with a space."
>
> It is just as I hoped. *His despair vanishes*. He becomes electric, turns to the keyboard and begins to type at a rapid speed. Now he is gone from me. He is disappearing into the code. (Ullman 2012, 8–9; italics added)

In this literary excerpt, an information is progressively being assembled—the narrator provides the very last inscription ("The parameters never start with a space")—and a location is, in turn, defined: let entities be enrolled, actions be delegated, and opponents be gotten around. And the joyful technical lightning strike soon unfolds.

Let the reader forgive me if I rave a little at this point of the chapter, but both technical and scientific practices as documented in these two sections provide such a refreshing perspective on computer programming that it is difficult for me to remain placid. We see indeed how the standard cognitive-behavioral framing of computer programming as a problem-solving process (cf. chapter 3) can be misleading. Programmers may never *solve* any problem; when confronted to a remote entity that refuses to generate electric pulses on data, they more or less collectively constitute a chain of reference that, if equipped enough, may indicate a problematic location, a location that, in turn, may trigger the enrollment of new actants and technical work-arounds of impasses. Nothing is solved; something is located, thus eventually triggering the drawing of a zigzag that will soon be forgotten. "Problem solving" and even the likable expression "debugging" may both miss the point: by amalgamating two different and equally important sets of practices, they may not adequately catch the subtle practical tempos a programmer goes through when defining *appropriate* lists of instructions.

Yet, despite my enthusiasm, this tentative model still lacks something crucial. Indeed, where does this "appropriateness" come from? Is it not something I surreptitiously invoke from outside, without defining its attributes? At this point, it surely is. Fortunately, this is precisely the topic of the next section of this chapter.

Attached to a Scenario

We have seen so far that programming can be viewed as the expression of two sets of intimately related practices. The first set implies the multiplication and alignment of inscriptions in order to assemble chains of reference that can provide information about remote entities whose trajectories are affected in undesirable ways. These practices echo well, to some degree, with some of the laboratory practices required for the construction of scientific facts. The second set of practices—much more difficult to capture— implies the inclusion of new actants in order to get around impasses. These practices of inclusion, delegation, and bypassing echo well, to some degree, with practices required for the running of technical projects. From this point, we may conjecture that during a computer programming episode, scientific and technical practices are intimately articulated, the programmer constantly shifting from one mode to the other. This tentative but empirical look at computer programming unfolds many crucial elements— inscriptions, chains of reference, impasses, detours—that most standard takes on computer programming do not stress.

At this point of the chapter though, something essential to computer programming is still taken for granted. While I keep on talking about "programming episodes," what defines the limits and the scope of such episodes? Where do these "meta-instructions" that establish the boundaries of the programming episodes come from? What is this wind that pushes programmers in the back, making them inquire into remote entities, enroll actants, and get around impasses? In the previous section of the chapter, readers may have noticed that I surreptitiously used the term "programming project" to speak about the technical skills DF was deploying for the composition of PROG. But where does this *projection* come from? At this point, this aspiration, this desire shall not be ignored anymore. It is time now to address the issues of *projection* and *attachment* without which there would simply be no programming practices.

Lucy Suchman thoroughly explored this relationship between projects and situated actions or, as she put it, "the utility of projecting future actions *and* the reliance of those projections on a further horizon of activity they do not exhaustively specify" (Suchman 2007, 19; emphasis in the original). Initially struggling against mid-1980s artificial intelligence experts who tended to consider the relationship between *plans* and *actions* as deterministic— the former rigorously defining the latter—she proposed an alternative view of plans as resources that set up horizons without specifying the actions required to reach them. To clarify her proposition, she used the example of canoe:

> In planning to run a series of rapids in a canoe, one is very likely to sit for a while above the falls and plan one's descent. The plan might go something like "I get as far over to the left as possible, try to make it around that next bunch." A great deal of deliberation, discussion, simulation, and reconstruction may go into such a plan. But however detailed, the plan stops short of the actual business of getting your canoe through the falls. When it really comes down to the details of responding to currents and handling a canoe, you effectively abandon the plan and fall back on whatever embodied skills are available to you. The purpose of the plan in this case is not to get your canoe through the rapids, but rather to orient you in such a way that you can obtain the best possible position from which to use those embodied skills on which, in the final analysis, your success depends. (Suchman 2007, 72)

Plans do not determine actions. Rather, by suggesting future orientations, plans help express skills in appreciable conditions. Moreover, building on Suchman's example, we can also assume that plans create something like another world, another *layer of existence*: by telling *stories*, plans express figures that could not exist without them. Before running the rapids, when I am expressing my plan above the fall, I am *projected* into another space ("into the rapids," "as far over to the left as possible"), another time ("later"), and toward other human and nonhuman actants ("me, alive, at the end of the rapids," "the canoe, struggling to get around the next bunch," "the powerful rapids I—hopefully—managed to run"). In this respect, by establishing a triple *shifting out* (Latour 2013, 234–257) into other space and time, and toward other actants, plans are also *narratives* that help us engage into desirable processes.

Yet this definition of plans as narratives establishing desirable horizons without specifying how to reach them is still quite loose. In what sense are these narratives different from, say, bedtime stories for children or Hollywood mega-productions? What specific transformations do plans-narratives institute? How do we address the modifications they suggest? To better understand the specificity of these narratives—or, as I will soon call them, these *scenarios*—we shall consider the narrative DF constructed for the completion of PROG. One point of departure could be two days before the programming episode we have followed in the last sections. At that time, I was struggling with the data I had previously collected from a crowdsourcing task. Unable to make sense of these data, I asked the director of the Lab (DIR) for some advice:

Thursday February 4, 2016, at the office of DIR

FJ: The thing is that I am still struggling to find measures that could make sense of the variations of the rectangles drawn by the workers [and] depending on the images.[11] Because at this point, I have this kind of result:

[FJ shows images on his laptop to DIR, see figure 4.33]

FJ: But the rectangles vary both in terms of size and alignment. That is, some rectangles are well aligned and small compared to the image; others are aligned but vary in terms of dimensions; others are aligned but in groups of different sizes; and others are just spread out everywhere.

DIR: Well, there's surely a way to measure how much overlap there is. But in any case, you should get other views than these. You can't see anything here.

...

There are many ways; but for example, you could go through each pixel and see how often they are in a rectangle. And once you get these graphs, we can help you find a measure that explains the variations.

FJ: You mean, something like getting for each pixel, the relative difference of the number of rectangles they are part of?

DIR: Yes. Or rather, I guess in your case, for each image, the proportion of pixels that are part of one rectangle, two rectangles, and so on. ... And then you can get gray-scale images, or graphs like histograms. For example, assume you're giving zero to every pixel that is labeled by no one, one for every pixel that is labeled by only one worker, etc. You add this up and you'll get a maximum or, like twenty. Then you can normalize between zero and one or do other things. But for now at least, you should get better matrices from these images.

DIR's advice was clear: if I wanted to find correlations between the pixel-values of the images and the rectangles drawn by the workers, the very first step was to *simplify* the collected results through the design of *better matrices*. But how should these matrices be designed? This issue was the raison d'être of PROG: in order to define simpler/better matrices whose values can be expressed by graphs, PROG should instruct my computer to transform the values of each image and its associated rectangles. In short, the graphs that could help me explain the dispersion/alignment of rectangles required matrices that still needed to be designed computationally by an instructed

Figure 4.33
Sample of labeled images shown to DIR.

computer. The first narrative—or plan—that further supported the formulation of PROG can thus be summarized as such: "FJ shall make a computer assemble matrices whose pixel-values correspond to the number of rectangles each pixel is part of."

I soon tried to write this program that could help me have a better grip on the data I had collected but was soon confronted to my incapacity to specify the problem with Matlab. What should be the first step? And the second step? Using the project's helping clause that allowed me to ask for help whenever I needed to (cf. above), I sent an email to DF:

Monday, January 15, 2016. Email from FJ to DF, header "Struggling with Matlab. ..."

Hi DF,

For my project I need to process each pixel of each image individually in order to count how many rectangles belong to each pixel. I got the idea, I think, but am still struggling with Matlab to write the script. Would you have some time to help me do it? That'd be great!

Have a great day,

FJ

Monday, January 15, 2016. Email from DF to FJ, header "Struggling with Matlab. ..."

Hi Florian,

No problem. What about this afternoon then? It should be quite easy. We'll check this together.

DF

Monday, January, 15, 2016. Email from DF to FJ, header "Struggling with Matlab. ..."

This afternoon is great. I'll be in my office. Come whenever you want.

See you then!

FJ

A couple of hours later, DF arrived at my office. Before starting to program, he told me what he intended to do:

DF: "Well, I think I know how to compute this. It shouldn't be difficult. So for each rectangle, we have the x and y coordinates right?"

FJ: "Well, a rectangle is defined by four values"

DF: "Yes so x and y [coordinates] and then the size, right?"

FJ: "Yes."

DF: "So basically we have this."

[DF starts to write in FJ's logbook]

DF: "And this, and then size. And all this defines the rectangle."

[DF draws figure 4.34 (A)]

DF: "Here [pointing at figure 4.34 (A)], you initialize all pixels of the matrix with the value 0. Then you iterate on all rectangles. So for the first rectangle of the image [starts to draw in FJ's logbook], you have the coordinates and you check what pixels of the matrix are in the rectangle."

[DF draws figure 4.34 (B1)]

DF: "And you add one for these pixels in the matrix. And then you do the same for the second rectangle [starts to draw in FJ's logbook] that might be here."

[DF draws figure 4.34 (B2)]

DF: "And you also add one for all these pixels. So here [pointing at figure 4.34 (B2)], some pixels in the matrix will have the value 0, some will have the value 1, and some others will have the value 2."

FJ: "OK, I see."

DF: "And you do this for all the rectangles. And once you have a script that works for one image, it's easy to adapt it [the script] to go through all the images."

FJ: "Sure."

DF: "And well, when you have these matrices with values 0, 1, 2, etc., you can make all the graphs you want like gray-scale images or histograms [draws in FJ's logbook] like this."

[DF draws figure 4.34 (C)]

DF: "Where x is the number of rectangles and y the number of pixels."

At this point, the narrative of PROG has thickened. From "FJ shall make a computer create matrices whose pixel values correspond to the number of rectangles they are part of," it has become "for every image, DF shall first make a computer use the dimension of the image to create an empty matrix,

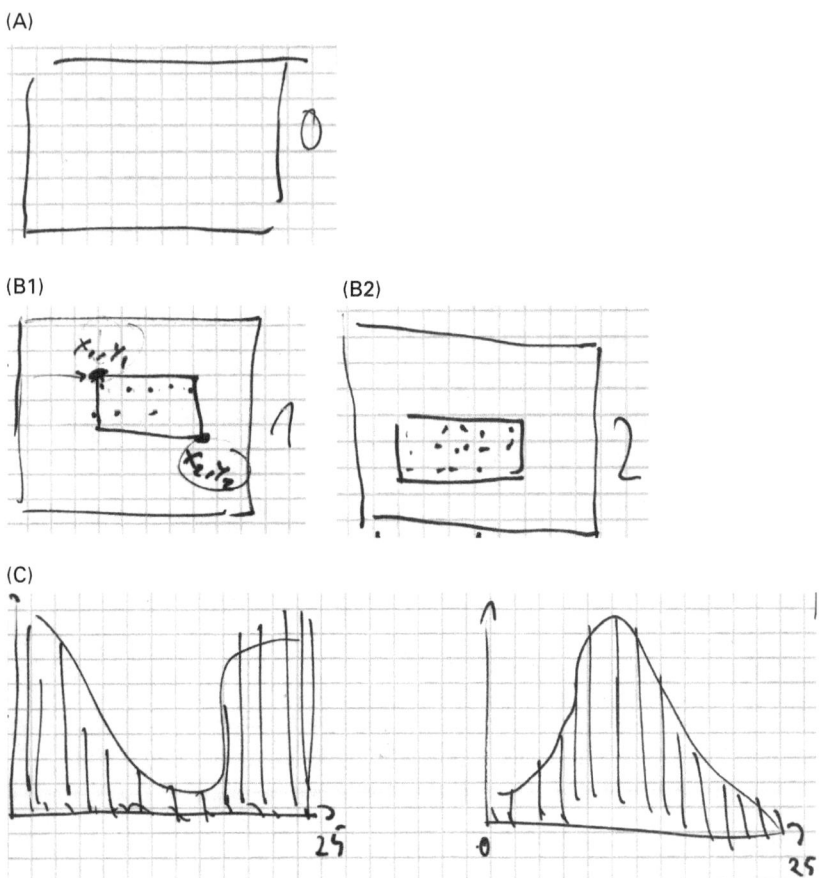

Figure 4.34
Drawings of DF in FJ's logbook.

then define the first rectangle of this image according to its coordinates as defined in its correlated .txt file, then add this rectangle to the matrix, then define the second rectangle, then add it to the matrix, and so on for every rectangle of the image." Even though the topic is slightly different from Suchman's (2007, 72) example of canoe, DF's narrative also works as a *resource* that sets up a horizon without specifying the actions required to reach it. Nothing is said about how to define the empty matrix, how to define a rectangle, and how to increment the matrix with these rectangles. Yet, altogether, the pileup of these steps institutes a desired future that the following actions should try to reach. Moreover, similar to Suchman's example, DF's narrative also creates another layer of existence. His story projects us into another time ("in a couple of minutes"), another space ("in front of the Matlab IDE"), and toward other actants ("incremented matrices," "gray-scale images," "histograms," "FJ being able to produce meaningful graphs thanks to the new program").

But DF's narrative—when considered in the light of the last two sections of this chapter—also suggests an important difference between narratives that institute desired futures and, say, bedtime stories for children or Hollywood mega-productions. When after the narrative has been expressed—that is, after having been projected into other times, other locations, and toward other actants—hopefully children fall into sleep and spectators leave the movie theater to carry on their occupations, DF's narrative *still has a hold on him*. More than just establishing a triple shifting out into other space and time and toward other actants, DF's narrative *engages* DF; it asks DF to do things. In this sense, as soon as DF expresses the narrative, he finds himself simultaneously in two positions: he is the writer of the narrative who can modify it any time he wants but also the actor who has to follow the narrative he has just expressed (Latour 2013, 391). Following Austin (1975) and recent works in STS (Barad 2007), we can consider these narratives as *performative* in the sense that they engage those who articulate them. In our case, DF holds the narrative but is also held by it.

To underline the literary and performative dimensions of these particular narratives that are crucial for computer programming—since they institute a desired horizon to be achieved, hence supporting both alignments of inscriptions and technical detours—I shall call them *scenarios*.[12] The cinematographic connotation is voluntary. Indeed, a scenario—in the case of cinema or computer programming—is a *narrative*: it tells a story and therefore instantiates a beginning, an end, a plot, and characters that all possess ontological weights. Second, in both cases, a scenario is *performative*: it has a hold on both the movie director who is asked to *transform* it into a movie as well as on the programmer who tries to make it *become* an actual computer program. Third, if a scenario roughly describes the successive scenes of a movie or the successive steps of a computer program, it says almost nothing about how to shoot these scenes or implement these steps. While in both cases, the scenario draws desirable horizons, almost everything still needs to be done in order to reach them. Fourth, if the plot, steps, characters, or variables are described by the scenario, nothing prevents the movie director, the programmer, movie stars, or recalcitrant instructions to *modify* some of its constitutive elements. In both computer programming and movie production, a scenario can be revisited to better take into account unpredictable contingencies.

While they are not sufficient to assemble computer programs, scenarios are nonetheless crucial for computer programming. These flexible yet performative narrative resources institute horizons on which programmers can hold—while being held by them—thus establishing the boundaries of computer programming episodes. Scenarios both trigger and are blended with alignments of inscriptions and technical detours; altogether, they form programming courses of action we can now consider in all their sinuosity.

But again, at this point, something is still missing. We are very close but are not there yet. If the notion of "scenario" is useful to better understand what helped DF shift between scientific and technical modes of practice, thus framing the programming sequences we have previously followed, it does not make us understand *why* DF wanted to engage himself in it. If scenarios provide the frame and the energy of programming episodes, where does this energy initially come from?

Something is definitely overflowing scenarios, making them "put into gear" more or less delightful *affects*: how do we consider them as well? If scenarios give horizons, they do not by themselves allow to grab what arises from programming episodes. INT's stubbornness, the multiple inclusions of actants, and the numerous work-arounds of impasses; all of this—in the middle of the action—is terribly uncertain. But when the program accomplishes what was hoped for at the beginning of the episode—or modified during the episode—something is happening that cannot be reduced to the consequence of what allowed it to happen. This is the important contribution of the sociology of *attachments* against the social science of taste: reducing beloved objects to the conditions—social or material—of their appreciations tells us *nothing* about the objects themselves (Hennion 2015, 2017). While an object—a painting, a piece of music, a computer program—is constructed, it also exists in its own right. Or perhaps even more; as it is constructed, it exists more intensely. But how do we grab this appreciation of the constituted object? In our case, how do we consider the upsurging of PROG? We may perhaps refer to what DF tells me at the end of the programming episode:

FJ: "Well, thanks. I'm always impressed by your patience."

DF: "You're welcome. It was quick. And you know, I love it so it's not a problem."

FJ: "You love spending time on these lines of code?"

DF: "Sure. It's fun. What I really like is that you should never lose the thread. And when the script does the thing, it means you didn't lose it."

What may this excerpt tell us about the *affects* of computer programming? The notion of scenario seems, by itself, unable to provide a clearer understanding of what PROG, once assembled, does to DF. But, following DF and using the scenario as stepping stone, it helps to make appear something *lovable*: being able to constantly evaluate what has been done against what still needs to be done. This is what DF steadily needs to grab, the thread he tries never to lose: this scenario suggests a path, a plot, but also says nothing about how to follow it. Following a story by tracing his own path: a curious experience of establishing something by reaching it. But this reach, this access to the horizon—one should not simply consider it as the satisfaction of realizing something that was previously projected. Taking DF seriously—but also other Lab collaborators who participated in other "helping sessions"—we may consider it as the asymptote of a constant evaluation. "This" had to be done, then "this," then "this," and now, *there is nothing else to do* until the next affect-bearing scenario, of course. The specificity of the affects of computer programming may lie in the recurrent upsurging of this temporary "nothing else."

This is only an adventurous proposition about the attachments that bind programmers to the scripts they may *instaure* (Latour 2013, 151–178; Souriau [1943] 2015). More systematic studies are obviously necessary to enrich the above speculations. But let the reader not forget, once again, that one goal of this chapter, besides its analytical ambitions, is also to point to innovative avenues of research on computer programming situated practices. In that sense, looking at the formation of scenarios and their complex relationships with the attachments they may suggest—but not strictly produce—could be a relevant way to inquire into what *moves* programmers, sometimes to the point of spending huge amount of unpaid (or detoured) hours on uncertain free and open-source software projects. In the light of programmers' attachments to scenarios, what Demazière, Horn, and Zune (2007, 35) called the "enigma of free software development"—the ability to produce coherent programming results from evanescent involvement—could, for example, be tackled in an alternative way. While entangled modes of regulations among these voluntary collectives are certainly important for the actual production of free and open-source software, these arrangements may also benefit from being considered in the light of the *passions* they make exist. What is indeed happening when a scenario is realized through a computer script? Can such an affective event only be reduced to the organizational processes (Demazière, Horn, and Zune 2007), individual incentives (Lerner and Tirole 2002), or ideologies (Elliott and Scacchi 2008) that made it possible? Is there not something in DF's emotive spark that may also contribute to the formation and maintenance of programmers' communities? It is the whole ecology of programming work—be it free, open-source, or corporate—that may deserve to be considered also in the light of what programmers are after when they are writing numbered lists of instructions.

*   *   *

Despite its lengthy and tortuous aspect, the point I wanted to make in this part II is quite simple. Once we inquire into computer programming courses of action, we see that they engage the alignment of inscriptions, the work-around of impasses, and the definition of scenarios. These three modes of practices are intimately related: Working around impasses implies the localization of a problematic phenomenon that itself requires a scenario to be considered problematic. DF and more generally, perhaps, *programmers* constantly shift from one mode to the other until temporally realizing their desired narratives.

The main difficulty lay in the preparatory work required to distinguish the process of programming from its result. For complicated reasons we covered in chapter 3, a confusing mix has progressively been established between human cognition and programmed computers. This confusion led, in turn, to important misunderstandings such as cognitive studies of programming that ended up being tautological as they supposed the existence of what they tried to account for. As I wanted to analyze the situated practice of programming, I had to distance myself from cognitivism and embrace very minimal, yet powerful, enactivism that considers cognition as the process by which we grasp affordances of local environments.

Unfortunately, I could play only at the edge of computer programming practices, and many questions were left unanswered. Regarding the alignment of inscriptions, it would be insightful to learn more about the different modalities, organizations, and even institutions that participate in a programmer's multiplications and articulations of inscriptions. Regarding the working around of impasses, what about exploring more thoroughly the equipment that supports the identification and enrollment of new actants? This may even lead to innovative programming devices and equipment. Concerning scenarios, I will soon document the formation of some specific, easily transposable ones. But in light of the fascination exerted by computer programming as well as its importance for contemporary societies, I wish there were more studies documenting the actions that sometimes make the *joy* of programming emerge. In these times of controversies over algorithms—entities that seem to rely on ground-truthing and programming activities—these are, I believe, crucial research directions.

III   Formulating

It is easy to study laboratory practices because they are so heavily equipped, so evidently collective, so obviously material, so clearly situated in specific times and spaces, so hesitant and costly. But the same is not true of mathematical practices: notions like … "calculating," "formalism," "abstraction" resist being shifted from the role of indisputable resources to that of inspectable and accountable topics. … We seem to be inevitably contaminated by [these notions], as if abstraction has rendered us abstract as well!

—Latour (2008, 444)

We are not out of the woods yet. We may have a clearer idea about the whys and wherefores of *ground-truthing* (part I) and *programming* (part II), yet we still lack, at this point of the inquiry, one activity that is sometimes crucial to the formation of algorithms in computer science laboratories. Without accounting for these practices, I could only propose an extremely partial constitution of algorithms.

One way to become sensitive to the "missing mass" of our inquiry could be to look at a recent academic paper in computer science. And why not choose the subfield of image processing since it is the empirical ground of this ethnographic venture? While browsing, for example, through a paper entitled "Learning Deep Features for Discriminative Localization" (Zhou et al. 2016), we would encounter many things we are now familiar with. We would read about a specific problem (localizing class-specific image regions) that, according to the paper's authors, is solved satisfactorily by means of a computer program they call CAM, which stands for "class activation mapping." We would see that the problem, CAM, and what this program should retrieve all derive from an already-assembled ground truth (in this case, ImageNet Large Scale Visual Recognition Challenge [ILSVRC] 2014) that has been split into two parts: a training set and an evaluation set. We would also feel, behind the printed words and numbers, the long and fastidious computer programming episodes that were necessary to provide and discuss the paper's results. After all, if the authors did not write lists of instructions capable of triggering electric pulses in meaningful ways, they could not have provided any statistical evaluations of their algorithm's performances.

However, while browsing through this academic paper that presents and tries to convince us about the relevance of a new image-processing algorithm, we would very quickly bump into cryptic passages such as this one:

By plugging $Fk=\sum x,y\, fk\,(x, y)$ into the class score, Sc, we obtain

$$Sc = \sum_k w_k^c \sum_{x,y} fk\left(x,y\right) = \sum_{x,y} \sum_k w_k^c fk\left(x,y\right) \tag{1}$$

We define Mc as the class activation map for class c, where each spatial element is given by

$$Mc\left(x,y\right) = \sum_k w_k^c fk\left(x,y\right) \tag{2}$$

Thus, $Sc=\sum_{x,y} Mc\,(x,y)$, and hence $Mc\,(x,y)$ directly indicates the importance of the activation at spatial grid (x,y) leading to the classification of an image to class c. (Zhou et al. 2016, 2923)

Such sentences that mix English words with combinations of Greek and Latin letters divided by equals signs are indeed widely used by computer scientists when they communicate about their algorithms in academic journals. Of course, as grown-up readers, we immediately understand that such an excerpt deals with *mathematics* and that (1) and (2) are proper *formulas* (or *equations* once their variables are replaced by numerical values). But if we only consider the descriptive system developed so far in this inquiry, we have no grip on these mathematical inscriptions. The conceptual apparatus of the inquiry enables us to deal with graphs and numeric values as they refer somehow to both data and targets as defined by ground truths. The inquiry's apparatus also enables us to deal with lines of code as they refer to numbered lists of instructions that trigger electric pulses in desired ways. But what about mathematical formulas? Where do they come from? Why do computer scientists need them, and how are they assembled? At this point, I do not have any other choice. In this last and important part III, I will have to consider the role of mathematics in the formation of algorithms.

The road I am about to take is dangerous; one second of inattention and my action-oriented method will be lost. For intricate reasons that I will cover, mathematical entities such as "theorems," "proofs," or "formulas" are indeed extremely resistant to empirical considerations; even though they certainly are the products of situated activities, they are often considered fundamental ingredients of thoughts. This tenacious habit is frequently the starting point of a downward spiral, itself leading to grand questions such as: "Are mathematics the expressions of abstract structures or individual consciousness?" So many innocent souls have been consumed by such floating interrogations! To avoid digging my own grave in this cemetery of practice, I will have to be extremely cautious and process one small step at a time. But with some patience, the construction of mathematical knowledge as well as its further enrollment in the formation of algorithms may be partially accounted for. Altogether, these efforts to define *formulating practices* will allow me to link both ground-truthing practices (necessary to establish the terms of solvable problems) and programming practices (necessary to make computers compute in desired ways). Within the present constituent effort, what we tend to call "algorithms" may then be described as uncertain products of (at least) these three interrelated activities.

As in part II—and largely for similar reasons—I will require operationalization efforts before diving into ethnographic materials. I will first have to put aside the vast majority of studies on mathematics. Too many topics, too many studies, too many methods; without preliminary cleaning efforts, dealing with mathematics in an action-oriented way is doomed to fail. As we shall see in chapter 5, the only way not to duck will be to start (almost) afresh, from very basic observations and hypotheses. Progressively, these hypotheses—well inspired by several STS on mathematics—will make us realize that mathematical entities such as "theorems," "proofs," or "formulas" are quite akin to more familiar scientific facts. If mathematical knowledge is often considered the expression of some superior reality, it might only be due to its extreme combinability. Once the vascularization of mathematics is put forward, we will realize that its indubitable power also comes from the humble instruments and actions that make nonmathematical topics *mathematicable*. This important point will, in turn, allow me to define *formulating practices* as the empirical process of merging networks that sustain given domains of activity with networks that sustain certified mathematical knowledge. In chapter 6, I will account for a small yet successful formulating effort that took place within the Lab. This third and last case study will underline the centrality of certified mathematical knowledge for the progressive formation of algorithms as it both forces the refinement of ground truths and unfolds scenarios for further programming episodes. It will also allow me to consider recent issues related to machine learning and artificial intelligence in an unconventional way. The last section of chapter 6 will be a brief summary.

5 Mathematics as a Science

This chapter aims to consider mathematical knowledge not as the expression of some superior reality but as a huge collection of scientific facts whose shaping necessitated a fair amount of practical work. As we will see, by considering mathematical knowledge to be one specific product (among many others) of scientific activity, we may provide a reasonable explanation of its capacity to make important differences in other scientific domains (neurology, geography, gambling, computer science, etc.). Once this operationalization exercise is over, I will come back to the main goal of this part III: understanding when, how, and why mathematical knowledge takes active part in the constitution of algorithms (chapter 6).

Where Is the Math?

If we want to better understand how mathematical entities (formulas, theorems, conjectures, equations) are manipulated and related to ground truths and programming languages, we first need to better understand where they come from. Such entities surely do not exist by themselves; they need to be assembled by people in specific designated places. Where are these places? Who are these people, and what do they do?

Such trivial questions lead to many, many heterogeneous answers. This is one reason why dealing with mathematics can be dangerous: Where shall we start? From the mathematics of ancient Greece (Heath 1981a, 1981b; Netz 2003)? From mathematics of medieval Islam (Berggren 1986; Netz 2004)? From baroque mathematics of continuous change (Bardi 2007; Boyer 1959)? But if we use the adjective "baroque," we already define the seventeenth century in quite an orientated way (Deleuze 1992). Shall we then focus on more contemporary mathematics such as set theory (Ferreirós 2007; Tiles 2004), Weierstrass functions (Bottazzini 1986), and the subsequent "crisis of foundations" that shook up mathematics at the beginning of the twentieth century (Ewald 2007; Ferreirós 2008; Hesseling 2004; Mancosu 1997)? But what do we mean by "mathematics" anyway? Do we mean mathematical *texts* (Rotman 1995, 2006; Sha 2005)? Do we mean *famous mathematicians* such as Leibniz (Antognazza 2011), Gauss (Tent 2006), or Cantor (Dauben 1990)? Do we mean *philosophies of mathematics* that try to define what mathematics *is* (Aspray and Kitcher 1988; Corfield 2006; Hacking 2014)? Our head is spinning and we start to feel dizzy. But it is not over yet! Indeed, are we talking about *arithmetic* (Husserl 2012), *algebra* (Everest 2007), *geometry* (Netz 2003; Serres 1995, 2002), or *logic* (Fisher 2007; Rosental 2003)? Maybe are we talking about the *evolution* from numbers to logic (Kline 1990a), from logic to geometry (Kline 1990b; Netz 2003), from geometry to algebra (Kline 1990c; Netz 2004)? And even within arithmetic, geometry, algebra, or logic, are we talking about *theorems* (Villani 2016), *proofs* (Lakatos 1976; MacKenzie 1999, 2004, 2006) or *conjectures* (O'Shea 2008)? We do not know. We are lost in questions whose only enunciation makes us want to do something else. But we cannot; we must find a way to address mathematics as it seems important for the constitution of algorithms. How can we do so?

One way to avoid this spiral of confusion could be to start from some very basic hypotheses. We would, of course, have to develop these hypotheses and justify them by using concrete examples. To do this, we may need to mobilize a tiny part of the gigantic mathematics literature that scares us. One step after the other, one hypothesis after the other—coupled with some STS assumptions—we may end up with an operative definition of mathematical knowledge that could suffice to achieve our specific task: accounting for the way that computer scientists, when they try to assemble new algorithms, are sometimes able to mobilize certified propositions previously shaped by their mathematician colleagues. We surely do not need to revolutionize our understanding of these powerful statements we sometimes call "theorems," "conjectures," or "formulas." If we just manage to shape one simple version of what mathematicians *do* (instead of what mathematics *is*), our last duty—accounting for formulating practices—will be greatly facilitated.

Written Claims of Relative Conviction Strengths

To initiate our operationalization exercise and shape our first hypotheses, let us start with three scenes that all gravitate around mathematical notions:[1]

Scene 1

January 1994. Charles Elkan is in turmoil: his theorem demonstrating that only two truth values can be expressed by a system of fuzzy logic is highly contested.[2] What went wrong? The initial presentation of his theorem at the Eleventh National Conference on Artificial Intelligence went very well. The paper that further appeared in the conference proceedings was even selected for the "Best Written Paper Award" (Elkan 1993). The program committee saluted the elegance of the proof as well as its significance for further developments in expert systems. Everything was in place for his theorem to be accepted. But many logician colleagues—who did not attend the conference but did read some of its proceedings published by MIT Press—are quite upset. Elkan can even follow their dissatisfaction on the newly established internet forum "comp.ai.fuzzy" that is dedicated to advanced discussions in fuzzy logic theories and systems. The critiques are harsh. Some say—and try to demonstrate—that Elkan's basic hypotheses are flawed. Others accuse him of deliberately weakening fuzzy logic as it is a threat to old, "dusty" classical logic. Some colleagues even suspect him to be a thick-headed Aristotelian! As one of his friends advises him, Elkan should now "cool things down" and publish a "smoother" version of his theorem that could include some of its soundest critiques.

Scene 2

Summer of 1890. Alfred Kempe is puzzled;[3] although not really because Percy Heawood recently managed to find a flaw in the proof of the four colors conjecture Kempe previously published in the *American Journal of Mathematics* (Heawood 1890; Kempe 1879). Heawood did a great job, and being refuted is part of the game anyway. No, it is more that even though his proof was shown to be erroneous, Kempe does not think that Francis Guthrie's 1852 candid proposition—that says that four colors suffice to color any map drawn on a plane in such a way that no neighboring countries have the same color—is wrong. But how could such a basic intuition lead to such great difficulties? Do mathematicians not have the tools to prove this conjecture and make it a theorem once and for all? "Poor Heawood," thinks Kempe. "He is now hooked on it, as I was fifteen years ago. He'd better drop it; this four colors thing is old hat."

Scene 3

November 8, 2013, 3 p.m. I sit at the back of the lecture hall.[4] Around three hundred undergraduate students are also attending this Friday afternoon "Information, Computing and Communication" class that aims to inculcate (communicate?) the foundational concepts of computer science to future civil and mechanical engineers. I see my younger brother and his friends—good students—in the second row. They've just started their academic curriculum; I've almost finished mine. But here we are in the same classroom, waiting for the same information (orders?). The professor adjusts his microphone: "All right. Hi, everyone. So, last week we talked about the Nyquist-Shannon sampling theorem. Today, we'll start with another contribution of Claude Shannon to the mathematical understanding of digital signals, which is the Shannon-Hartley theorem. It is quite a powerful theorem that can be summarized with this formula here:

$$C = B \log_2(1 + \frac{S}{N}).$$

Of course, we'll go through it together."

At this point, we do not need to make any a priori distinction between "theorems" (scenes 1 and 3), "conjectures" (scene 2), "proofs" (scene 1 and 2), and "formulas" (scene 3). We just need to notice that all three scenes, while presumably concerning mathematics, deal with *claims that attract more or less adherence*. In scene 1, Elkan's claim about fuzzy logic first attracts the adherence of the Eleventh National Conference on Artificial Intelligence's program committee. But then, in January 1994, his claim repulses many logician colleagues who do not hesitate to publish "counterclaims" on the web forum "comp.ai.fuzzy." In scene 2, Kempe's claim about the veracity of Francis Guthrie's claim (the "four colors conjecture") also first attracts the adherence of the editorial board of the *American Journal of Mathematics*. But then, in the summer of 1890, Kempe dissociates himself from his own claim and adheres to that of Heawood. However, Guthrie's 1852 "candid" claim has not lost all of its conviction strength yet, which makes Kempe puzzled about the fate of Heawood. Scene 3 is quite straightforward: Shannon and Hartley's claim—and its correlated formula projected on the lecture hall's whiteboard—is about to be taught to a crowd of undergraduate students in engineering. There is little room for doubt here: in November 2013, Shannon and Hartley's claim attracts the adherence of quite a lot of people. In fact, their claim is so strong that a well-known pedagogical device—the exam—will soon verify that all students properly adhere to it.

These basic but fair observations are all we need to start our operationalization exercise. Mathematicians certainly do a lot of things, but among these things, they make claims that attract the adherence of more or fewer individuals. Let us assume then that the grand notions of "theorems," "conjectures," "formulas," or "proofs" can all be grasped in a down-to-earth manner; let us assume that, to a certain extent, they are claims that convince more or fewer individuals.

This way to consider mathematical knowledge—theorems, conjectures, proofs, formulas—as the product of some rhetoric may sound odd at first. Many grand narratives have indeed chanted the abstract power of mathematical truths that, by themselves, supposedly describe some superior reality.[5] But this is precisely the road we do not want to take, at least not yet. If we do not want to crash on the sharp rocks of epistemological accounts of mathematics, we need to plug our ears and, for the moment, ignore the sirens of necessity. Fortunately for us, our first operational hypothesis—mathematicians make claims that convince more or fewer individuals—echoes well the central thesis of Lakatos's (1976) important book on mathematics. As he showed, instead of an accumulation of self-evident discoveries, mathematics should be considered a creative process during which concurrent claims are subjected to criticism and improvement. But how are such claims criticized or improved? How do they gain or lose their relative conviction strength? Shannon and Hartley's claim in scene 3 seems much stronger than Elkan's claim in scene 1. Similarly, in 1890, the claim Kempe made in 1879 is now powerless in front of Heawood's claim (scene 2). How do such differences come about?

To better understand how (mathematical) claims gain or lose conviction strength, we need to make another basic observation about scenes 1, 2, and 3. If more or fewer individuals could adhere to the scenes' claims, it means that they could *access* these claims. What medium allowed such access? Some claims are oral, but we are obviously not dealing with them here. The claims in scenes 1, 2, and 3 are all *written*. This important characteristic allows individuals to read them and eventually—very rarely—adhere to them. In scene 1, it is Elkan's written claim as it appears in the conference's proceedings that makes the program committee adhere to it. But in January 1994, it is the multiplication of written counterclaims on the web forum "comp.ai.fuzzy" that begins tormenting Elkan. In scene 2, both Kempe and Heawood access their respective claims by reading mathematical journals. Finally, the engineering students in scene 3 are asked to adhere to Shannon and Hartley's claim projected on the classroom's whiteboard. Of course, Shannon and Hartley did not write their claim on the projected document; many individuals intervened to carry their claim further through time and space until reaching this specific lecture hall. But this translation process does not change the overall shape of the claim; it is still something that is written down on a flat surface. At this point, we can therefore slightly refresh our first hypothesis: mathematicians surely do a lot of things, but among these things, they *write* claims that attract the adherence of more or fewer individuals.

It is also fair to assume that the written claims in the above scenes did not appear ex nihilo. In order to be published in proceedings, specialized web forums, mathematical journals, or the slides of a computer science professor, they all had to overcome a series of tests, *trials* upon which their existence as written claims depended. I agree that this hypothesis flirts with the metaphysics of subsistence—close to "process thought" (cf. introduction)—as proposed by influential, yet contested, thinkers. Let us then consider it an assumption we need for our operationalization exercise. "Whatever resists trials is real" (Latour 1993a). The above (mathematical) written claims are real; they thus resisted trials. But what trials?

Resisting Trials, Becoming Facts

The first kind of trial we can consider regarding the conviction strengths of (mathematical) written claims such as those in scenes 1, 2, and 3 are the trials they must endure *before* their actual publication. Examining what we often call the "sources" of claims is indeed a common way to evaluate their seriousness.

For example, we can make the fair assumption that, all things being equal, a claim published in the journal *Nature* will generally have more conviction strength than a claim posted on some social media platform with very little monitoring. Without even considering their respective content, both claims will have different capabilities. Why is that? We must immediately put aside the question of prestige or symbolic power; these are shortcuts our sociological method of inquiry forbids us to manipulate. A more empirical grip on this topic would quickly point to the number of individuals who could prevent the publication of a claim. Very few people—or bots—can prevent me from publishing a claim on, say, Facebook. Conversely, many individuals can prevent me from publishing a claim in the journal *Nature*. Taking into account those who have to be convinced by claims in order for them to circulate and reach a broader audience is crucial as it somewhat calibrates the cost of disagreement. If someone disagrees with a claim I publish on Facebook, they can just shrug their shoulders and move on to something else.[6] But if the same person disagrees with a claim I publish in *Nature*, they will have to disagree with me, my institution, the funding agencies that supported my research, *Nature*'s editorial board, those responsible for the nomination of this board, and so on. Compared with a claim I publish on Facebook, a claim I publish in *Nature* is initially supported by a far bigger team of *external allies* (Latour 1987, 31–33).

But if we consider our three scenes, we quickly realize that surviving *publication trials*—and thus enrolling external allies—is not enough to assure any durable conviction strength of (mathematical) claims. Although this lecture, in terms of convinced *gatekeepers,* may be enough to quickly account for the conviction strength of Shannon and Hartley's claim within the lecture hall—the students being literally crushed by all its external allies (their professor, their manuals, all those responsible for the engineering curriculum of their university, the exam they will soon have to pass)—it does not help us understand the relative strengths of Kempe's, Heawood's, and Elkan's claims (scene 1 and scene 2). In scene 2, both Kempe's and Heawood's claims survived similar publication trials; both propositions were initially supported by roughly the same number of individuals.[7] Yet Kempe's claim became distrusted as Heawood's appeared certified. The situation is even more confusing in scene 1: even though Elkan's claim successively resisted the scrutiny of the sixty-eight individuals responsible for the publication of the proceedings and the selection of the "Best Written

Paper,"[8] his claim is seriously shaken up by posts on a web forum with almost no monitoring (Rosental 2003, 81–86). Again, these counterclaims must have survived other kinds of trials in order to gain such strength.

Another kind of trial that may provide strength to written claims is one that consists in successively enrolling *internal allies* by means of citations and references (Latour 1987, 33–45). Equipping one's claim with previously published claims is indeed an important conviction strategy that has even become a whole field of study.[9] In addition to allies outside of the written document, a claim with references and citations is now supported by allies inside of it. Or is it? While often necessary, augmenting the conviction strength of a claim by means of references and citations can be a risky endeavor. What if the references do not match the claim, or worse, what if some unmentioned references contradict the presented claim? In some cases, this *citation trial* is overcome. One example is Shannon's initial paper that presented the basic elements of what would later be called the "Shannon-Hartley theorem" (Shannon 1948). In this paper, Shannon enrolls previously "solidified" claims made by Ralph Hartley (hence his later inclusion in the theorem's name) and thirteen other important mathematicians. As far as I know, no serious disagreements about the use of these references emerged after Shannon's initial publication. But the same was not true of Elkan's publication. Although he mobilized thirty-nine internal allies to strengthen his claim about the limitations of fuzzy logic, his contradictors managed to find and publish many strong "counter references" on the specialized web forum. Elkan soon appeared as someone unaware of many recent uses of fuzzy logic in advanced expert systems (Rosental 2003, 157–168). Although they were at first certainly useful to convince the program committee of the Eleventh National Conference on Artificial Intelligence, the internal allies of Elkan's paper ended up working as stepping stones for his contradictors.

However, surviving or not surviving citation trials is, again, not enough to account for the relative conviction strengths of the claims in all of our scenes. Indeed, in scene 2, Kempe's 1879 paper makes only three references to former mathematical propositions, the first two being loose statements made by Augustus De Morgan and Arthur Cayley to the London Mathematical Society (Kempe 1879, 193–194) and the third one being a more important claim made by Augustin-Louis Cauchy about polyhedrons (Kempe 1879, 198). Yet this scarcity of references did not prevent his claim—the proof that Guthrie's 1852 proposition was correct—from convincing his mathematician colleagues for eleven years. The same is even truer of Heawood's claim, for his 1890 paper includes no references other than Kempe's 1879 paper. Again, this scarcity did not prevent his claim from attracting the adherence of the chief person concerned: Kempe himself (MacKenzie 1999, 22). There must be something else in published (mathematical) claims that makes them gain, sometimes, in persuasion strength.

Some potential objectors of published (mathematical) claims will not be impressed by lists of convinced gatekeepers nor by the references invoked by the author. To be convinced by a claim, these skeptical readers want to *see* the thing the author asks them to believe in. This strategy that consists of presenting the thing in question to the reader was precisely the one used by Heawood in his paper against Kempe. He did not only rely on external allies; he also showed a *figure* (see figure 5.1) that, according to Kempe's 1871 claim, was impossible to draw:

> Mr. Kempe says—the transmission of colours throughout E's red-green and B's red-yellow regions will each remove a red, and what is required is done. If this were so, it would at once lead to a proof of the proposition in question [the four-colours conjecture]. … But, unfortunately, it is conceivable that though either transposition would remove a red, both may not remove both reds. Fig [below] is an actual exemplification of this possibility. (Heawood 1890, 337–338)

We do not need to spend too much time on the specificities of Heawood's figure[10] nor on the role of drawings in published mathematical claims.[11] Here, the important thing to notice is the conviction strategy; just as scientists engaged in many other fields—biology (Rheinberger 1997), chemistry (Bensaude-Vincent 1995), climatology (Edwards 2013)—mathematicians try to gain in persuasion strength by adding the referent of what they write about. At this point, then, "this is not a question any more of belief: this is *seeing*" (Latour 1987, 48). If, until now, I put the adjective "mathematical" in parenthesis, it was not to grant too much specificity to mathematical claims; they too are part of the scientific genre that tries to silence potential objectors by gathering more and more supporters. Scientific as well as mathematical texts can indeed be compared with bobsled tracks allowing very little room for maneuver while implying high level of skills. In both cases, readers must start at point A, pass through checkpoints $B_{1,2,...,n}$, and finally finish at point C, the claim that tries to be established as a *fact*.

If scientific literature can be described as texts gathering many external and internal allies in order to isolate their readers and force them to take

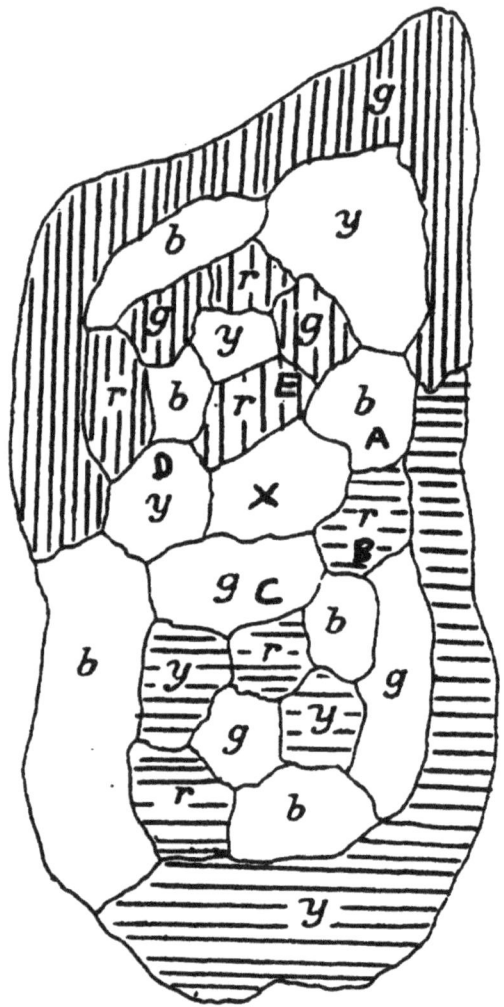

Figure 5.1
Reproduction of Heawood's figure showing that Kempe's proof does not hold. *Source:*
MacKenzie (1999). Reproduced with permission from Sage Publications.

only one path, different scientific domains progressively shaped their own specific rhetorical habits.[12] In the case of mathematics, this whole *captation trial* (Latour 1987, 56–61) that consists in subtly controlling the movements of potential objectors has been finely analyzed by Rotman (1995, 2006). As he showed, mathematical publications are full of verbs in the imperative form, such as "construct," "define," "connect," or "compute." But a close analysis of these imperative forms reveals that they are in fact split into two distinctive types: *inclusive imperative* to establish premises—often equipped with references—and *exclusive imperative* to present lists of actions an imaginary reader should perform to reach the claimed result:

> Inclusive command marked by the verbs "consider," "define," "prove" and their synonyms—demand that speaker and hearer institute and inhabit a common world or that they share some specific argued conviction about an item in such a world; and exclusive commands—essentially the mathematical actions denoted by all other verbs—dictate that certain operations meaningful in an already shared world be executed. (Rotman 2006, 104)

These elements are crucial for our operationalization exercise as they indicate the felicity conditions of captation trials within mathematical texts. If skeptical readers, thanks to all the allies mobilized by the writer, have no other choice than to accept the premises and follow *one* specific path in order to reach one necessary conclusion, a mathematical text and its concomitant claim have, at least temporally, overcome their captation trial. In this respect, Kempe's 1879 paper on the four colors conjecture is quite illustrative. Remember that Kempe wanted to prove that four colors suffice to color any map drawn on a plane in such a way that no neighboring countries have the same color. How did he enjoin his readers to reach this conclusion? With a succession of inclusive commands, both Kempe and his imaginary skeptical reader start by defining a perfectly four-colored "singly connected surface" divided into many "districts" (Kempe 1879, 193). Once this basic common world has been instituted, they then consider two sets of "detached regions" either colored in red and green or in yellow and blue (Kempe 1879, 194). These premises allow Kempe and his reader to further define the properties of "points of concourse" (points where boundaries and districts meet) that themselves permit the definition of six classes of districts with different characteristics: "island districts," "island regions," "peninsula districts," "peninsula regions," "complex districts," and "simple districts" (Kempe 1879, 195–196). Once this quite complex common world has been instituted, Kempe then switches to exclusive commands and asks his reader to execute a series of operations:

> Now, *take* a piece of paper and *cut it out* to the same shape as any simple-island or peninsula-district, but larger, so as just to overlap the boundaries when laid on the district. *Fasten this patch* (as I shall term it) to the surface and *produce* all the boundaries which meet the patch ... to meet at a point, (a point of concourse) within the patch. If only two boundaries meet the patch, which will happen if the district be a peninsula, *join them* across the patch, no point of concourse being necessary. The map will then have one district less, and the number of boundaries will also be reduced. (Kempe 1879, 196–197; italics added)

By asking the reader to reiterate this patching process, the whole imagined map is progressively reduced to one single district with no boundaries or points of concourse. Kempe then asks the reader to reverse the process; that is, to "strip off the patches in reverse order, taking off first that which was put on last. As each patch is stripped off it discloses a new district and the map is developed by degrees" (Kempe 1879, 197). At this precise point, Kempe switches to inclusive command again, thus instituting a *second common world* based on the first one that has just been modified. The author and the reader, together again, define the progressive reconstitution of all districts, boundaries, and points of concourse. Little by little, they soon realize that their recombination of districts, boundaries, and points of concourse is equivalent to, respectively, faces, edges, and points of polyhedrons as already defined by Augustin-Louis Cauchy in 1813 (Kempe 1879, 198). Once this polyhedron world has been instituted, Kempe switches one last time to exclusive command and makes the reader reach the claimed result: obviously—look, we have just done it together!—four colors suffice to color any map drawn on a plane in such a way that no neighboring countries have the same color.

We do not need to understand every little step of Kempe's paper. We just need to appreciate how Kempe manages to control the movements of his reader; from the initial premises to the conclusion, the reader is literally carried through Kempe's line of argument. His allies are quite numerous—"single connected surface," "districts," "detached regions," Cauchy's "polyhedrons"—and his transitions are smooth enough to transport the reader through the flow of necessity. But as we saw, Kempe's *captatio* was only temporary, for as eleven years later, Heawood managed to escape from Kempe's line of argument and propose a figure that dismantled the whole rhetorical edifice (see figure 5.1).

Publication, citation, and captation trials—just as any other claim trying to gain conviction strength and become a fact, mathematical claims must survive many jeopardies. Yet this is still not enough. A claim published in an important journal, with well-arrayed references and a smooth line of argument, may still vanish if it is not carried further by later claims. This is a sine qua non condition as there is no such thing as a solitary scientific fact: "Fact construction is so much a collective process that an isolated person builds only dreams, claims and feelings, not facts" (Latour 1987, 41). The fate of a claim, its progressive transformation into a solidified fact, depends ultimately on how it is used by later claims. We saw that Kempe's claim, despite its captation strength, ended up being refuted by Heawood. From the status of mathematical fact, it turned into mere fiction. What about Heawood's claim? It is difficult to call it a fact as it only concerned Kempe's fiction; it successively refuted Kempe's claim but did not provide any confirmable, or refutable, proposition. What about Elkan's claim, then? Despite Elkan's efforts to make it stronger—especially via the inclusion of many coauthors, better arrayed references, and smoother transitions (Elkan et al. 1994; Rosental 2003, 282–331)—it ended up being known for the doubtful reactions it gave rise to; that is, precisely, *for not being a fact*. Among our arbitrary mathematical examples, only Shannon's claim survived this important *posterity trial*, as scene 3 already suggested it. In fact, Shannon's claim survived the posterity trial so well that it progressively became part of a very small number of facts that are constantly used as resources in later claims. As it became more and more enrolled without any skeptical modalities, it became a black box with certified content presented in a clear-cut form. This *stylization process* (Latour 1987, 42) is typical of scientific facts that are much enrolled in later claims. Although Shannon went through several demonstrations in his initial paper, only the results of these demonstrations were progressively retained. These results were later concatenated, polished, and linked with former results established by Hartley until reaching a stylized form expressed by the formula presented in scene 3. Soon, perhaps, this strong mathematical fact may even become a "single sentence statement" (Latour 1987, 43): a scientific fact that is so accepted that it no longer needs any reference. If this happens, Shannon and Hartley's theorem will be part of tacit, undisputable, and necessary knowledge.

These last elements about blackboxed polished facts that may become part of tacit knowledge allow us to respond to an important objection:

Objection of a skeptical reader

But is not mathematics different from all the other scientific disciplines in that it deals with fundamental truths? We could feel it when you presented Kempe's paper: in order to overcome the captation trial, he followed the timeless laws of deduction, did he not?

Not so long ago, it would have been very difficult to respond to this classical objection.[13] But thanks to the philological efforts made by Reviel Netz (2003, 2004), we now know that what we call "deduction" and "logical relations" are themselves blackboxed polished facts that were initially published around the middle of the fifth century BCE in Greece and southern Italy.[14] At that time, several self-educated amateurs who, presumably, tried to distance themselves from ancient Greece's highly polemical culture,[15] were surprised to discover that when they wrote *only* about the properties of lettered diagrams drawn on wax tablets, they could, step by step, express indisputable propositions. More precisely, by starting with some lettered parts of a diagram—say, two segments—they could, in turn, compare them with another lettered part of the same diagram. This very basic operation, made possible by the combination of drawings and letters on a flat surface, can be reconstituted as such: "This segment A here is equal to that segment B there. *And* that segment B there is equal to that segment C over there." In turn, thanks to the lettered diagram, Greek geometers could surreptitiously use conjunctive adverbs in a necessary way: "*Therefore* this segment A here is equal to that segment C over there." The shift seems trivial but is in fact crucial. Indeed, this first necessary result could be used to compare other parts of the diagram: "And that segment D over there is two times segment C. *Therefore*, segment A is half segment D." Progressively, by comparing more and more parts of the diagram, using more and more conjunctive adverbs and cumulating more and more intermediary results such as "A is half segment D," the Greek geometer could end up with a complicated yet necessary true proposition—the written list of indexical steps going from his first basic assertion to his last complicated one being the *proof* of the veracity of his claim.

For the sake of this section that only tries to present mathematical claims as part of the broader family of scientific claims, we do not need to dig further into the fascinating work done by Netz. Suffice it here to say that thanks to his efforts, we can now assert with some confidence that even deduction is the solidified product of past accepted claims. These constructed-yet-fully-logical laws of necessity must certainly have been surprising in ancient Greece.[16] But after centuries of enrollments in further claims, this style of reasoning—that obviously overcame its posterity trial—was progressively blackboxed, polished, and stylized until acquiring the status of indisputable knowledge.[17] Who would now quote Aristotle when using the inference rule of *modus ponens*? Yet even these principles of logic—dear to the formalist school of mathematics[18]—went through a process similar to that of Shannon and Hartley's theorem that very few mathematicians in signal processing would now try to contest. Just as the theorem they helped to shape, deductive laws were themselves shaped a long time ago by people equipped with specific instruments (in this case, lettered diagrams drawn on wax tablets and indexed to small Greek sentences).

Flat Laboratories

In the previous sections, we spent some time trying to stress the similarities between mathematical and scientific claims. It appeared that both need to survive similar trials to become, eventually, indisputable facts. No superior necessity helps mathematical claims to become certified facts; they too need to convince their readers in order to be enrolled in later claims and become, very rarely, polished black boxes.

However, so far, we have only considered one side of the coin. Although looking at mathematical published claims helps us realize that successful mathematical propositions could be considered genuine certified knowledge, we can legitimately assume that mathematicians do not prepare, write, and read papers all their working time. They must also spend time and energy on the *things* they write about. All the claims we considered in the last sections were indeed about things: limitations of fuzzy logic systems for Elkan, the four colors conjecture for Kempe, Kempe's claim about the four colors conjecture for Heawood, and maximum rate of information transmission over noisy channels for Shannon (and later, Hartley). But how are these things assembled? What practices lead to the presentation of these mathematical things—or objects—in published materials? Are these practices different from laboratory practices in other scientific communities?

As we prepare to look inside the locations in which mathematical objects are shaped, we immediately face a difficulty: there are very few empirical studies of such locations. Although there are robust studies about controversies within mathematical domains (Warwick 1992, 1993; MacKenzie 1999, 2000, 2004, 2006; Rosental 2003, 2004) and historical reconstructions of the shaping of mathematical objects from famous mathematicians' logbooks (Lakatos 1976; Pickering and Stephanides 1992), there are very few laboratory studies of mathematics.[19] It is thus with limited means that I will now try to stress the scientific aspect of mathematics a little bit more:

Scene 4

Salk Institute for Biological Studies at La Jolla (California), winter of 1972.[20] Paul Brazeau is on edge. His boss, Professor Roger Guillemin, is after him, casting doubts on his ability to handle the lab's brand new—and very expansive—radioimmunoassay. It is true that the graphs recently printed by the massive bioelectronic instrument are surprising; instead of showing that Guillemin's newly purified peptide triggers the growth hormone, it shows that it decreases it. This drives Guillemin crazy. But Brazeau and his technicians retro-inspected the whole experimental procedure a dozen times: there were no mistakes. The right amount of purified peptide was injected in the carefully assembled rat pituitary cell culture, and no mishandling occurred during the operationalization of the radioimmunoassay. "It's terribly simple," thinks Brazeau. "Either I am no conscientious professional or, for the last three years, we were all wrong about this peptide."

Scene 5

Dublin, fall of 1843. William Rowan Hamilton is in a challenging mood: even though he bumps into another impasse in his attempt to extend complex number theory to a three-dimensional space, he is obviously making important progress.[21] He is particularly proud of his new starting point; what a mistake it was to start his previous experiments from tiring algebraic models! As he now starts geometrically by moving from $x+iy$ to $x+iy+jz$, he possesses a three-dimensional line segment that is far easier to test (even though it adds a second imaginary number j right from the start). His first experiment was, in that sense, very conclusive. Thanks to the advice of his German colleague Gotthold Eisenstein, he could reach an equivalence between algebraic and geometrical definitions of the square of his three-dimensional segment by abandoning the assumption of commutation between i and j. He could then further test his model by multiplying two arbitrary coplanar triplets according to his new noncommutative rule for ij. Although he struggled at first to define the orientation of his new product, he realized—after several attempts—that Pythagoras's theorem could nicely do the trick. Here again, an encouraging achievement. Yet this last move led him to another problem: the algebraic and geometrical representations of this coplanar multiplication differ by a factor of $(bz—cy)^2$. "I must find a way to remove this superfluous term," he thinks. "I don't want to start the whole thing over again!"

Despite their cryptic aspects, what do these two scenes tell us about laboratory practices? Can we draw similarities between what takes place within Guillemin's laboratory of endocrinology (scene 4) and what takes place within Hamilton's laboratory of mathematics (scene 5)?

We can first notice that both scenes deal with *experiments*; they both put something to the test in order to evaluate its reactions. The peptide in scene 4 is, in 1973, still undefined. Guillemin—in line with recent claims about this class of amino acid polymer—is convinced that it should trigger the rat's growth hormone.[22] But how much is such growth hormone triggered? And under what circumstances? To have a clearer view on the capacities of this peptide, he puts Brazeau in charge of implementing an experiment he recently designed. In scene 5, a complex three-dimensional line segment $x + iy + jz$ is, in 1843, still undefined.[23] Hamilton hopes that this "triplet"—as he calls it—will allow him to extend the geometrical representation of complex number theory.[24] But at this point, nothing is certain. To better understand the capacities of his complex three-dimensional line segment, he puts it through two successive experiments: he first *squares* it and then *multiplies* it with another arbitrary coplanar triplet.

In both scenes then, experiments are run to test undefined entities. Yet experiments do not happen by themselves; in both scenes, *instruments* are used by scientists in order to help them probe their undefined entities. In scene 4, the delicately assembled rat pituitary cell culture and the very expansive radioimmunoassay are the two principal tools used to test the peptide. It is worth noting that both instruments are highly visible and take up a lot of space. The instruments in scene 5 are a priori less impressive but equally important. The first instrument is, obviously, the algebraic apparatus as progressively defined by medieval Islamic mathematicians; without any means to express relationships among variables in a condensed and succinct manner, Hamilton could not juggle his triplet.[25] But he also needs a coordinate space to express his triplet geometrically. In that sense, without the efforts of seventeenth-century mathematicians such as Descartes, de Fermat, Newton, and Leibniz, Hamilton would have no means to consider the transformations of his triplet. He further requires some insight from noncommutative algebra, as then recently proposed by Gotthold Eisenstein, to handle the complex product *ij* (Hankins 1980). Finally, he needs good old Pythagoras's theorem to multiply his initial triplet with another arbitrary coplanar triplet.[26]

At this point, we need to make another down-to-earth observation: although both laboratories have instruments to conduct experiments on undefined entities, the shapes of these instruments differ from each other. On the one hand, there is a bioelectronic assemblage that gathers peptides, Brazeau, rat cells, laboratory technicians, and an imposing metal box full of electronic parts; on the other hand, there are books, paper, Hamilton, and a pencil. There is little room for doubt here: the instruments do not take up the same amount of space. Hamilton's instruments appear *dryer* and *thinner* whereas Guillemin's instruments appear *wetter* and *thicker*. One could say— and that is the terminology I will use for the remainder of this section—that Hamilton's laboratory is *flat* whereas Guillemin's laboratory is *bulky*. Both laboratories are engaged in the same process—testing the reactions of an undefined entity—but they use instruments that are different in terms of occupied space.[27]

Can we in turn say that Guillemin's laboratory is more *expansive* than Hamilton's laboratory? If we only consider the relative price of their instruments, it seems indeed to be the case: paper is cheaper than laboratory technicians, most books (even in nineteenth-century Ireland) are cheaper than a radioimmunoassay from the 1970s, and pencils are cheaper than a rat pituitary cell culture. Yet if one considers the relative networks of both laboratory apparatuses, the question appears trickier. Indeed, how many efforts were needed to cultivate and sell standardized rat cells? Many, indubitably. But how many efforts were required to establish coordinate spaces? Many, indubitably. And what about algebra? As Netz (1998, 2004) showed, without centuries of commentaries on Greek geometrical writings, without

Byzantine libraries, and without the classification efforts of Bagdad mathematicians, no algebraic system of notation could have come into existence. The same is true of Pythagoras's theorem; many long-standing efforts were required to gather, compile, and preserve Pythagorean propositions from early antiquity to nineteenth-century Ireland. Let us then stick to the topological difference between our two laboratories: Hamilton's laboratory is *flatter* than Guillemin's.

If we continue to analyze both scenes, we can see that despite their topological differences, both bulky and flat instruments end up producing comparable *inscriptions*; that is, readable traces on documents. Indeed, the bulky bioelectronic experimental assemblage of scene 4 ends up producing graphs whose curves indicate that the rat's hormone decreases. The results of the experiment on the undefined peptide conducted by Brazeau are pieces of paper anxiously examined by Guillemin.[28] Similarly, the flat experimental assemblage of scene 5 ends up producing a series of coupled algebraic and geometrical equations; at first, both equations appeared equivalent (which was good news for Hamilton), but in the second step of the experiment, both appeared dissimilar (which was bad news for Hamilton). Yet, just as for Brazeau and Guillemin, the results of Hamilton's flat experiments are readable traces on documents he examines with his eyes.[29]

At this point then, we can tentatively say that both scenes deal with experiments, instruments (of different topologies), and series of inscriptions. But where does all this work lead to? At this stage, it certainly cannot lead to any published claim that may later become a scientific fact. Within these two laboratories, scientists impose tests on undefined entities, but how can these practices lead to the formation of *objects* capable of being described in academic papers?

Scene 6

Salk Institute for Biological Studies at La Jolla (California), January 1973.[30] There is nothing to do about it; even after two other meticulous experiments, the graphs printed by the radioimmunoassay still show that the rat's hormone decreases when put in contact with Guillemin's peptide. The rat pituitary cell culture is indisputable as are the composition of Guillemin's peptide, the radioimmunoassay, and Brazeau's professionalism

(Guillemin quickly admits it). The only way to escape from this impasse is to cast doubt on what the peptide does. Leading figures in endocrinology—including Guillemin—thought that this class of peptide triggered the growth hormone; obviously, it does the opposite. After being in contact with rat pituitary cell culture for a certain amount of time and after having gone through the radioimmunoassay with some consistent parameters, this *new thing* significantly decreases the rat's growth hormone. As it is certain that there have been no mistakes during the experimental procedures, a paper is now being prepared to convince skeptical readers about the existence of this new scientific object Guillemin starts to call *somatostatin* (literally, "that which blocks the body").

Scene 7

Dublin, fall of 1843.[31] There is nothing to do about it: the superfluous term $(bz—cy)^2$ within the geometrical expression of the length of a complex line segment cannot be removed without adding a new imaginary quantity. The rules of algebra—including noncommutativity—are indisputable, as are Pythagoras's theorem and Hamilton's scriptural operations (he ran the whole experiment several times). The only way to escape from this impasse is to cast doubt on the premises of the experiment: What if the extension of the geometrical representation of complex number theory required not three but four dimensions? Indeed, only the inclusion of a third imaginary quantity k as the product of i and j can make the superfluous term $(bz—cy)^2$ disappear. It is true that this new imaginary quantity needs in turn a fourth axis in order to be geometrically represented, but who cares? After the introduction of k as either an imaginary quantity (in the algebraic representation) or a fourth dimensional axis (in the geometrical representation), this *new thing* can be squared and multiplied while producing equivalent equations, hence effectively extending the geometrical representation of complex number theory. If Hamilton now manages to define the quantities k^2, ik, kj, and i^2—almost a formality at this stage—he will be able to completely define the behavior of this new mathematical object he starts to call *quaternion* (literally, "that which is made of four").

Again, beyond their cryptic aspects, what do these two scenes tell us about the formation of new objects within scientific laboratories? Can we draw some similarities between the progressive shaping of *somatostatin* (scene 6) and *quaternions* (scene 7)?

We can first see that in both scenes, inscriptions printed out by instruments begin by expressing singular phenomena. In scene 6, the graphs printed by the radioimmunoassay indicate confidently that after the peptide is injected in the rat pituitary cell culture over a specific period of time and after it goes through the radioimmunoassay with specific parameters, the growth hormone decreases significantly. This is what is inscribed within the graphs Guillemin can read; the whole experimental process ends up decreasing the rat's growth hormone. Trustful graphs become flatter; *therefore* the growth hormone decreases.

Similarly, in scene 7, the inscriptions produced by the hands of Hamilton indicate that after a fourth dimension is added to the triplet in order to geometrically express the new imaginary quantity k—itself required to make the superfluous term $(bz—cy)^2$ disappear—both algebraic and geometrical representations of complex number theory become equivalent. Again, this is the phenomenon described by the inscriptions Hamilton can read on a sheet of paper; the whole experimental process ends up expressing an extension of the equivalence between geometrical and algebraic representation of complex number theory. A trustful geometrical equation becomes equivalent to another algebraic equation; *therefore*, the geometrical representation of complex number theory is extended.

However, and this is the crucial point, by virtue of the experimental setting, the origins of these two phenomena—"quantifiable inhibition of the growth hormone" and "extension of the equivalence between geometry and complex number theory"—can be attributed to specific *things*. In scene 6, the only element whose actions were undefined at the beginning of the experimental process was the peptide. The actions of rat pituitary cell cultures, radioimmunoassay, Brazeau, and the technicians were all predictable; the unpredictable phenomenon—the graphs becoming flatter—must thus result from the action of this peptide-thing that "blocks the body." Similarly, in scene 7, the only element whose actions were undefined at this stage of the experimental setting was the third imaginary quantity k geometrically expressed by a fourth dimensional axis. The actions of noncommutative algebra, Pythagoras's theorem, and Hamilton's pencil and paper operations were all predictable; the unpredictable, yet anticipated, phenomenon—geometrical and algebraic equations becoming equivalent—can only be

attributed to this four-dimensional thing that "groups together four numbers." In both scenes, new things emerge from the same attribution process;
scriptural traces of a new phenomenon are imputed to the behavior of a
previously undefined entity.

At the end of both scenes, this attribution process that imputes a behavior to a previously undefined entity by virtue of an experimental setting
ends up being summarized by a term that encapsulates what the now
defined thing does: "that which blocks the body" becomes *somatostatin* and
"that which groups four numbers" becomes *quaternion*. New objects come
into existence, but there has been no miracle; in both cases, the shape of
the new object was progressively defined as scientists made it "grow" from a
list of actions to the name of a thing. In scene 6, *somatostatin* was first "the
graphs become flatter," *then* "under these experimental conditions, there
is a diminution of the growth hormone," *then* "our new peptide decreases
rat's growth hormone," and *finally* "somatostatin decreases rat's growth
hormone." The same *reification process* (Latour 1987, 86–100) happened in
scene 7: *quaternion* was first "two equations become equivalent," *then* "there
is an extension of geometrical representation of complex number theory,"
then "four-dimensional representation allows the extension of geometrical
representation of complex number theory," and *finally* "quaternions express
geometrically complex number theory in a four-dimensional space." In
both cases, experiments, instruments, and alignments of inscriptions—in
short, *laboratory practices* (Latour and Woolgar 1986)—progressively led to
the shaping of scientific objects whose properties and contours could, in
turn, become the topics of papers claiming their existence.[32]

However, as we saw in the previous section, both somatostatin and quaternions as presented in papers that can be read by skeptical colleagues still
need to overcome many trials to become certified scientific facts capable
of being blackboxed, stylized, polished, and enrolled in further claims and
experimental settings. Although both objects came into existence within
their respective bulky and flat laboratories, they still need to attract the
adherence of a wider community. But when the doubts of skeptical readers are removed, when the veracity of both claims are certified by the
scientific institution, we can in turn confidently say that Guillemin *discovered* somatostatin and that Hamilton *discovered* quaternions. Or can we?
We saw indeed that both objects were the results of laboratory practices
that progressively shaped them. Can scientists discover objects they were

previously constructing? Were somatostatin and quaternions already part of "nature" even though they had to be shaped in well-equipped (yet topologically different) laboratories? This is where the story starts to become tricky. If STS has long shown that scientific objects need to be manufactured in laboratories, the heavy apparatus of these locations as well as the practical work needed to make them operative tend to vanish as soon as written claims about scientific objects become certified facts. Once there are no more controversies or disagreements about a new scientific object, nature tends to be invoked as the realm that always already contained this constructed scientific object. Here, we encounter something we discussed in chapter 4 where we were dealing with computer programming practices: when facts are certified and enrolled in further studies, the experiments, instruments, communities, and practices that allowed their progressive formation are generally put aside (Latour and Woolgar 1986, 105–155). This is what makes the history and sociology of sciences (including mathematics) so difficult to conduct; as established facts are purified from the artificial setting that supported their formation, the temptation is great to start from these established facts and extrapolate backward (Collins 1975).[33]

However, if one is not interested in the history or sociology of sciences, if one "just" wants to speak about *objective* facts and eventually enroll them in further claims, the reference to nature appears completely justified. In that sense, one may of course say—as a kind of convenient shortcut—that Hamilton "discovered" quaternions or that Guillemin "discovered" somatostatin, but only because these objects ended up being accepted as certified facts, put in black boxes, translated, polished, and enrolled in later claims. As both initially manufactured objects presented in written claims successively resisted trials, the conditions of their production within dedicated laboratories can be, temporarily, neglected; nature can take over and support their raison d'être. In this respect, Latour's funny analogy is quite instructive:

> Nature, in scientists' hands, is a constitutional monarch, much like Queen Elizabeth the Second. From the throne she reads with the same tone, majesty and conviction, a speech written by Conservative or Labour prime ministers depending on the election outcome. Indeed she *adds* something to the dispute, but only after the dispute has ended; as long as the election is going on she does nothing but wait. (Latour 1987, 98)

The notion of "nature" is thus convenient to speak about noncontroversial scientific facts—why not?—but as soon as one speaks about scientific *controversies* or about scientific objects *in the making,* one needs to consider nature as the uncertain result of scientific practices.[34] This cautious position toward nature applies to "conventional" bulky scientific objects such as somatostatin as well as to "unconventional" flat scientific objects such as quaternions. Again, no superior reality makes mathematical objects appear to mathematicians. They too need to be shaped within (flat) laboratories equipped with instruments that print inscriptions.

Mathematic*able*

A good thing has been taken care of: it seems indeed that the construction process of scientific facts is quite similar to the construction process of mathematical facts. Theorems (cf. scenes 1 and 3), mathematical systems (cf. scenes 5 and 7), conjectures (cf. scene 2), and even formulas (cf. scene 3) may all be considered genuine scientific claims that try to convince colleagues of the existence of objects previously shaped within (flat) laboratories. If the vast majority of these claims do not overcome the trials that can make them become certified facts, some of them (e.g., Shannon-Hartley's theorem, Hamilton's theory of quaternions) may become stylized and polished black boxes that are used as instruments in further experimental settings. It is this huge—and changing—repository of certified mathematical facts that we may call "mathematical knowledge." Moreover, several elements of this certified body of knowledge may, sometimes, become part of tacit, indisputable, and necessary knowledge (e.g., the logical laws of deduction).

However, despite the striking similarities between their respective construction processes, certified scientific and mathematical facts—and their correlated objects—still seem to differ significantly:

Objection of a skeptical reader

All right, let's assume that both facts—and correlated objects—go through similar construction processes, as you obviously believe (while only relying on small, incomplete examples). An important difference subsists: mathematical objects never stop being used for the constitution of nonmathematical objects! We could even see it in the laboratory of endocrinology you used to illustrate your point. The graphs printed by the radioimmunoassay, which quantify how much the growth hormone is decreased by the peptide, are importations of solidified mathematical facts (in this case, basic analytical geometry). The same is certainly true of the inner mechanisms of the radioimmunoassay; complex mathematical theories must have been used to develop this costly instrument. Similar processes happen all the time in demography, climatology, political science, biology, and so on. Mathematical objects such as logarithms, Gaussian functions, or probabilities infiltrate all domains of "hard" science, helping scientists to shape new objects and facts. Yet the inverse is not true: how could peptides or radioimmunoassay help mathematicians shape new objects? Mathematicians have to do things by themselves, without the help of the other sciences. This is why mathematics is the queen of all sciences: without the work of mathematicians in their "flat laboratories"—we may keep that—there would simply be no exact sciences. Mathematical objects are so powerful; they must be of some superior nature. How could it be otherwise?

There are two glitches in this classical objection. First, it is not tenable to say that the practice of mathematics is self-sufficient, for many disciplines intervene in the construction process of mathematical objects and facts. Netz (1998, 2004) showed, for example, how *archiving* and *standardization* were central to overcome the stagnation of Greek geometry.[35] Thanks to the assembling of well-arrayed corpora of papyruses and parchments—especially in Byzantium—late antiquity commentators such as Eutocius became able to compare, annotate, and complete the entangled multiplicities of Greek geometrical writings. Progressively, these systematic standardization efforts made early antiquity's geometrical propositions commensurable; unlike Greek geometers,[36] medieval mathematicians—especially in Bagdad's *House of Wisdom* (Netz 2004, 131–186)—could *see* what Greek geometry was. Equipped with "intellectual technologies" (Goody 1977)—here, collections of standardized Greek geometrical treatises—mathematicians such as al-Khwarizmi and Khayyam could systematize and classify the geometrical problems solved by the Greeks. These systematic comparisons progressively led, according to Netz, to the formation of the algebraic language: "Al-Khwarizmi's algebra was, ultimately, a fairly unambitious ambition, translated into major transformations. Without himself doing anything beyond classifying the results of the past, Al-Khwarizmi, effectively, created the equation" (Netz 2004, 143).

Since archiving and standardization were, and are,[37] central to the formation of mathematical objects, do we have to say that these two respectable disciplines are the queens of the queen of all sciences? To me, a more reasonable position would be to accept that hierarchal classification of disciplines is misleading. When something allows something else to come into existence, it may not be a matter of vertical hierarchy but of *horizontal arrangement*.

This leads us to the second objection regarding the usability of mathematical objects for the assembling of nonmathematical objects. It is true that the combinational capabilities of mathematical facts are surprising. In every scientific discipline, recent or ancient mathematical discoveries are used to conduct experiments, organize inscriptions, express new phenomena, and eventually define new objects. I would go even further than our skeptical reader and expand this extreme combinability of mathematical objects to everyday life. For example, how many times a day do we use the basic precepts of arithmetic? Obviously, mathematics is everywhere, from laboratories of high energy physics to cashiers' desks. This capacity to infiltrate heterogeneous domains of activity is very impressive. But does it necessarily mean that mathematical objects come from a different *nature*? Does their plasticity necessarily manifest a supernatural *essence*?

Let us consider Guillemin's laboratory of endocrinology since it is the example used by our skeptical reader. It is true that the results printed by the computer of the radioimmunoassay required the application of elementary mathematical theories in order to indicate a diminution of the growth hormone. Was there some magic? Not if we consider more precisely the process by which the rat pituitary cell culture was "flattened" to become representable as a graph with numerical values varying through time. What happened indeed within the radioimmunoassay? Schematically, the very small radioactive waves emitted by the rat pituitary cell culture were captured and, after a series of translations, counted by the costly equipment. Radioactive waves became signals that, in turn, became discrete values varying through time. This transubstantiation process—or, more succinctly, *translation process*—that made a cell culture go from the state of complex liquid to the state of a writable list of (radioactive) values spread over time is precisely what allowed the enrollment of the elementary mathematical notion of "ratio" and the further calculation of the growth hormone's decreasing. How did the ancestral theory of ratios as developed by the Pythagoreans become applicable to the world of endocrinology? The concrete efforts to form differently (trans-form) the cell culture into quantifiable inscriptions, thus making it become a geometrical graph, allowed the *connection* between ratios and Guillemin's peptide. It was by flattening the cell culture and adapting it to the flat ecology of ratios that these mathematical objects became applicable to the cell culture. Nothing mysterious happened; by progressively translating a complex entity into a scriptural form, it became possible to link it with certified mathematical facts.

Another—better—example of such an empirical process that makes non-mathematical entities become mathematicable is provided by Michal Lynch (1985) in his book *Art and Artifact in Laboratory Science*. During the 1970s, an important topic in neurology was the plasticity of the brain; that is—briefly stated—its capacity to recover lost functions through the reorganization of some of its tissues. How this reorganization occurs was a controversial topic at the time of Lynch's laboratory study. Two major conjectures were in competition. The first one considered that the reorganization occurred through the densification of the synapses—the structures that allow interneuronal communication between axons and dendrites—*within* the damaged brain territory.[38] The second theory, labeled "axon sprouting," considered that the reorganization was due to the extension of axons *adjacent* to the damaged territory. For many reasons encompassing results of then recent laboratory experiments as well as promising industrial applications, the director of the laboratory studied by Lynch believed that axon sprouting was the main ingredient for the brain's reorganizational capacity (Lynch 1985, 32–33). But how could he demonstrate it? Many pitfalls got in his way. First, neurons are very small. Observing their (re)organization required powerful zooms. Fortunately, the advent of electron microscopy—a technology recently purchased by the laboratory—allowed him to make ultrastructural observations. But this led to another issue: at that time, these observations could only be made on tiny *slides* whose flat topology was different from the bulky topology of neurons. Fortunately, a "methodic series of renderings of laboratory rats" (Lynch 1985, 37) could be organized to properly slice brains and adapt them to ultrastructural visibility. But this extraction of brain slides led to another issue as a reorganizational brain process can only happen within a living brain. How could it then be possible to observe brain plasticity on dead sliced samples? Fortunately, the availability of *many* standardized laboratory rats with almost identical brains allowed the organization of a "chain of sacrifices" (Lynch 1985, 38). Although it was not possible to observe the reorganization of one living damaged brain, it progressively became possible to observe the reorganization of "same" damaged brains killed at different time intervals. A regular series of discrete—and meticulously referenced—dead slices permitted the reconstitution of the evolution of one living brain trying to palliate its damages. Yet the scientists followed by Lynch still needed to discern specific events within the mess of every single slide. They were indeed trying to account for axon fibers that were expanding their territories to damage zones. But how could they define territories of axons as well as their potential expansions? Fortunately—and this greatly contributed to designing the whole project—one interesting characteristic of the "dorsal hippocampus" helped them to establish points of reference common to all electron microscopic observable sections. It had indeed been demonstrated—and accepted—that the structure of the dorsal hippocampus looks like a grid, the dendrites of its cell bodies regularly intersecting axons indexed to different brain regions (Ramón y Cajal 1968). Therefore, if the brain researchers managed to produce electron microscopic observable slices of dorsal hippocampus extracted from similarly damaged rats' brains (killed at different time intervals), the "natural" grid structure produced by the intersections of the dendrites of dorsal hippocampus's body cells with axons indexed to different brain regions could constitute an initial empirical base for further measurements (Lynch 1985, 35–39). In other words, as it was certified that one specific part of the dorsal hippocampus contained cell bodies whose dendrites *always* intersected regularly with axons indexed to two different brain regions, which I call here α and β, it became possible to damage the β brain regions of all rats and then check if the axons indexed to α "sprouted" to infiltrate the territory of the axons previously indexed to β. But again, a new problem arose: how to go from specific electron microscopic views on slices to a *panorama* of many slices distributed over time? At the time of Lynch's study, the easiest way to operate this translation was first to take analogical photographs of electron microscopic dorsal hippocampus displays. Brain scientists then had to develop these photographs in high definition and equip them with a coordinate system scaled according to the ultrastructural levels of observation (between 2,160 and 24,000 times, depending on the photographs). How did Lynch's scientists concretely manage to equip these high-definition photographs? They pinned down the photographs on a cardboard sheet, hence creating a chronological *montage* of the microscopic displays. As Lynch put it, "these successions of photographs provided the visible configuration of brain ultrastructure that was addressed in the analytical phase of the study" (Lynch 1985, 38). But here again, it was not enough to measure an extension of axons indexed to α. Even though the dendrites of dorsal hippocampus's cell bodies regularly intersected axons indexed to α and β, it remained necessary to affix a *referential* common to all photographs. How did the brain scientists do this? It is difficult here not to quote Lynch's account:

> As each montage was constructed, it was analytically addressed in the following manner: a clear plastic sheet was laid over the surface of the photographs, and a linear scale was drawn over the surface of the sheet running in a vertical direction which paralleled the edge of the columnar montage of photographs. ... A scale of "microns" (computed with reference to the magnificational power of the photographs) was plotted for the drawn-line, where the "zero" point was set at a horizontal line that approximated the alignment of the granule cell body layer. ... *Measurement along this scale was used to estimate linear distance along the "vertical" alignment of granule cell dendrites as they arose from the cell bodies and coursed "upward."* (Lynch 1985, 38; italics added)

Flat linear distances are a priori far removed from neurons and the potential sprouting of their axons. Yet, once enlarged photographs of tiny little slices of standardized rats' dorsal hippocampus are mounted on cardboard and equipped with a linear scale drawn on clear plastic sheets whose "zero" point corresponds to the cell body of each slice, this venerable mathematical theory and its correlated objects become very, very close (Latour 1987, 244). The experimental setting of the laboratory and all of its instruments producing "alignable" inscriptions—standardized rats; tiny, carefully washed (and stained) slices of rats' dorsal hippocampus; montages of enlarged photographs; linear scales drawn on clear plastic sheets—end up conferring to rats' dorsal hippocampus the same *form* as graphs on which linear distances can be estimated. At the end of this measurement process, ratios of intact/dead terminals—junctions between axons and dendrites—plotted in terms of days *post* the lesion could even be computed by the scientists, thus demonstrating *statistically* the phenomenon of axon sprouting: "Measurement of this expansion showed a consistent reoccupancy of the lower 25 per cent of the region of the granule cell dendrites formerly occupied by the [damaged] layer of axons" (Lynch 1985, 35).

Again, as Lynch demonstrated, no magic intervened; laboratory practices made the relationships between axons and dendrites become *mathematicable*. Standardized rats became dorsal hippocampus, tiny slices became
enlarged photographs, and a montage of cardboard became one regular
geometrical space whose occupancy evolved through time. If some polished mathematical facts—computation of surfaces progressively occupied
by intact terminals—did help demonstrate the existence of a nonmathematical phenomenon (axon sprouting), this *event* necessitated a succession
of translations in order to connect the wet and bulky ecology of the brain
with the dry and flat ecology of mathematics.

Formulating: A Definition

Mathematics does not apply to the world. A cascade of translations is required
to connect nonmathematical entities with certified mathematical facts. But
at this point of our operationalization exercise, one question remains: if
the rats' dorsal hippocampus of the brain research laboratory we have just
considered and the rat pituitary cell culture of Guillemin's laboratory both
end up being trans-formed in order to fit with the networks sustaining
solidified mathematical objects (themselves formerly described by claims
that progressively became certified facts and even, sometimes, single sentence statements part of tacit undisputable knowledge), do they not *lose*
many properties on the road? After all, from a rich and complex region of
the brain, the dorsal hippocampus becomes a tinkered montage of gridded
photographs; from a rich and complex soup of cells, the rat pituitary cell
culture becomes a simple graph. To make both entities mathematicable,
they must endure important *reductions*. But is it worth it? What justifies
such flattening and drying?

In these specific situations, the gains of these reductions are important
because the properties of the mathematical objects as formerly defined
by mathematicians within their flat laboratories are progressively "lent"
to the pituitary cell culture and the dorsal hippocampus. First, both entities become *easier to handle*. After the translation process from a cell soup
to a graph, Guillemin does not need the cell soup anymore. He certainly
conserves it for potential verifications, but whenever he needs to see or
show the rat pituitary cell culture, he can now use the graph printed by the
radioimmunoassay that expresses only the tiny important part of the soup's

properties. The same is true of the brain research laboratory studied by Lynch: instead of handling tiny slices of hippocampus, brain scientists can now consider gridded photographs. One direct consequence of this ergonomic gain is that the reduced entities become also *more sharable*. Although it is impossible to e-mail—or, in these cases, fax—wet and bulky dorsal hippocampus, after their translation into a succession of photographs, trustful brain scientist colleagues based on the other side of the world are also able to scrutinize them. Transforming the hippocampus into gridded pieces of paper allows it to invest extended—yet expansive and fragile—communication networks. Such a reduced and flattened hippocampus therefore also becomes *more comparable*; if the brain scientists based on the other side of the world also manage to operate similar reductions on the dorsal hippocampus, they may be able to compare both successions of gridded photographs. The same is also true of Guillemin's graphs: instead of comparing cell soups, endocrinologists can compare graphs, a far easier endeavor.

Another gain of reducing entities and making them fit with the flat network of certified mathematical knowledge is that reduced entities become much more *malleable*; new takes appear that, in turn, suggest new instruments, tests, and inscriptions. For example, when active junctions between axons and dendrites become points within a uniform geometrical space, the instruments already defined by mathematicians for this geometrical space can be used to further probe the still undefined phenomenon of axon sprouting, thus producing new inscriptions that will precisely help to define it. Within this geometrical space, new tests can be made, such as *measuring* surfaces, *counting* terminals, and *calculating* ratios of occupancy. These tests and their correlated instruments will, in turn, produce readable inscriptions—here, lists of numbers—that will help further characterize the phenomenon under scrutiny. The same is true of Guillemin's rat pituitary cell culture: once complex biochemical reactions become discrete values varying through time, all the instruments that become available through this graphic form can be used to further probe the cell soup. What is the *slope* of the graph? What is the *speed* of the growth hormone's decreasing? Again, a flat reduced form enables the use of new instruments and the production of new readable inscriptions that help with the characterization of a new phenomenon.

This leads us to one last gain of these crucial reduction processes, perhaps the consequence of all the other gains:[39] when an entity is made compatible with mathematical facts, it also becomes *enrollable* within the written claim that will try to attest to its reified existence. This element is crucial if we want to understand the full additional strength these reduction processes may give to undefined entities. How indeed to include axons within a text claiming their ability to sprout? How to include Guillemin's new peptide within a paper attesting to its decreasing effect on the growth hormone? Reducing them until they reach the same *form* as certified "flat" mathematical facts allows them to become the *referents* of the prose that presents them to their respective scientific communities. In addition to making both axons and peptide easier to handle, more shareable, more comparable, and more malleable, reducing them to make them compatible with the flat ecology of mathematical facts allows them to be included *inside* the texts that talk about them. The reified object "axon sprouting," more than just being described in a paper, is also present within the paper in the flat and dry form that precisely allowed its mathematization (in this case, according to Lynch [1985, 40–49], as a succession of gridded photographs whose points move "upward"). Similarly, the reified object "somatostatin," more than just being described in a paper, is also within the paper in the form of a graph summarizing its behavior (Brazeau et al. 1973). The attentive reader may have noticed that we have now come full circle from the beginning of this operationalization exercise where we were talking about written claims of relative conviction strengths. The end results of laboratories, experiments, instruments, and inscriptions are indeed the formulation of claims that try to attract the adherence of individuals. In this respect, we should now be in a position to better understand the fascinating power of mathematical objects and facts; they may go through construction processes that are similar to other scientific facts, but their particular flat and dry ecology makes them relevant for the formation of nonmathematical objects and facts. They make undefined entities easier to handle, more shareable, more comparable, more malleable, and more enrollable within claims they precisely help to *formulate*.

It is not mathematical facts and their correlated objects that give, by themselves, some additional strength to the transformed entities they sometimes encounter. Rather, it is the flat ecology within which mathematical knowledge deploys itself that, sometimes, provides advantages to the entities that acquire the same *form*. This last element allows me to finally define the activity of *formulating* more technically; for the remainder of this part III, I shall call *formulating* the empirical process of translating an undefined entity until it acquires the same form as already defined mathematical object. The encounter between a "made-flat" entity and a mathematical object—that previously had to be constructed in a laboratory and presented in a claim whose conviction strength made it a polished fact—will, in turn, help scientists to further characterize the behavior of the entity and present its reified version in a written claim. Just as any scientific claim (including those formulated by mathematicians), this written claim will still have to overcome publication, citation, captation, and posterity trials to become, eventually, a certified fact. A circle has been drawn; we are now back to where we started. With all these elements in mind, it is high time to return to computer science in the making and engage with ethnographic materials.

6 A Third Case Study

As in part II when we were dealing with computer programming, the journey was long and full of zigzags. But we did not have any other choice: in order not to get lost in our further explorations of the role of mathematics in the formation of algorithms, we needed to understand where certified mathematical facts come from; how they solidify; and how, sometimes—very rarely—they become part of tacit necessary knowledge. Thanks to STS works on mathematics as well as heterogeneous examples taken from nineteenth-century protograph theory, contemporary controversies in fuzzy logic, a well-accepted theorem in theoretical signal processing, and the laboratory practices that led to the shaping/discovery of quaternions, we progressively realized that mathematical objects—and the certified facts that describe them—need academic papers, trials, laboratories, instruments, and inscriptions to come into existence. Moreover, when nonmathematical disciplines, such as endocrinology or brain research, need to borrow the heuristic and ergonomic strength of certified mathematical objects and facts to qualify bulky and wet entities (e.g., a new peptide, axons of dorsal hippocampus), a cascade of translations is required in order to make these entities compatible with the flat ecology of certified mathematical facts. Consequently, we saw that the indubitable power of mathematics should be understood in the light of the mundane practices that allow nonmathematical entities to become "mathematicable." These mundane yet often ignored practices aiming to connect undefined entities to certified mathematical knowledge are what I call "formulating."

But how do formulating practices express themselves within computer science laboratories? What is their role in the construction of algorithms? In light of the previous parts of this book, how does *formulating* articulate with *ground-truthing* and *programming* activities? This is what we are going to consider in this third case study.

Presentation of the Empirical Materials

This case study is taken from the saliency-detection project we already encountered in chapter 2. Just to refresh the memory of the reader, this saliency-detection project included two PhD students and a postdoc—BJ, GY, and CL—that I shall keep on referring to as a single entity: "the Group." In a nutshell, the Group's argument that framed the project was that saliency detection in image processing may become industrially more interesting if saliency-detection algorithms could detect, segment, and evaluate the varying importance of salient objects and human faces within complex digital photographs. This new problematization of the saliency problem called for the construction of a new ground-truth database gathering unlabeled complex digital images and their manually labeled counterparts, the "targets." The new ground truth was central to the formation of the Group's algorithm as this database materially established the terms of the problem to be solved computationally. To effectively shape its algorithm, the Group divided its new ground-truth database into two sets: a training set and an evaluation set. The training set was used to study the relationships between input-data and their targets. Once these relationships were defined and expressed in a computational model, the Group translated this model into numbered lists of machine-readable instructions, thus assembling a genuine computer program. The performances of this program could then be evaluated on the evaluation set of the ground truth by means of standard statistical measures. The new ground-truth database, the principles of the computational model, and the processing performances of the correlated computer program were later presented in an academic paper that was rejected by the committee of an important conference in image processing. Yet one year later, a revised version of the article won the "Best Short Paper Award" at a smaller conference.

In the following sections, I will mainly focus on the training set and the practices that led to the *formulation* of the relationships between input-images and their targets that was then translated into lines of code. As the targets of the Group's new ground truth were quite complex, I will focus exclusively on one of the targets' component: the relative importance values of the detected and segmented faces (see figure 6.1). My goal is to account for the formulating practices that led to the characterization of a way to automatically calculate the relative importance values of detected faces, thus retrieving one—small—part of the ground truth's targets. Accounting

Figure 6.1

Montage assembled from the data of Group's ground truth. On the left, an "input-image" of the Group's new ground-truth database. In the middle, the same image as labeled by the workers of the crowdsourcing task. The crowdworkers did not all agree on the salient features of the image. If all of them labeled the whole body of the woman, then some others also labeled her face, the face in the middle of the image, and the face on the right-hand side of the image. The gray-scale image on the right is based on the labeled image in the middle. It was post-processed within the Lab after the crowdsourcing experiment. Each gray-scale zone corresponds to one target of the unlabeled image on the left. These zones are what the computer program, as defined by the computational model, should retrieve in the best possible way. The relative saliency values of the targets—expressed by different gray-scale values— were defined as the ratios of the number of rectangles that surround them over the number of workers who performed the labeling task on the image. In this case, four-teen workers performed the labeling task. Fourteen rectangles surrounded the whole woman, which makes the shape of her body have the maximum value 1. But thirteen rectangles also specifically surrounded the face of the woman, making it have the value 0.93. Twelve rectangles surrounded the face in the middle (value 0.85), and ten rectangles surrounded the face on the right (value 0.71). The background of the gray-scale image—everything that is not labeled—has the value zero. All these values and zones have been defined with the help of the labels drawn by the workers. At this point, the goal of the Group's project was to find a way to automatically transform the image on the left into the image on the right without the help of the labels. In this case study, we will only examine how the Group found a way to automatically retrieve the relative saliency values of faces. We will not deal with nonface elements nor with any sort of segmentation. Following the Group, the question we will have to answer is thus the following: How do we retrieve face importance values (e.g., 0.93, 0.85, 0.71) from input-images such as the one on the left?

for these practices will allow me to link this part III with part I (ground-truthing) and part II (programming). This case study will also serve as stepping stone to touch on the now widely discussed topics of machine learning and artificial intelligence.

To better understand the practices that lead to the definition of a computational model for face importance, we will have to closely examine the Group's training set and the progressive reorganization of its data. Yet, as a Matlab

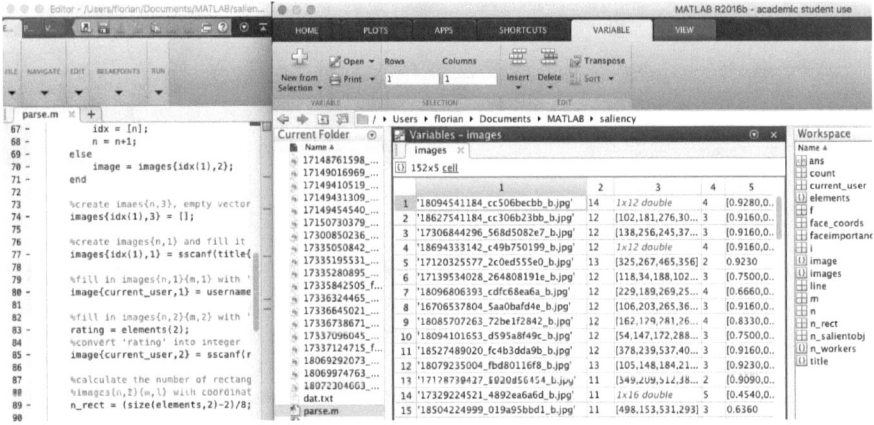

Figure 6.2

Screenshot of the Group's training set used for the modeling of face importance values as it appeared in the Matlab software environment. On the right, the Workspace of Matlab IDE indicates all the variables used to create the database. In the center of the screenshot, a spreadsheet that summarizes the organization of the database. The first column of the spreadsheet gathers the IDs of the input-images of the training set. The second column indicates the number of crowdworkers who performed the labeling task on the input-image of the same row. The third column gathers the coordinates of the face-detection rectangles as provided by BJ's algorithm when run on the input-image of the same row (more on this below, in the main text). Each group of four coordinates refers to (a) the point on the *x* axis of the input-image where the rectangle starts; (b) the point on the *y* axis where the rectangle starts; (c) the point on the *x* axis where the rectangle ends; and (d) the point on the *y* axis where the rectangle ends. The fourth column indicates the number of salient feature within the input-image according to the crowdworkers. This value can be different from the number of groups of four coordinates in column 3. The fifth column refers to the importance values of the faces as the Group computed them based on the labels of the crowdworkers. On the left of the spreadsheet, the window Current Folder indicates the folder currently accessed by Matlab IDE. On the far left, the Editor shows a small part of the Matlab script that was required to parse the data of the crowdsourcing task and organize it as a Matlab database. The computer programming practices that were needed for the completion of this Matlab script were similar to those I described in chapter 4.

training set is quite confusing (see figure 6.2), I will not be able to base my analysis on "real" screenshots. Just like in chapter 4 when I was accounting for programming practices, I will have to simplify the Group's training set and retain only the elements that are relevant for the present analysis. The simplified version of the Group's training set will thus be presented as in table 6.1. As we are going to follow a succession of translations, the first translation of the Group's training set will be counted as one, the second translation as two, and so on. The initial form of the training set will be counted as translation 0.

This case study is organized as follows. I will first start by illustrating how the anticipation of formulating practices may sometimes impact on the design of ground truths. It seems indeed that translating undefined

Table 6.1

Translation 0: Simplified Matlab IDE as it will be presented for the remainder of the analysis

Input-images ID	Coordinates of labeled faces (BJ's model)	Face importance values of labeled faces
image1.jpg	[52; 131; 211; 295] [479; 99; 565; 166] [763; 114; 826; 168]	[0.928] [0.857] [0.714]
image2.jpg	[102; 181; 276; 306] [501; 224; 581; 304]	[0.916] [0.818]
image3.jpg	[138; 256; 245; 379] [367; 142; 406; 202]	[0.916] [0.636]
...	...	...
image152.jpg	[396; 151; 542; 280]	[0.928]

Note: The term "Translation 0" indicates that it is the "initial" state of the training set. This "Translation 0" is of course relative to the sequence we will follow: many other translations were necessary to give this dataset its "initial" form. The first column refers to the input-images' IDs. For this case study, we will only need to consider the first three and the very last input-images. For the sake of clarity, I simplified their IDs. All the rows between image3 and image152 are summarized by the ellipsis "...". The second column indicates the coordinates of the labeled faces in the input-images. These coordinates were provided by BJ's face-detection algorithm (more on this in the main text). The last column gathers the importance values of these faces as provided by the crowdworkers. These are the only data we need in order to follow the group as it tried to define the relationship between input-images and the varying importance values of their faces.

data-target relationships to make them fit with certified mathematical knowledge requires, sometimes, preparatory efforts. In the subsequent section, I will account for the formulating practices that led to the characterization of a computational model that could satisfactorily retrieve face importance values from input-images. As we shall see, many parallels can be drawn between what the Group did to its data-target relationships and what other scientists do to the undefined entities they try to characterize. In that sense, apart from the fact that they often rely on ground-truth databases, the formulating practices that sometimes take place within computer science laboratories may not be very different from formulating practices that take place within laboratories of biology, anthropology, or physics. In the next section of the chapter, I will link formulating practices with programming practices as defined in chapter 4. As we shall see, formulating data-target relationships can make appear polished mathematical facts that operate as *scenarios* for further programming episodes. Finally, I will consider machine-learning techniques as audacious attempts at automating formulating practices at the cost of more ground-truthing and programming efforts. This last element will make me tentatively deal with what is nowadays called (often indiscriminately) "artificial intelligence."

But first things first; for the moment, let us go back to November 2013 at the Lab's cafeteria.

Ground-Truthing—Formulating

November 2013, at the Lab's cafeteria: I meet the Group for the very first time. As I know almost nothing about image processing, ground truths, and saliency detection, this first Group meeting is for me difficult to follow. But during the presentation of the project, the Group soon shares with me one important assumption:

Group meeting, the Lab's cafeteria, November 7, 2013

CL: "Experiments have shown that saliency of faces varies according to their size and number. Basically, one large face is considered more important than many small faces."

GY: "And when there are many faces, each face 'loses' some saliency, so to speak."

FJ: "But when there are many faces, they are also smaller, no?"

GY: "Well, not necessary. You can have one large face on the foreground and many faces in the background."

FJ: "I see. And the other algorithms don't do that?"

SL: "No, they don't pay attention to faces. At least in saliency. And that's precisely the point of including faces to saliency."

As I will find out a few days later, the experiments CL mentions at the beginning of the above transcription come from papers in gaze prediction (Cerf, Frady, and Koch 2009), cognitive psychology (Little, Jones, and DeBruine 2011), and neurobiology (Dekowska, Kuniecki, and Jaśkowski 2008) published in peer-reviewed journals. These papers claim that the relative size and number of faces within a given scene tend to affect their attraction strength. Roughly stated, in a given scene, one large face will generally attract more attention than one small face that itself will attract more attention than many small faces but less attention than, for example, two larger faces. That the importance of faces is somehow related to their size and number within a given image is an important assumption for the Group as it further contributes to defining the selection criteria of the images of the new ground truth:

Group meeting, the Lab's cafeteria, November 7, 2013

CL: "So if it's OK for you, you can start downloading images. Meanwhile, we'll keep working on the code [of the experiment]."

FJ: "Sure."

CL: "But again, it has to be complex images. And most of them must also contain faces."

BJ: "And faces of different sizes and number."

FJ: "You mean, images with many faces as well?"

BJ: "Yes because it impacts on their importance. Otherwise everybody will agree and we won't have continuous values."

How could crowdworkers disagree if the dataset only includes simple images with one centered face or object? As one goal of the Group's project is to refine saliency and make it become more flexible, the images the workers will be asked to label should also give interpretative opportunities. In that sense, the recent findings in gaze prediction and neurology are decisive: gathering images with more or less faces of different sizes may guarantee some healthy disagreement among workers.

Still dazed by all these new stories about ground truths and models, I soon started downloading images on the Lab's server. At the second Group meeting, on November 14, 2013, I showed the Group sample images just to be sure I understood the instructions correctly. As the feedback was positive I continued to download photos. On November 16, 2013, nine hundred carefully selected complex images were available on the Lab's server. But the day after, I received an email from BJ:

Friday, November 17, 2013. Email from BJ to FJ, header "About the distribution of faces"

Hey FJ,

I've quickly processed the faces in the images you selected and binned the x axis. Here is the distribution of our database over number of faces and face size so far.

[see figure 6.3]

We'll try to model things later so we need to equalize a little with more images with two or more large faces. So if you can keep on digging for such images (say two hundred), that'd be great.

Best,

BJ

Many questions immediately arose. First, how did BJ manage to count the number of faces and calculate their respective sizes for every image I put on the server? It turned out that BJ had previously worked on a face-detection algorithm that does precisely this: detecting, counting, and measuring the size of faces within images.[1] Capitalizing on BJ's previous work on face detection was even a reason why this saliency project was launched in the first place (see chapter 2). But why would the current distribution impact the model the Group will have to shape *after* the crowdsourcing task that was not even submitted? This is precisely the question I asked BJ:

Friday, November 17, 2013. Email from FJ to BJ, header "About the distribution of faces"

Sure, no problem. But, if I may, why is it so important to equalize at this stage of the project?

Best,

FJ

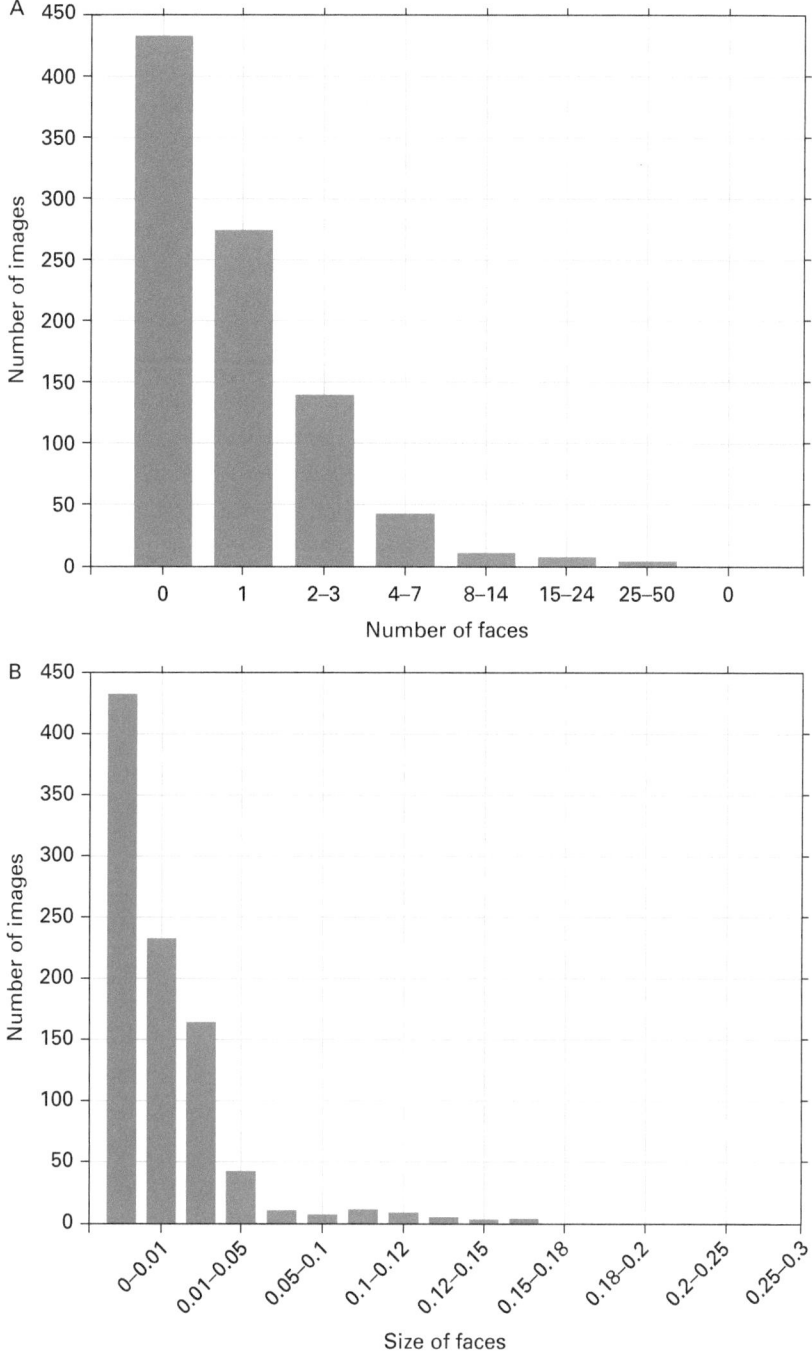

Figure 6.3

Two graphs sent by BJ illustrating the distribution of the database on November 17, 2013.

Saturday, November 18, 2013. Email from BJ to FJ, header "About the distribution of faces"

Great if you can do it.

It's just that if face importance really varies with size and number, we'll surely need a bigger range of cases to fit the data.

Best,

BJ

At this stage of the chapter, we do not need to understand what "fit the data" means (we will cover this in the next section). Suffice here to notice the *projection* BJ makes toward the Group's forthcoming analysis of the relationship between input-images and the importance values of faces, the one small aspect of the output-targets I decided to cover in this case study. In November 2013, the Group does not possess any ground-truth database yet: the web application is not finished; the crowdworkers have not labeled any images; no coordinates of rectangles have been stored in the Lab's server; no multilevel targets have been post-processed. At this stage, there is nothing. Or is there? We saw indeed that the Group has an assumption based on papers it considered trustworthy: the perceived importance of faces is somehow correlated to their size and number. This assumption suffices to make BJ foresee a convenient way to connect the output-target relationship of face values with—hopefully—some certified mathematical claim that will, in turn, help to qualify it. It is indeed not the first time that BJ and the other members of the Group have embarked on the construction of a new algorithm. They have done it before—especially the postdoc CL—and know what to expect. It is perhaps this habit that pushes them to be on the safe side. If equalizing face data can facilitate the future work that will consist in automating the passage from input-images to output-targets that still need to be constructed, it is indeed important to do it.

At the end of chapter 1, I suggested two complementary analytical perspectives on algorithms: a "problem-oriented perspective" that should inquire into the problematization processes leading to the formation of ground truths and an "axiomatic perspective" that should inquire into the numerical procedures extracted from already constituted ground truths. The distinction between these two perspectives was motivated by the need to better understand the formation of the ground truths from which algorithms ultimately derive—hence the "problem-oriented" perspective—while not completely reducing algorithms to these ground truths—hence the "axiomatic" perspective. But I also stipulated, though quite loosely, that both perspectives should be intimately articulated as *ground-truthing* and what I now call *formulating* activities may sometimes overlap, specific numerical features being suggested by ground truths (and vice versa). We see here concretely how these two processes can overlap; the uncertainty related to the construction of a ground truth relying on anonymous and scattered crowdworkers certainly encourages the development of equalizing habits that can further help connect with certified mathematical facts capable of specifying a new phenomenon.

Reaching a Gaussian Function

March 2014: the post-processing of the crowdworkers' rectangular labels is now over. The Group finally possesses a new ground-truth database gathering input-images and their corresponding multilevel targets (see chapter 2, figure 2.8). At this stage, one can say that the Group effectively managed to redefine the terms of the saliency problem, at least at the "laboratory level" (Fujimura 1987). The task of the not yet fully designed algorithm is now clear: from the input-images of the ground truth, it will have to retrieve their corresponding targets in the best possible way. The ground-truth database is thus the material base that will allow both the shaping of the algorithm as well as its evaluation in terms of precision and recall statistical measures.

The next move of the Group is to split the ground truth into two subsets: a training set and an evaluation set. Only the training set containing two hundred images and targets is used to design the computational model. The remaining six hundred images and targets are stored in the Lab's server and will only be used to test the accuracy of the model's program and compare it with other models' programs already proposed by concurrent laboratories (cf. figure 2.9).[2] Within the training set, 152 images contain faces. It is thus this subset of the training set that is used to define a way to automatically retrieve face importance values from input-images without the help of the workers' labels.

Let us have a closer look on this subset of the training set. What does it look like? For the case that interests us here—the definition of the relationship between input-images and face importance values—the training set

Table 6.2
Translation 0 of the Group's training set

Input-images ID	Coordinates of labeled faces (BJ's model)	Face importance values of labeled faces
image1.jpg	[52; 131; 211; 295] [479; 99; 565; 166] [763; 114; 826; 168]	[0.928] [0.857] [0.714]
image2.jpg	[102; 181; 276; 306] [501; 224; 581; 304]	[0.916] [0.818]
image3.jpg	[138; 256; 245; 379] [367; 142; 406; 202]	[0.916] [0.636]
...	...	...
image152.jpg	[396; 151; 542; 280]	[0.928]

concretely looks like a spreadsheet of 152 rows and five columns (only the first three columns are represented in the simplified table 6.2).[3]

The first column of table 6.2 refers to the IDs of the input-images, the second column refers to groups of four coordinates—each group providing information about one face of the input-image (more on this below)—and the third column refers to the importance values attributed by the crowd-workers to each labeled face of the input-images. The data of this Matlab spreadsheet—actually, a genuine database—is crucial as it is the material base of the still to be defined model that will have to retrieve face impor-tance values as provided by the labels of the crowdworkers *without the help of these labels*. But arranged in such a spreadsheet, these data remain quite confusing. How indeed to discern the relationship between the faces of input-images and their correlated face importance values in such an austere classification? Something needs to be done to better appreciate what this relationship looks like.

A convenient way to get a better grip on this relationship between faces of input-images and their importance values—the still-undefined *entity* the Group tries, precisely, to define—is to make it seeable all at once. But how to see faces and their importance values within one legible document? Importance values are numbers so they can be represented as dots within a readable drawing—for example, a *graph*—rather easily. But what about faces? What *are* they? Technically, within the training database—thanks to BJ's face-detection algorithm—the faces of input-images are *groups of four coordinates linked to one image ID*. But how then do we make these groups commensurable with face importance values? One necessary operation is to *reduce* these groups and *translate* them into something else, hopefully comparable to the face importance numerical values. In line with its documented initial assumption regarding the size and number of faces—an assumption that participated in the collection of the data in the first place (cf. above)—the Group decided to summarize every group of coordinates with only two numerical values: a "number-value" and a "size-value." The number-value is provided by BJ's face-detection algorithm. It refers to the absolute number of faces within each input-image. This value can sometimes be superior to the number of labeled faces as crowdworkers have not always labeled as salient all the faces within the input-images. The "size-value" refers to the size of the faces labeled as salient by the crowdworkers. Again, BJ's face-detection algorithm helped to produce these values as it computed the faces' sizes as the ratio of the area of the face-detection rectangle over the size of the image. After the Group wrote the appropriate scripts in the Matlab Editor to compute these values with the help of BJ's face-detection algorithm, the spreadsheet of its training set is reorganized as in table 6.3.

If this first translation successively reduces each labeled face of input-images to two numerical values—a "number-value" (column 2) and a "size-value" (column 3)—it remains difficult to compare them with their importance values deriving from the workers' labels. Indeed, how would it be possible to represent such different orders of magnitude on the same scale? We saw that face importance values can vary between zero and one. But what about "number-values" and "size-values"? Number-values can be problematic as they can vary from one to ninety-eight. But the real issue comes from the size-values that can vary from 0.0003 (smallest labeled face of the training set) to 0.7500 (the biggest labeled face of the training set): four orders of magnitude separate the smallest size-value from the highest. And six orders of magnitude separate the smallest size-value (0.0003) from the highest number-value (98). With such differences of scale, it is extremely difficult to gather all these values in one readable document.

Yet all these numerical values possess an important property: they are numerical values and can thus be written down, studied, and tested in *flat laboratories* by researchers called mathematicians (as we saw in chapter 5). In fact, a whole subfield of mathematics—*number theory*—daily dedicates itself to the study of these flat and dry entities. An important proto number

Table 6.3

Translation 1 of the Group's training set

Input-images ID	number-values	size-values of labeled faces	Face importance values of labeled faces
image1.jpg	3	[0.065] [0.014] [0.008]	[0.928] [0.857] [0.714]
image2.jpg	2	[0.042] [0.012]	[0.916] [0.818]
image3.jpg	3	[0.030] [0.0054]	[0.916] [0.636]
...	...	...	...
image152.jpg	1	[0.053]	[0.928]

theorist, John Napier, even shaped/discovered what he called, in 1614, "logarithm": the inverse of exponentiation.[4] Thanks to this mathematical fact that is now a "single sentence statement" (Latour 1987, 21–62), it is nowadays easy to translate values of different orders of magnitude and represent them on one same readable drawing. Thanks to the instrument of logarithm, both number-values and size-values referring to the faces of input-images can be further translated by the Group into *logarithmic values*. Thanks to this basic operation—imbedded in Matlab—the initial problem of scale vanishes, and a whole set of comparable integers now appears in the Group's dataset (see table 6.4). And the undefined entity "relationship between faces of input images and their importance values" the Group tries to describe becomes a little bit more characterizable.

But still, at this stage, the training set remains hard to read. Whereas the Group is mainly interested in the faces of its training set, the database keeps being organized around the IDs of the input-images. This organization of the data was important at the beginning of the translation process as it helped to indicate what BJ's face-detection algorithm was to look at. But at this stage, this image-centered organization is cumbersome. It is then time for the Group, once again, to reorganize its spreadsheet to center it around its face-related data: log(number-values), log(size-values), and face importance values. When put together, these "triplets" of values give a unique "signature" to each of the 266 labeled faces of the training set (see table 6.5).

After this third translation, the training set has become a list of signatures gathering triplets of relatively close values. Though quite common and mundane, the efforts undertook by the Group from Translation 0

Table 6.4
Translation 2 of the Group's training set

Input-images ID	log(number-values)	log(size-values)	Face importance values
image1.jpg	0.477	[-1.187] [-1.853] [-2.096]	[0.928] [0.857] [0.714]
Image2.jpg	0.301	[-1.376] [-1.920]	[0.916] [0.818]
Image3.jpg	0.477	[-1.522] [-2.267]	[0.916] [0.636]
...	...	...	...
image152.jpg	0	[-1.275]	[0.928]

Table 6.5
Translation 3 of the Group's training set

	Face signatures
1	[0.477; -1.187; 0.928]
2	[0.477; -1.853; 0.857]
3	[0.477; -2.096; 0.714]
4	[0.301; -1.376; 0.916]
5	[0.301; -1.920; 0.818]
6	[0.301; -1.522; 0.916]
7	[0.301; -2.267; 0.636]
...	
266	[0; -1.275; 0.928]

start to pay off: every labeled face is now described by a unique combination of numbers. But still, in this list form, it remains hard for the Group to discern a relationship among the values of these triplets: how do face importance values interact with both number-values and size-values? Even though this list well simplifies the initial spreadsheet, it still has an important inconvenience: it looks like any other list—from shopping lists to lists of bond prices. The values within these lists may differ, but the lists themselves have always roughly the same shape: they remain successions of lines (Goody 1977, 78–108). How then to grasp the particularity of the undefined entity the Group tries to characterize? How to define its shape, its unique behavior?

If the forms of lists of numbers are difficult to differentiate, these lists have nonetheless a crucial quality: they can—at least since the second half of the seventeenth century—give form to the values they contain. Indeed, when coupled with an appropriate *coordinate space*, the numbers contained by lists can be transformed into points that draw distinguishable shapes. As the transformation of lists of values into graphs is nowadays a "single sentence statement" part of tacit and necessary knowledge, the Group just needs to write the Matlab instruction "`scatter(data(:,1), data(:,2), data(:,3))`" to create the *scatterplot* of figure 6.4.

Every labeled face of the training set is re-presented in this Matlab scatter-plot of log(number-values)—x axis—and log(size-values)—y axis—against importance values—z axis, ψ in the plot. At this point, the undefined entity the Group tries to characterize starts to get a shape. Its behavior begins to appear; a genuine *phenomenon* is being drawn that has specific characteristics. It starts "slowly" with low ψ values before drawing a steep slope. This slope then stops to form a kind of ridge before abruptly dropping again. The bell shape of this phenomenon might not talk to everyone. Yet to the Group's members, who are used to encountering mathematical objects, it soon reminds them of a *Gaussian function*:

Friday April 14, 2014. The terrace of CSF's cafeteria, discussion with BJ

FJ: But how did you know that face importance was a Gaussian?[5]

BJ: Well, once we got the plot, it was sure that it was a Gaussian.

FJ: I mean, it could have been something else?

BJ: Sure, but here, the data *drew* a Gaussian.

FJ: But *you* juggled the data in the first place!

BJ: Yes, but it's just to make something appear. You have to do these things; otherwise you have nothing to model.

Thanks to this fourth translation of the training set, the Group has a strong intuition: the relationship between faces of input images and their importance values is surely close to some kind of Gaussian function, a polished certified mathematical object whose behavior is now decently understood and documented. But how could the Group be certain that the phenomenon its experiment created really behaves like a Gaussian function? After all, a Gaussian function is something smooth while the scatterplot the Group asked Matlab to draw is quite discontinuous. From a distance, this heap of

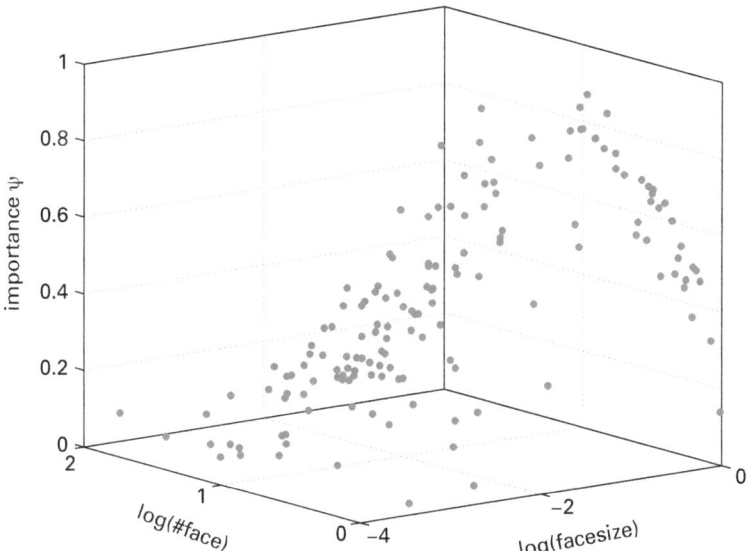

Figure 6.4
Translation 4 of the Group's training set.

points may look like a Gaussian function but when one looks closer, its shape appears rough and uneven.

This is where Matlab, as a huge repository of certified mathematical knowledge, is again crucial as the simple instruction "fit(x.', y.', 'gauss2')" allows the Group to verify its intuition by producing other graphs and captions (see figure 6.5).

Once again, the training set is translated, trans-formed. Its shape is now smooth and homogeneous; it *becomes* an actual function. This new translation of the training set also produces a series of new inscriptions describing the junction between the previous rough heap of points and its smooth counterpart. Let us have a look at these inscriptions: What do they refer to? The last piece of inscription—"R^2 = 0.8567"—indicates that more than 85 percent of the variability in the z data points that constitute the phenomenon the Group tries to qualify can be described by this mathematical function. The inscriptions "$\mu1$ = -1.172" and "$\mu2$ = 0.4308" refer to the peak of the function. They assert that the xy point [−1.72; 0.4308] corresponds to the function's highest z value. Finally, the inscriptions "$\sigma1$ = 0.9701" and "$\sigma2$ = 0.7799" indicate the standard deviation of the function

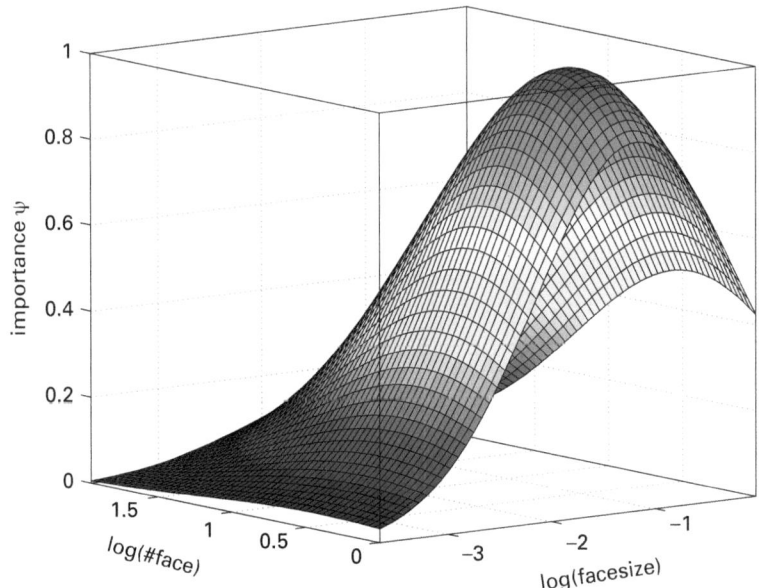

Figure 6.5
Translation 5 of the Group's training set: Gaussian function fitted on the distribution and normalized between 0 and 1. Function's information: `General model Gauss2:` `f(x,y) = exp(-((x-μ1)^2/2σ1^2)-((y-μ2)^2/ 2σ2^2))`. Coefficients: μ1 = -1.172 ; μ2 = 0.4308 ; σ1 = 0.9701 ; σ2 = 0.7799 ; R2 = 0.8567.

along the x axis and y axis, respectively. Altogether, "μ1," "μ2," "σ1," and "σ2" form the *parameters* of the Gaussian function.

In this chapter, I try to account for the formulating practices required for the shaping of an image-processing algorithm (and potentially many others). As a consequence, we do not need to understand every subtlety of these mathematical objects called Gaussian functions. All we need to understand is, first, that Gaussian functions do not come from some superior reality: just as any other mathematical object, Gaussian functions had to be shaped within flat laboratories and described in written claims that had to overcome many trials to become polished certified facts (see chapter 5). Second, we need to understand that thanks to the parameters provided by Matlab—themselves relying on the training set as transformed into a list of coordinates (see table 6.5)—the Group becomes able to *deduce* face importance values as provided by crowdworkers *from* log(number-values) and log(size-values) as provided by the input-images after being processed by BJ's algorithm. In other words, the Group can now decently retrieve face importance values without any labels. This is the consequence of a certified mathematical fact about Gaussian functions. As Matlab reminds the Group after the fifth translation, any z value of this Gaussian function at any point (x,y) can be expressed by the following *formula*:

```
z=f(x,y)=exp(-((x-μ1)^2/2σ1^2)- ((y-μ2)^2/2σ2^2)).
```

When reorganized more elegantly, this formula provided by the certified mathematical knowledge embedded in Matlab gives us:

$$z = f\left(x_i, y_i\right) = \exp\left(-\frac{\left(x_i - \mu_1\right)^2}{2\sigma_1^2} - \frac{\left(y_i - \mu_2\right)^2}{2\sigma_2^2}\right).$$

A connection has been made with the flat ecology of mathematics; thanks to this fifth translation and its correlated inscriptions, the Group now possesses all the elements it needs to *compute* face importance values. With the fourth translation, the undefined entity "relationship between face importance values and faces" became an observable phenomenon. With this fifth translation and the connection it creates with a certified mathematical fact, the behavior of this phenomenon is describable: for any duets (x, y) with coordinates (log[number-value],log[size-value]), there is a z coordinate described by the following *equation*:

$$z = f\left(x_i, y_i\right) = \exp\left(-\frac{\left(x_i - (-1.172)\right)^2}{2(0.9701)^2} - \frac{\left(y_i - 0.4308\right)^2}{2(0.7799)^2}\right).$$

But how does the parametrized equation of the formula that describes the Gaussian function work concretely? How does this equation effectively output face importance values close to those provided by the crowdworkers? Let us consider the first input-image of the training set—the one we used to introduce the topic of the case study in figure 6.1. We saw that, thanks to BJ's face-detection algorithms, the faces of this input-image can be described as [0.065; 3], [0.014; 3], and [0.008; 3], the first values of these duets representing the size-value of the face, the second value representing its number-value. Now, by plugging the log values of these three duets (x_1, y_1), (x_2, y_2), and (x_3, y_3) into the formula provided by the certified mathematical knowledge embedded in Matlab (itself deriving from the Group's translations of the training set), one obtains the three following equations:

$$f(x_1, y_1) = \exp\left(-\frac{(\log(0.065) - (-1.172))^2}{2(0.9701)^2} - \frac{(\log(3) - 0.4308)^2}{2(0.7799)^2}\right) = 0.998$$

$$f(x_2, y_2) = \exp\left(-\frac{(\log(0.014) - (-1.172))^2}{2(0.9701)^2} - \frac{(\log(3) - 0.4308)^2}{2(0.7799)^2}\right) = 0.779$$

$$f(x_3, y_3) = \exp\left(-\frac{(\log(0.008) - (-1.172))^2}{2(0.9701)^2} - \frac{(\log(3) - 0.4308)^2}{2(0.7799)^2}\right) = 0.633$$

The values [0.998], [0.779], and [0.633] are the three face-importance values of the three faces of input-image1 as computed by the Group's *computational model*. We can see that these values are close but not similar to the "original" values [0.928], [0.857], and [0.714] as computed from the crowdworkers coordinates. This is the cost but also the benefit of the whole formulation as the Group now possesses a face importance model that can retrieve different, yet close, face importance values without the help of the crowdworkers' labels.

But the translation process is not over yet. After the statistical evaluation of the whole algorithm on the evaluation set (see chapter 2), one last operation needs to be done; the Group still has to present its reified *object* within the claim that attests for its existence. This is another advantage of formulating practices—more than connecting undefined entities with certified mathematical facts that help to characterize them, it also allows the inclusion of the characterized object inside the text that presents it to the peers. At this point, I must then quote the passage of the Group's initially rejected manuscript where the computational model for face importance is presented:

> We use the following function, denoted as G in Eqn. 2, as a model for varying importance of faces in our saliency algorithm.
>
> $$\psi_i^f \approx G(s_i^f, n_i) = \exp\left(-\frac{(\log(s_i^f) - \mu_1)^2}{2\sigma_1^2} - \frac{(\log(n_i) - \mu_2)^2}{2\sigma_2^2}\right) \qquad (2)$$
>
> Here, ψ_i^f is the importance values of fth face in i^{th} image. s_i^f and n_i are the size of the fth face in i^{th} image and the number of faces in i^{th} image, respectively. Note that s_i^f is the relative size compared to the size of the image, therefore it is between 0 and 1. The parameters of the Gaussian fit are $\mu_1 = -1.172$, $\mu_2 = 0.4308$, $\sigma_1 = 0.9701$. $\sigma_2 = 0.7799$, and the base of the logarithm is equal to 10.

Our efforts paid off: we finally managed to account for these sentences that mix English words with combinations of Greek and Latin letters divided by equal signs that are widely used by computer scientists when they communicate about their algorithms in academic journals. We first had to better understand how mathematical facts and objects come into existence. We then had to accept that the power of these facts and objects does not come from a superior reality but from the mundane formulating practices that progressively translate and reduce undefined nonmathematical entities—peptides, axons, relationships between values of Matlab databases—in order to, eventually, connect them to the flat ecology of mathematical knowledge. We also had to better appreciate the extra strength these connections provide to undefined entities: formulating practices—and the reductions that go with them—make undefined entities easier to handle, more sharable, comparable, malleable, and enrollable within texts claiming for their existence and behavior. With all these elements of chapter 5 in mind, we further had to account for how formulating practices are expressed in the construction of new image-processing algorithms (and potentially many others). We first saw that the anticipation of these practices may sometimes impact on the shaping of ground truths. We then saw how these practices—and all the translations they call for—progressively make an undefined entity become a mathematical object capable of being described by a formula. These connections with the flat ecology of mathematics—in fact, genuine transformations into well-documented mathematical objects—participate in the assemblage of computational models that further appear in academic publications. To paraphrase Latour (1999a, 55), we saw in this section that mathematics has never crossed the great abyss between ideas and things. Yet it often crosses the tiny gap between the already geometrical graph of Translation 4 (figure 6.4) and the solid formula as provided by Translation 5 (figure 6.5). Once this tiny gap is crossed—and this requires many preparatory small gaps—mathematics provides full additional strength to the object under scrutiny.

Yet despite this small victory, something remains mysterious. Indeed, a mathematical formula such as the one summarizing the (very small part of) the Group's model within its academic paper is surely powerful as it allows us to retrieve face importance values without the data provided by the crowdworkers. In that sense, this formula decently describes the behavior of the phenomenon "relationship between faces of input images and their importance values" that was still an undefined entity at the beginning of the formulating process. But in this "formula state," such a computational model cannot make any computer compute anything. In this written form, within the Group's manuscript, the model might be understandable to human beings, but it is not able to trigger electric pulses capable of making computers compute. Yet it somehow needs to; as the performances of the Group's model will also be evaluated on the evaluation set of the ground truth, the model must also take the shape of an actual program. What is then the relationship between the mathematical inscriptions that describe computational models and the actual computer programs that effectively compute data by means of electric pulses?

Formulating—Programming

The point I want to make in this section is quite simple: if mathematical inscriptions that describe computational models in academic papers cannot, of course, trigger electrical pulses capable of making computers compute actual data, they nonetheless work, sometimes, as *transposable scenarios* for computer programming episodes.

In chapter 4, we saw that computer programming practices imply the alignment of inscriptions to produce knowledge about a remote entity (e.g., a compiler, an interpreter, a microprocessor) that is negatively affected in its trajectory. We also saw that programmers constantly need to enroll new actants to get around impasses. More importantly for the case that interests us here, we also found that both aligning and contouring actions needed to be "triggered" by special narratives that engage those who enunciate them. Building on Lucy Suchman and Bruno Latour, I decided to call these performative narratives "scenarios."

Scenarios are crucial as they provide the boundaries of programming episodes while enabling them to unfold. But their irritating drawback is that while they constitute indispensable resources that set up desirable programming horizons, they often tell little about the actions required to reach these horizons. We experienced this when we were following DF in his small computer programming venture. Even though his scenario stipulated the need for the incrementation of an empty matrix with rectangles defined by coordinates stored in .txt files, the scenario said almost nothing about how to do this incrementation. The lines of code had to be progressively assembled as this process was required to align inscriptions and to get around impasses.

Yet some scenarios might be more transposable than others. Let us imagine the following programming scenario: "FJ shall make a computer compute the square root of 485,692." Though quite short, this imaginary example can be considered a genuine scenario as it operates a triple shifting out into other space (at my desk) and time (later) and toward other actants (the Matlab Editor, my having completed the script, etc.) while also engaging me, the one who enunciated it. How could I reach the horizon I am projecting? If I am using Matlab or many other high-level programming languages, the program would be the single instruction "sqrt(485692)." The passage from the scenario to its completion would thus seem quite direct. Let us imagine a trickier scenario: "FJ shall make a computer compute k-means of five clusters over dataset δ." How could I reach this horizon? For the case of Matlab and several other high-level programming languages, the program will, once again, be the single instruction "kmeans(δ,5)"— another straightforward accomplishment.[6] Both imaginary scenarios thus appear quickly transposable into lines of code; the horizon they establish can be reached without many tedious alignments of inscriptions and workarounds of impasses.

Are both imaginary scenarios *simpler* that the scenario defined by DF in chapter 4? It is difficult to say as both square roots of large numbers and k-means of five clusters are not so trivial operations.[7] Rather, it seems that there is a difference of *density*: while our imaginary scenarios can be translated into code almost as they stand, DF's scenario needs to be completed, patched, and refreshed. If nothing seems to stand in between the terms of the statements "square root of 485,692" and "k-means of five clusters," many gaps surely separate each term of the statement "empty matrix incremented with coordinates of rectangles."

The issue is trickier that it seems. One may indeed think that these differences of density within programming scenarios come from scenarios themselves. One may, for example, think that if DF's scenario is less transposable than our two examples, it is because it is less *precise*. But it is actually the opposite: whereas "square root of 485,692" and "k-means of five clusters" tell us almost nothing about how to perform such tasks, DF's scenario takes the trouble to specify a succession of actions. Yes, there are differences of density, but no, they are not necessarily related to what is inside scenarios. So where do these differences come from? I believe these differences of density might be linked to the diffusion of the operations necessary to realize a scenario. My hypothesis, which still needs to be further verified, is that the more an operation is *common* to the community of users and designers of programming languages, the less it will need to be decomposed, translated, and completed. The most striking example of such diffusion-related difference of density within a programming scenario is certainly arithmetic operations. What can be more common to users and designers of programming languages than adding, subtracting, dividing, and multiplying elements? Electronic computers themselves have been progressively designed around these widely distributed operations (Lévy 1995). The terms "add," "subtract," "multiply," or "divide"—when part of a scenario—will thus be immediately translated into their well-known mathematical symbols "+," "/," "−," and "*." The same is true of many other widely diffused calculating operations. "Sine," "cosine," "greatest common divisor," "logarithms," and even sometimes "*k*-means clustering" are all operations that can be straightly transposed from scenarios to programs.

Though quite wild, these propositions will allow us to better understand how the Group's computational model can be almost directly transposed into an actual computer program. Let us first consider once again the formula describing the model shaped by the Group. We saw that the phenomenon observed by the Group was a particular Gaussian function that could be described as

$$z_i = f(x_i, y_i) = \exp(-\frac{(\log(x_i) - \mu_1)^2}{2\sigma_1^2} - \frac{(\log(y_i) - \mu_2)^2}{2\sigma_2^2}),$$

where x_i is the size-value of the ith face, y_i is the number-value of the ith face, and μ_1, μ_2, σ_1, σ_2 are the parameters of the Gaussian fit. When all the parameters of this formula are replaced by the numerical values provided by Matlab, the model becomes the following equation:

$$z_i = f(x_i, y_i) = \exp(-\frac{(\log(x_i) + 1.172)^2}{1.88218802} - \frac{(\log(y_i) - 0.4308)^2}{1.21648802}).$$

From that point, the Group just needs to transpose this mathematical scenario almost as it is within Matlab Editor. This translation gives us the following line of code:

```
z = exp(-((log10(x)+1.172)^2/1.88218802)-((log10(y)-0.4308)
   ^2/1.21648802));
```

As we can see, there is an almost one-to-one correspondence among the mathematical operations as expressed within the equation and the mathematical operations as expressed within the program of this equation: "exp," "–," "log," and "+" all keep the same shape. Only the squaring and dividing operations had to be slightly modified.

Yet in this state, the Group's program of the model will not do anything; it still needs to become iterative to process the changing values of $x_{1,2,...,266}$ and $y_{1,2,...,266}$. Here again, the scenario as defined by the computational model is quickly transposable. We saw in the last section that the training set could be reorganized as needed, as long as the Group manages to write the appropriate Matlab scripts to instruct the training set's reorganization. To operationalize its computational model, the Group just needs to organize the faces of its training set according to their size-values and number-values. Expressed within the Matlab software environment, this reorganization takes the (simplified) form of table 6.6.

This reorganized Matlab spreadsheet will allow the program to know what data it should process. With Matlab programming language, the data of every cell of such spreadsheets can be accessed by inscribing a duet of values in between curly brackets. For our case, the instruction "cell{1,1}" will ask INT to consider the value [0.065]; the instruction "cell{1,2}" will ask INT to consider the value [3]; and so on.[8] Thanks to this referential system, it is possible to ask INT to go through all the cells of the spreadsheet and iteratively plug their values inside the equation. Moreover, the

Table 6.6
Simplified view on the Group's reorganization of the training set

	1	2
1	[0.065]	[3]
2	[0.0143]	[3]
3	[0.008]	[3]
4	[0.042]	[2]
5	[0.012]	[2]
6	[0.030]	[3]
7	[0.0054]	[3]
...	...	...
266	[0.053]	[1]

spreadsheet has a finite length of [266]. This easily accessible information—
it is the number of rows of the spreadsheet—can be used to instruct INT
to start at line 1 of the spreadsheet and stop at its end. When all the size-
values and number-values are processed, they will finally be integrated in
the spreadsheet for their further use in the definition of the remainder of
the Group's algorithm (remember that we only considered one tiny part
of the Group's whole algorithm). The small yet crucial script that permits to
operationalize the Group's computational model for face importance takes
the form of figure 6.6. When run, this small script outputs something close
to table 6.7.

At this point, we can say that the Group managed to assemble a model
that effectively computes data. The deal is now changed: every digital
image can now—potentially—be processed by the Group's model program
for face importance evaluation. Of course, it only forms one small aspect
of the Group's saliency-detection project that ended up being rejected by

```
1. for i = 1:length(cell)
2.     x = cell{i,1};
3.     y = cell{i,2};
4.     z = exp(-((log10(x)+1.172)^2/1.88218802)-((log10(y)-0.4308)^2/1.21648802));
5.     cell{i,3} = z;
6. end
```

Figure 6.6
Operational script for the computation of face importance values.

Table 6.7
Simplified view on the results of the Matlab script as instructed by the Group's
mathematical model

	1	2	3
1	[0.065]	[3]	[0.998]
2	[0.0143]	[3]	[0.779]
3	[0.008]	[3]	[0.633]
4	[0.042]	[2]	[0.964]
5	[0.012]	[2]	[0.732]
6	[0.030]	[3]	[0.935]
7	[0.0054]	[3]	[0.527]
...	...	...	...
266	[0.053]	[1]	[0.853]

the reviewers of the conference (before being awarded the "Best Short Paper Award" at a smaller conference one year later). But still, some existence must be granted to this tiny entity we carefully followed. For three torturous parts divided into six chapters, we have looked for these things we like to call "algorithms"; now we finally glimpse one. And in such a prototypical state, this small piece of algorithm *is* the uncertain product of accountable courses of action.

The (Varying) Reality of Machine Learning

So far in this case study, we saw that although ground-truthing activities—in their capacity as producers of training and evaluation sets and enablers of performance measures—influence formulating activities, expectations regarding future formulating requirements may also influence the initial generation of ground truths. We then saw how formulating courses of action unfold in situ. As we continued to follow the Group in its algorithm project, we saw that many practical translations were necessary to make a training set acquire the same form as a mathematical object. Moreover, we saw how the results of formulating activities—in this case, a mathematical formula—relate to programming activities, the former providing transposable scenarios to the latter.

When we combine these empirical elements with those of part I and part II, we get a quite unusual action-oriented conception of algorithms (see figure 6.7). Indeed, it seems that sometimes what we tend to call an algorithm may be the result of three interrelated activities that I call ground-truthing, programming, and formulating. Of course, these activities may not be the only ones partaking in the constitution of algorithms (hence the interest in launching other ethnographic inquiries). At least, however, in these days of controversies, we can now realistically account for some of the constitutive associations of algorithms.

Yet this action-oriented conception of algorithms remains unduly narrow. Nowadays, is there such a thing as a solitary algorithm? As we have seen throughout the chapters of this book, the constitution of one algorithm undertakes the enrollment of many other algorithms. This was noticeable when we were dealing with ground-truthing practices; whether the selection of images on the Flickr website, their uploading onto the Lab's server, the administration of the crowdsourcing task, or the subsequent pixel-level segmentation of multilayered salient elements, these moments were all supported by additional

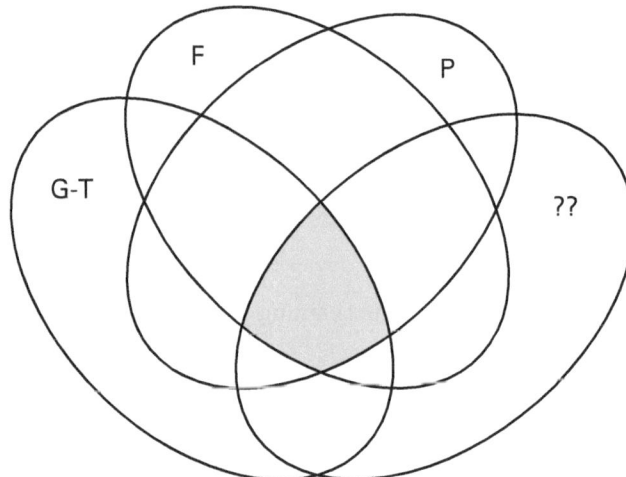

Figure 6.7
Schematic of the interpolation of ground-truthing (G-T), programming (P), and formulating (F) activities. The gray area in the middle of the figure is where algorithms sometimes come into existence. The fourth ellipse tagged "??" stands for other potential activities my inquiry has not managed to account for.

algorithms, among many other things. The same is true of computer programming. Even though this specialized activity currently contributes significantly to the constitution of new algorithms, it goes itself through numerous algorithms, many of which operate close to the computer's hardware to help interpreters, compilers, or processors compute digital data in appreciable ways. Moreover, as we just saw in this chapter, formulating practices are also irrigated by algorithms, an especially visible example being BJ's algorithm that reliably counted the number of faces in an image and calculated their respective sizes. During the constitution of algorithms, algorithms are everywhere, actively contributing to the expression of ground-truthing, programming, and formulating activities. Yet we may reasonably assume that, one way or another, these other algorithms also had to be constituted in specific times and places, being themselves—if my proposition is right—the products of, at least, the same three activities (see figure 6.8).

This conception of algorithms as the joint product of ground-truthing, programming, and formulating activities—themselves often supported by other algorithms that may have undergone analogue constituting

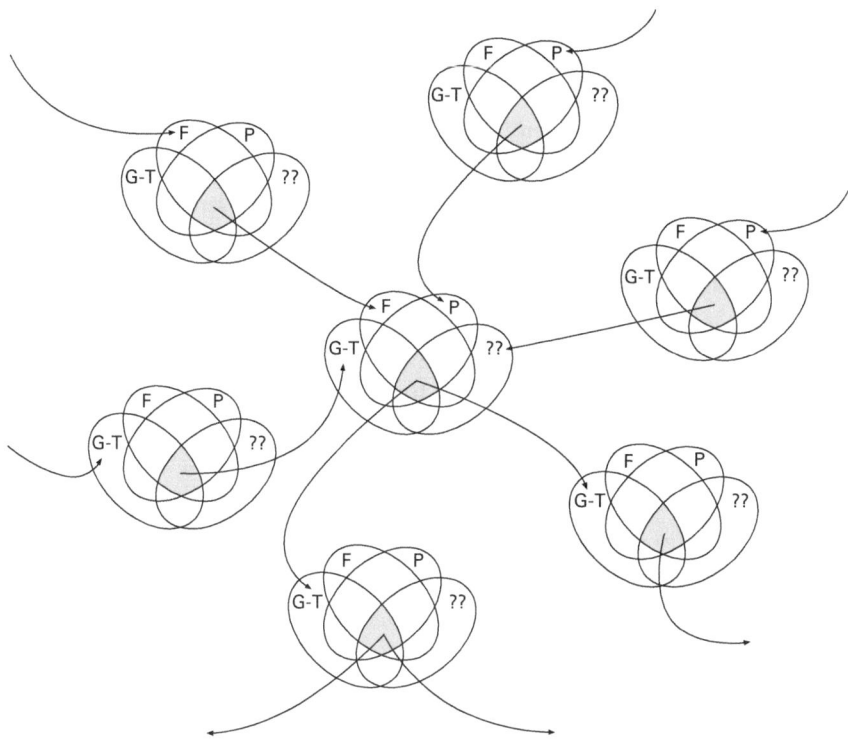

Figure 6.8
Complementary schematic of constituted algorithms partaking in the constitutive activities of other algorithms.

processes—complicates the overall picture while making it more intelligible. Indeed, whenever controversies arise over the effect of an algorithm, disputants may now refer to this basic mapping and collectively consider questions such as: How was the algorithm's ground truth produced? Which formulas operated the transformation of the input-data into output-targets? What programming efforts did all this necessitate? And, if deeper reflections are required, disputants may excavate another layer: Which algorithms contributed to these ground-truthing, programming, and formulating processes? And how were these second-order algorithms constituted in the first place? These are the kinds of empowering questions the present book aims to suggest to fuel constructive disputes about algorithms—a political argument I will develop further in the next, and concluding, chapter.

Again, however, something is still missing. Although the inquiry may sharpen the overall picture, it still fails to address a massive issue—an issue that may even be the most discussed algorithm-related topic at present among the press and academia: *machine learning*. Machine learning is an extremely sensitive topic, sometimes considered in itself (Alpaydin 2010), other times in relation to closely related, yet evolving, terms such as "big data" (Bhattacharyya et al. 2018) or "artificial intelligence" (Michalski, Carbonell, and Mitchell 2014); it is sometimes presented as industrially well established (Finlay 2017) and at others, as still in its infancy (Domingos 2015); it is sometimes praised for its performance (Jordan and Mitchell 2015), and other times criticized for the danger it (but what is *it*?) seems likely to represent to the collective world (Müller 2015). As soon as it is articulated, the term "machine learning" triggers warring feelings of familiarity and ignorance, hopes and fears, utopia and dystopia; a strange madness that seems very incompatible with the down-to-earth vision I am trying to constitute here. In these difficult conditions, how do we address, even superficially, iterations of machine learning as expressions of lived *courses of action*?

One way to scratch the very surface of machine learning, in the light of our empirical and theoretical equipment, may be to make the following observation: during the formulating process accounted for in the section entitled "Reaching a Gaussian Function," something crucial happened just after the Group wrote and ran the Matlab instruction "`fit (x', y', 'gauss2')`." Before this quick Matlab computation—which took only a few seconds—face values (x), size-values (y), and importance values (z) were simply *put* in the same three-dimensional coordinate space. As we saw, putting this together required several translations of the training set, but at a certain point, it was possible to arrange variables x, y, and z together within the same vector space (figure 6.4). At this point, these values were attached to different desires (themselves progressively shaped during ground-truthing processes); x and y values were the Group's *desired inputs,* and z values were its *desired outputs.* But their respective *antecedence* and *posteriority*—there are *first* inputs that should *then* become outputs—were not operationalized; x, y, and z values coexisted simultaneously in one mathematical world. But after INT had computed the translated training set by means of the instruction "`fit (x', y', 'gauss2')`" and printed the correlated graph, formula, and parameters (figure 6.5), number-values and size-values became mathematical *in*puts, and face importance values became mathematical *out*-puts. The Gaussian fit, as the Group happened to call it, made x and y values become operands, just as it made z values become the results of an operation. From the Group's perspective, temporality shifted, it was now possible to *start* with input values and *end* with output values. An operation has been implemented to allow sequential transformations; dimensionality has been reduced by extracting a before and an after.

This turning point, a shift in temporality, was enabled by the enrollment of and delegation to another algorithm. Indeed, when the Group wrote the Matlab instruction "`fit`," it asked INT to estimate the parameters of a function—in this case, a Gaussian one—from a series of coordinate points. At this precise point for the Group, this was a routine intuitive action that required only a handful of characters in the Editor of the Matlab IDE. For INT, however, which effectively computed this estimation of parameters, this was a not so trivial endeavor. How did INT do it?

If we refer to MathWorks' official 2017 documentation, the instruction "`fit (… 'gauss2')`" uses a nonlinear least square computerized method of calculation to estimate the optimal parameters of a Gaussian function from coordinate points.[9] It can thus be inferred that INT does something not so dissimilar to, first, defining the error associated with each point and then defining a function that is the sum of the squares of these errors before taking the partial derivative of the function's equation—with respect to the four parameters—thereby establishing four nonlinear equations that can in turn be solved by using, for example, the Newton-Gauss method. Though contested by several researchers in the field of statistical signal processing (e.g., Hagen and Dereniak 2008; Guo 2011)—thereby making it a genuine research topic—the nonlinear least square algorithm is currently a standard way of estimating parameters of Gaussian functions. Further, by writing this Matlab-imbedded instruction, the Group deployed another computerized method of calculation—one with its own shaping history—to take an important step toward formulating the relationships between the data and the targets of its training set.

That the Group used another algorithm to formulate its new algorithm should not surprise us; ground-truthing, programming, and formulating activities are full of moments where past algorithms contribute to the constitution of a new algorithm (see figure 6.8). What should beg our attention, however, is the decisive temporal shift provoked by the nonlinear least square algorithm subtending the Matlab "fit" instruction during the formulating process. Before the appearance of the Gaussian fit's parameters in the Command Window, the Group had no means to effectively compute the face importance values without the labels of the crowdworkers; its appearance, however, furnished the Group with such an operative ability. Can this specific algorithmically based predictive capacity for the constitution of the Group's algorithm be our entry point to the topic of machine learning?

It is tempting to assert that the algorithm invoked by the Group to help formulate its model *found* the Gaussian function. In fact, it would be more appropriate to say that the algorithm *found an approximation* of the initial function that already underlined the reorganized training set. In other words, given the ground-truth function $f(x,y)$ that, presumably, structured the relationship among size-values, number-values, and face importance values within the translated training set, the algorithm found a useful estimate $f'(x,y)$ that further allowed the production of prediction with an admittedly low probability of errors (hence its usefulness). According to Adrian Mackenzie (2017, 75–102), it is this very specific action that fundamentally consists of processing—some authors even say "torturing" (Domingos 2015, 73)—data to generate an approximation of an initially assumed function that is the main goal of machine learning algorithms, whether they are simple linear regressions or complex deep convolutional neural networks. As Mackenzie, building on the authoritative literature on this now widely discussed topic, astutely summarized it:

> Whether they are seen as forms of artificial intelligence or statistical models, machine learners are directed to build "a good and useful approximation to the desired output" (Alpaydin 2010, 41) or, put more statistically, "to use the sample to find the function from the set of admissible functions that minimizes the probability of errors (Vapnik 1999, 31)." (Mackenzie 2017, 82)

It seems, then, that machine learning algorithms—or "machine learners," as Mackenzie calls them—may be regarded as computerized methods of calculation that aspire to find approximations of functions that presumably organize training and evaluation sets' desired inputs and outputs, themselves deriving from ground-truthing practices (that are still sometimes oriented toward future-formulating practices, as we saw in a previous section of this chapter). This general argument allows us to better grasp the role played by the Gaussian fit during the Group's formulating process. By virtue of Mackenzie's proposition, the Matlab-embedded algorithm enrolled by the Group during its formulating process worked as a machine learner, building the mathematical approximation of the ground-truth function and its related formula (itself working as an easily transposable programming scenario).

Yet if the Matlab least square algorithm can be considered a machine learner, is it reasonable to say that there *was* machine learning during the Group's formulating episode? From Mackenzie's point of view as well as the perspective of the specialized literature, it may appear so; as soon as the Group ran the "fit" instruction, the project became a machine-learning project as its model relied on a statistical learning method that found a useful approximation of the desired output. However, from the Group's perspective, the story is more intricate than that as GY and BJ suggested to me after I shared some of my thoughts:

Wednesday, April 12, 2014. Terrace of the CSF's cafeteria. Discussion with GY

FJ: I'm still holding on to the Gaussian fit moment. ... To find the parameters, there was some kind of machine learning underneath in Matlab, was there not?[10]

GY: Huh, yes perhaps. Some kind of regression, I guess.

FJ: Which is a kind of machine-learning technique, no?

GY: Maybe, technically. But I wouldn't say that. You know, we saw it was a Gaussian anyway, *so it was no real machine learning.*

FJ: Real machine learning?

GY: Yes. For example, like when you do deep-learning things, you first have no idea about the function. You just have many data, and you let the machine do its things. And there, the machine *really learns.*

Friday, April 14, 2014. Terrace of the CSF's cafeteria. Discussion with BJ

FJ: So, machine learning is not what you've done with the Gaussian fit?[11]

BJ: No, no. I mean, there was a fit, yes. But it was so obvious, and Matlab does that very quickly, right? It's nothing compared to machine learning. If you look at what people do now with convolutional neural networks, it's very very different! Or with what NK is doing here with deep learning [for handwritten recognition]. There you need GPUs [graphical processing units], parallelization, etc. And you process again and again a lot of raw data.

There seems to be some uncertainty surrounding the status of the Gaussian fit. If it "technically" can be qualified as machine learning, it is also opposed to "real" machine learning, such as "deep learning" or "convolutional neural networks," where the machine "really learns." It seems that, for GY and BJ—and also for CL, as I learned later on—regarding the Gaussian fit moment as machine learning would misunderstand something constitutive of it. How should we qualify this uncertainty? How should we seek to grasp what, at least for the Group, gives machine learning its specific expression?

An element that, for the Group, seems to subtend the distinction between real and less real machine learning is the visual component that puts the instruction "fit" into gear: "We saw it was a Gaussian anyway, so it was no real machine learning." The visual component was indeed decisive in qualifying the phenomenon the Group tried to formulate; after several translations/reductions of the training set, the scatterplot of figure 6.4 literally looked like a Gaussian, and this similarity, in turn, suggested the use of the "fit" instruction to the Group. The dependent variables—size-values and number-values—were hypothesized before the formulating episode (they even contributed to the construction of the ground truth), and these were parsimonious enough to be visualized in an understandable graph. The group may well have used a machine learner made by others, in other places and at other times; this delegation was minimal, in the sense that most of the work involved in approximating the function had already been undertaken. This is evidenced by the instruction "gauss2" within the instruction "fit," which oriented INT's work toward a 2D Gaussian function with four parameters.

What about deep learning? Why do GY and BJ use it to distinguish between real and less real machine learning? It is important to note that in the spring of 2014—at the time of our discussions at the CSF's cafeteria—deep learning was becoming a popular trend among image-processing communities that specialized in classification and recognition tasks. This popularity was closely related to an important event that occurred during a workshop at the 2012 European Conference on Computer Vision, where Alex Krizhevsky presented a model he had developed with Ilya Sutskever and Geoffrey Hinton—one of the founding fathers of the revival of neural networks (more on this later)—for classifying objects in natural images. This model had partaken in the 2012 ImageNet challenge (more on this later) and won by a large margin, surpassing the error rate of competing algorithms by more than 10 percent (Krizhevsky, Sutskever, and Hinton 2012). The method Krizhevsky, Sutskever, and Hinton used to design their algorithm was initially called "deep convolutional neural networks" before receiving the more generic label of "deep learning" (LeCun, Bengio, and Hinton 2015; Schmidhuber 2015), pursuant to the terminology proposed by Bengio (2009). While this statistical learning method had already been used for handwritten digit recognition (LeCun et al. 1989), natural language processing (Bengio et al. 2003), and traffic sign classification (Nagi et al. 2011), this was its first time being used for "natural" object classification and localization. And in view of its impressive results, a new momentum began to flow through the image-processing community as deep learning started to become more and more discussed in the academic literature, modularized within high-level computer programming languages, and adapted for industrial applications.

In the Lab, NK was the member most familiar with the then latest advances in deep learning as suggested in the above excerpts. He was indeed conducting his PhD research on the application of deep learning for handwritten recognition of fiction writers, and it was through his work—and through communications during Lab meetings—that the topic progressively infiltrated the Lab. As a sign of the growing popularity of these formulating techniques, five doctoral students were moving toward deep learning when I left the field in February 2016, compared with only one—NK—when I arrived. Unfortunately, despite the growing interest in these techniques within the Lab, I did not have the opportunity to explore in detail a deep learning formulating episode. However, based on Krizhevsky's paper, which marked the rise of deep learning within digital image processing, it may be possible to dig further into—or rather, speculate on—the difference suggested by the Group between "real" and "less real" machine learning (despite the dangers that such an approach, based on a "purified account," represents; On this topic, see this book's introduction).

Let us start with the ground truth Krizhevsky, Sutskever, and Hinton used to develop their algorithm. If, to a certain extent, *we get the algorithms of our ground truths* (see chapter 2), then what was theirs? Krizhevsky,

Sutskever, and Hinton used a ground truth called ImageNet to train and evaluate their deep-learning algorithm. ImageNet was an ambitious project, initially conceived in 2006 by Fei-Fei Li, who was at that time a professor of computer science at the University of Illinois Urbana-Champaign.[12] Even though the detailed history of ImageNet—an endeavor that would represent an important step toward problem-oriented studies of algorithms (see chapter 2)—has yet to be undertaken, several academic papers (Deng et al. 2009, 2014; Russakovsky et al. 2015), journalist reports (Gershgorn 2017; Markoff 2012), and a section of Gray and Suri's (2019, 6–8) book *Ghost Work* nonetheless allow us to make informed assumptions about its genealogy.

It seems then that Fei-Fei Li, at least since 2006, was fully aware of something that we realized in chapter 2: better ground truths may lead to better algorithms. Just like the Group, who was not satisfied with ground truths for saliency detection, Li regarded the use of ground truths for the classification of natural images as too simplistic.[13] Through exchanges with Christine Fellbaum, who, since the 1990s, has been building WordNet—a lexical database of English adjectives, verbs, nouns, and adverbs, organized according to sets of synonyms called *synsets* (Fellbaum 1998)—the idea of associating digital images with each word of this gigantic database for computational linguistics progressively emerged. In 2007, when Fei-Fei Li joined the faculty of Princeton University, she officially started the ImageNet project by recruiting a professor, Kai Li, and a PhD student, Jia Deng. After several unsuccessful attempts,[14] Fei-Fei Li, Kai Li, and Jia Deng turned to the new possibilities offered by the crowdsourcing platform Amazon Mechanical Turk (MTurk). Indeed, while images could be quickly scrapped via a keyword search engine such as Google or, at that time, Yahoo, reliably annotating the objects in these images required time-consuming human work. And Amazon MTurk, as a provider of large-scale on-demand microlabor, effectively provided such valuable operations at an unbeatable price. Using ingenious quality control mechanisms, Li's team managed to construct, in two and a half years, a ground-truth database that gathered 3.2 million labeled images, organized into twelve subtrees (e.g., mammal, vehicle, reptile), with 5,247 synsets (e.g., carnivore, trimaran, snake).[15] Despite difficult beginnings,[16] ImageNet has made its way into computer vision research not only through the publicization efforts of Fei-Fei Li, Jia Deng, Kai Li, and Alexander Berg (Deng et al. 2010, 2011b; Deng, Berg, and Li 2011a) but also through its association with a well-respected European image-recognition competition called

PASCAL VOC that has now been followed by ILSVRC.[17] And it was in the context of the 2012 ILSVRC competition that Alex Krizhevsky, Ilya Sutskever, and Geoffrey Hinton developed their deep-learning method that surpassed, by far, all their competitors, initiating a wave of enthusiasm that we are still experiencing today.[18]

But what about the machinery implemented by Krizhevsky, Sutskever, and Hinton to develop their deep convolutional neural network algorithm? How did they formulate the relationship between the input-data (here, raw RGB pixel-values) and the output-targets (here, words referring to objects present in natural images) of the ImageNet ground truth? Let us start with the term "neural networks." We have already encountered it in chapter 3 when we were inquiring into the progressive invisibilization of computer programming practices. As we saw, the term neural network came from McCulloch and Pitts's 1943 paper, which was itself made visible by its instrumental role in von Neumann's *First Draft of a Report on the EDVAC* (von Neumann 1945). McCulluch and Pitts's main argument was that a simplified conception of "all-or-non" neurons could act, depending on their inputs, as logical operators OR, AND, and NOT and thus, when organized into interrelated networks, could be compared to a Turing machine. This analogy between logic gates and the inner constituent of the human brain was then used by von Neumann in his *Draft,* in which he was prompted to use unusual terms such as "organs" instead of "modules" and "memory" instead of "storage" (surprising analogies that must, crucially, be put into the 1945 context when military projects such as the ENIAC and the EDVAC were still classified). Yet, as intriguing as they were, McCulloch and Pitts's neural networks, in their role as logic gates, could not learn; that is, they could not adjust the weight of their "synaptic" interconnections according to measurable errors. It is a merit of Frank Rosenblatt's *perceptron* to have integrated a potential for repetition and modification of logic gates based on algorithmic comparisons between actual and desired outputs (Domingos 2015, 97; Rosenblatt 1958, 1962). But the perceptron algorithm that allows neural networks to modify their synaptic weight according error signals could only learn to draw linear boundaries among vectorized data, making it vulnerable to much criticism.[19] Nearly twenty years later, physicist John Hopfield, as part of his work on spin glasses, proposed an information storage algorithm that allowed neural networks to effectively perform pattern recognition, an achievement that finally brought to light this so-called connectionist approach to learning (Domingos 2015, 102–104; Hopfield 1982). Shortly thereafter, David Ackley, Geoffrey Hinton, and Terrence Sejnokwski built on Hopfield's insights and adapted his deterministic neurons into probabilistic ones, by proposing a learning algorithm for Boltzmann's machines (Ackley, Hinton, and Sejnowski 1985; Hinton, Sejnowski, and Ackley 1984).[20] Then came the real tipping point of this neural network revival, with the design of a stochastic gradient retropropagation algorithm (called "backprop") that could calculate the derivative of the network loss function and back-propagate the error to correct the coefficients in the lower layers, ultimately allowing it to learn nonlinear functions (Rumelhart, Hinton, and Williams 1986).[21] This was followed by a difficult period for this inventive and cohesive research community, who was once again gradually marginalized.[22] But this did not include the increasing computerization of the collective world from the 2000s and the development of web services, both of which led to an explosion of neural network*able* data (yet often at the expense of invisibilized on-demand microlabor). Krizhevsky, Sutskever, and Hinton's (2012) paper is one expression, among many others, of this renewed interest in neural networks, which goes hand in hand with the provision of large ground truths such as ImageNet. Yet besides big data-based labeled data, Krizhevsky, Sutskever, and Hinton could also rely on a stack of well-discussed algorithms (e.g., perceptron, learning for Boltzmann machines, backprop) to build their model; they were able to delegate a significant part of their formulating work to other neural network-related algorithms considered standard by the connectionist community in 2012.

What about the term "convolutional"? In this specific context, it is largely derived from a successful application of the backpropagation algorithm for optimizing neural networks to address an industrial issue: the recognition of handwritten postal codes. It was developed by LeCun et al. (1989) and aimed to exploit the potential of data expressed as multiple arrays—such as RGB digital images "composed of three colour 2D arrays containing pixel intensities in the three colour channels" (LeCun, Bengio, and Hinton 2015)—to minimize the number of neural network parameters as well as the time and cost of learning. In a nutshell, the operation consists of reducing the matrix image into a matrix of lower dimension using a convolution product—a classical operator in functional analysis dating back, at least, to the work of Laplace, Fourier, and Poisson. These convolutional layers are then followed by pooling layers, aimed to "merge semantically similar features into one" (LeCun, Bengio, and Hinton 2015, 439)—a typical way of doing this operation being, at the time of Krizhevsky, Sutskever, and Hinton's study, to use an algorithm called "max-pooling" (Nagi et al. 2011). And when Krizhevsky, Sutskever, and Hinton used convolutional neural networks, they effectively mobilized these convolution and pooling methods—integral parts of the standard algorithm "library"—to be used at their disposal.

Finally, what about the term "deep"? When convolutional layers, activation functions, and max-pooling layers are repeated several times to form a network of networks, this qualifies as "deep." In this case, AlexNet—as the algorithm presented in Krizhevsky, Sutskever, and Hinton ended up being called—was the very first neural network to integrate five convolutional layers in conjunction with three fully connected layers (Krizhevsky, Sutskever, and Hinton 2012, 2).

Though important, the technical features of the algorithm developed by Krizhevsky, Sutskever, and Hinton are not central to the proposition I wish to make here. It is more important to grasp the overall algorithmic machinery that they mobilized to formulate the relationships between their input-data and output-targets. Consider Boltzmann machines, backpropagation, convolutional networks, and max-pooling: although these algorithms were not mainstream in the image-processing and recognition community—as they came from an often marginalized connectionist tradition—they nonetheless constituted a relatively stable infrastructure that could be mobilized to find approximations of functions within large, yet reliable, training sets. The work of Krizhevsky, Sutskever, and Hinton was undoubtedly impressive in many respects. Nonetheless, they were able to capitalize on a modular *algorithmic infrastructure* capable of operating, at least theoretically, as a formulating machine (see figure 6.9).

Yet one important question remains: How did Krizhevsky, Sutskever, and Hinton actually get their input-data processed by their audacious yet standard algorithmic machinery? How did they effectively produce a function approximation? This is where another crucial ingredient emerges (in addition to the ImageNet ground truth and the more or less ready-to-use package of connectionist algorithms): Graphics Processing Units (GPUs). Indeed, the machinery of deep convolutional neural networks requires a lot of computing power. However, as Krizhevsky, Sutskever, and Hinton were processing images—that is, arrays containing pixel intensities—they were able

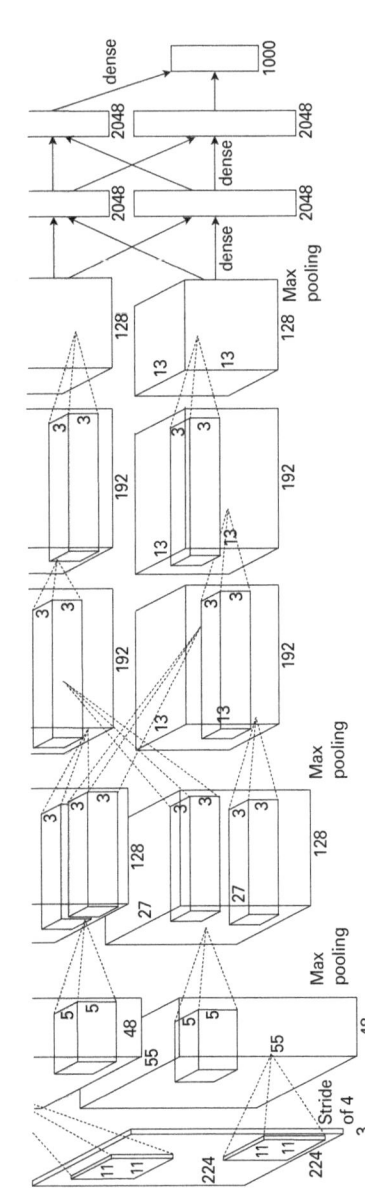

Figure 2: An illustration of the architecture of our CNN, explicitly showing the delineation of responsibilities between the two GPUs. One GPU runs the layer-parts at the top of the figure while the other runs the layer-parts at the bottom. The GPUs communicate only at certain layers. The network's input is 150,528-dimensional, and the number of neurons in the network's remaining layers is given by 253,440–186,624–64,896–64,896–43,264–4096–4096–1000.

Figure 6.9

Schematics of the algorithmic machinery that automatically formulated the relationship between the input-data and the output-targets of the ImageNet ground truth. *Source:* Krizhevsky, Sutskever, and Hinton (2012, 5). Courtesy of Ilya Sutskever.

to get some help from specially designed integrated circuits called GPUs (in this case, two NVIDIA GTX 580 3GB GPUs). It was necessary, however, to interact with these computing systems in such a way that allowed them to adequately express convolutional neural networks (and their whole algorithmic apparatus). This may be Krizhevsky, Sutskever, and Hinton's most impressive achievement, and it should not be underestimated. They may have had a large and trustworthy ground truth made by others, and they may also have had a rich and modulatory algorithmic infrastructure progressively designed by a vivid and supportive community of connectionists; all of these elements had yet to be rendered compatible with the ascetic environment of computers. And, if we refer to Cardon, Cointet, and Mazières's interview of a well-respected researcher in computer vision:

> [Alex Krizhevsky] ran huge machines, which had GPUs that at the time were not great, but that he made communicate with each other to boost them. It was a completely crazy machinery thing. Otherwise, it would never have worked, a geek's skill, a programming skill that is amazing (Cardon, Cointet, and Mazières 2018; my translation).

Besides the ground-truthing efforts made by Fei-Fei Li's team and the algorithmic infrastructure implemented by previous connectionist researchers, Krizhevsky, Sutskever, and Hinton also had to engage themselves in tremendous programming efforts to propose their deep learning algorithm: an "amazing" venture. Yet, after these efforts, and probably many retrofitting operations, they did manage to formulate a monster function with sixty million parameters (Krizhevsky, Sutskever, and Hinton 2012, 5).

When we compare the not quite machine learning of the Group's Gaussian fit with the real machine learning of Krizhevsky, Sutskever, and Hinton's deep convolutional neural networks, what do we see? Beyond the obvious differences, notably in terms of algorithmic complexity, an important similarity stands out: both lead to a roughly similar result; that is, an approximation of their respective assumed ground-truth functions. The function produced by the machine learner invoked by the Group may only have four small parameters, but it ends up transforming inputs into operands and outputs into results of an operation, just like Krizhevsky, Sutskever, and Hinton's sixty-million-parameter function does. Both machine learners approximate the assumed function organizing the data of their respective ground truths, thus remaining subordinate to them.

However, despite this important similarity, the two machine learners differ in that they emanate from differentiated processes; while the Gaussian fit takes over for only a brief moment, following manual translations that can be followed and accounted for, the machinery of Krizhevsky, Sutskever, and Hinton takes over much of the formulation of the training set. Whereas the Group must assume dependent variables, then translate/reduce its training sets according to these assumptions to progressively access a certified mathematical statement—here, a 2D Gaussian—Krizhevsky, Sutskever, and Hinton can delegate this formulating work to an algorithmic infrastructure. Yet again, if there has been *automation* of a significant part of the formulating activities, it is crucial to remember that this was at the cost of a symmetrical *heteromation* of the ground-truthing and programming activities. More than five years of ground-truthing ventures by Fei-Fei Li and her team as well as countless hours of programming work undertaken by Alex Krizhevsky (according to Cardon, Cointet, and Mazières 2018) have made it possible to automate the formulation of the relationship between input-data and output-targets, thereby rendering the former operands and the latter the results of an operation.

Speculating on these elements, we might be tempted to address machine learning—despite its great diversity—as unfolding along a *continuum* (figure 6.10). Machine learners make approximations of functions, but perhaps, the more their invocation relies on the stacking of other algorithms—operating as an infrastructure that automates the formulating activities—the more they constitute machine learning. According to this perspective, the term "machine learning" no longer refers only to a class of statistical techniques but now also includes a practice (and perhaps, sometimes, a habit) of delegation, requiring an appropriate infrastructure that itself touches on ground-truthing and programming issues.

This tentative requalification of machine learning, as a particular instance of formulating activities, may allow us to appreciate the issue of inscrutability in an innovative way. Instead of regarding the growing difficulty in accounting for the processes that have led to the formation of a machine-learned approximation of a ground-truth function as a limit, this conception of machine learning may see it as consubstantial with real machine learning: the more machine learning, the more delegation, and the more difficult it becomes to inspect what has led to the formation of the mathematical operation allowing the transformation of inputs into outputs. Yet—and this

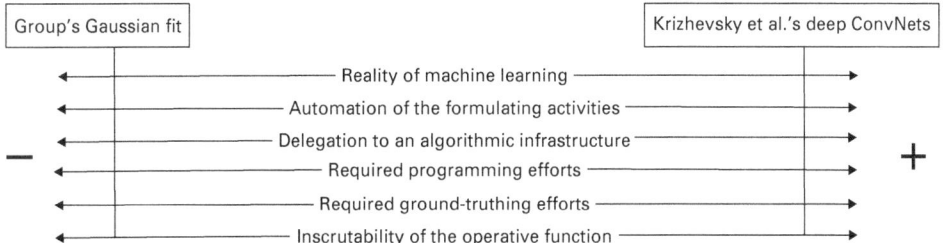

Figure 6.10
Schematic of machine learning considered a continuous phenomenon.

is the real promise of my speculative proposition—real machine learning's native inscrutability may have to be paid for by more ground-truthing and programming efforts, both of which are scrutable activities (as we saw in part I and part II).

I certainly do not here aspire to enunciate general facts; these tentative propositions are mainly intended to suggest further inquiries. This is even truer given that machine learning is both much discussed and very little studied, at least historically and sociologically. Yet as suggested by Jones (2018) and Plasek (2018), given machine learning's growing importance in the formation of algorithms, it is more crucial than ever to investigate the historical and contemporary drivers of this latest expression of formulating activities.

*   *   *

Here in part III, I tried to document the progressive shaping of a computational model in the light of the elements presented in part I and part II. Given that what I ended up calling "formulating practices" dealt with the manipulation of mathematical propositions, we first had to better understand *mathematical facts* and their correlated *objects*. Where do they come from? How are they assembled, and why do computer scientists need them? To answer these preliminary questions, we had to temporarily distance ourselves from many accounts of mathematics: our tribulations in chapters 3 and 4 taught us indeed to be suspicious of terms such as "thoughts," "mind," or "abstraction." In chapter 5, inspired by several STS on mathematics, we privileged a down-to-earth starting point: at some point in their existence, mathematical propositions can be regarded as written claims that try to convince readers. This initial assumption allowed us to consider the striking similarity between mathematics and the other sciences; the written claims made by both mathematicians and scientists must overcome many *trials* to become, eventually, accepted facts. Instead of existing as some fundamental ingredient of thought, mathematical knowledge progressively emerged as a huge, honorable, and evolving body of certified propositions.

We then had to consider the objects that these certified mathematical propositions deal with: Are they similar to scientific objects? By fictitiously comparing the work carried out in a laboratory for biomedicine with the work carried out in a laboratory for algebraic geometry, we realized that, yes, scientific and mathematical objects can be considered quite similar. In both cases, despite topological differences (the mathematical laboratory being often "flatter" and "dryer" than the biomedical one), experiments, instruments, and alignments of inscriptions—in short, *laboratory practices*—progressively led to the shaping of scientific objects, the properties and contours of which became, in turn, topics of papers aimed to convince skeptical readers.

The striking similitude between scientific and mathematical objects prompted us, in turn, to consider why mathematical objects often participate in the shaping of nonmathematical scientific objects. Still supported by STS works on mathematics, we realized that the combinatorial strength of mathematics derives largely from mundane translation practices that progressively reduce entities to make them fit with the flat and dry ecology of mathematical knowledge. By means of such reductions, scientists render the entities they try to characterize as *easier to handle, more sharable, more comparable, more malleable,* and *more enrollable* within written claims trying to convince colleagues of their reified existence. These elements finally allowed us to define *formulating practices* as the empirical process of translating undefined entities to assign them the same form as already defined mathematical objects.

We then tried to use these introductory elements to analyze a formulating episode that took place within the Lab. We started by considering how ground-truthing practices—especially the initial collection of the dataset—may sometimes function as a *preparatory step* for forthcoming formulating practices. This first element made us appreciate the need for a close articulation between the "problem-oriented perspective on algorithms" we initiated in chapter 2 and the "axiomatic perspective on algorithms" we expanded on in chapter 6.

We then inquired into the formation of one of the Group's computational models. We first documented the many translations and reductions of the Group's training set; from a messy Matlab database, the training set progressively evolved into a list of single values that the Group could translate into a scatterplot whose shape expressed a singular phenomenon. The Group's strong intuition that this phenomenon *looked like* a Gaussian function supported the further translation of the scatterplot into a graph that could, in turn, be expressed as a parametrized formula, thanks to centuries of certified mathematical propositions, among many other things.

We then saw that, although mathematical inscriptions describing computational models in academic papers cannot, of course, trigger electric pulses capable of making computers compute actual data, these mathematical inscriptions can nonetheless institute *transposable scenarios* for computer programming episodes. This element was crucial as it completed the connections among the three gerund-parts of this inquiry. Indeed, it seems that formulating practices rely on, and sometimes influence, ground-truthing practices that themselves are supported by programming practices that are themselves, sometimes, irrigated by the results of formulating practices. A whole action-oriented conception of algorithms started to unfold; what we like to call an algorithm may sometimes be the result of these three interrelated activities I here call *ground-truthing*, *programming*, and *formulating*.

Speculating on this, we finally addressed the widely discussed yet sociologically little-investigated topic of machine learning. Based on some (few) empirical clues regarding the varying reality of machine learning, I made the following, tentative, proposition: it may be that machine learning, once considered a *lived experience*, consists of the audacious capacity to automate formulating processes. However, this recently acquired habit may rely on increasing ground-truthing and programming efforts, the springs of which would benefit from further sociological studies.

Conclusion

If you want to understand the big issues, you need to understand the everyday practices that constitute them.
—Suchman, Gerst, and Krämer (2019, 32)

Constituent power thus requires understanding constitution not as a noun but a verb, not an immutable structure but an open procedure that is never brought to an end.
—Hardt (1999, xii)

There was a follow-up of the work required to ground the veracity of a computational model for digital image processing whose academic article was provisionally rejected (chapter 2), a description of the actions deployed to write a short Matlab program (chapter 4), and an analysis of the shaping of a four-parameter formula abstracted from a small training dataset (chapter 6). These empirical elements might seem quite tenuous when compared with the ogre to whom this book is explicitly addressed: *algorithms* and their growing contribution to the shaping of the collective world.

And yet, this book is nonetheless driven by a certain confidence. If I did not believe in its convenience, I simply would not have written (or at least published) it. What justifies such confidence? Which way of thinking supports such a presumption of relevance? In this conclusion, it is time to consider this inquiry's half-hidden assumptions regarding the political significance of its results, however provisional they may be.

Catching a Glimpse, Inflating the Unknown

In the introduction, I mentioned some of the many contemporary sociological works on the effects of algorithms, and I assumed these works progressively contributed to making algorithms become matters of public concern. I then suggested that the current controversies over algorithms call for composition attempts. As algorithms are now central to our computerized societies *while* engaging in moral and ethical issues, their very existence entails constructive negotiations. I then suggested that the ground for these contentious compromises needs to be somewhat prepared or, at least, equipped. As it stands, the *negative invisibility* (Star and Strauss 1999) of the practices underlying the constitution of algorithms prevents from grasping these entities in a comprehensive way; it is difficult, indeed, to make changes on processes that have no material thickness. I then suggested that one way—among other possible ones—to propose refreshing theoretical equipment was to conduct sociological inquiries in collaboration with computer scientists and engineers in order to document their work activities. This may lead to a better understanding of their needs, attachments, issues, and values that could help disputing parties to start negotiate, as Walter Lippmann (1982, 91) said, "under their own colors."

This was an unprecedented effort. While I could build on several STS authors dealing, among other things, with scientific and mathematical practices, I have most often, to be fair, been left to my own devices. However, it was a formative exercise that forced me, beyond the general framework proposed by the "laboratory study" genre, to propose methodologies and concepts—especially in chapters 1, 3, and 5—that I believe are well adapted to the analysis of computer science work. The careful and fastidious unfolding of courses of action allowed me to document the progressive formation of entities—ground truths, programs, and formulas—aggregating choices, habits, objects, and desires. Moreover, it seemed that the congruence of these entities and the practices involved in their shaping form, at least sometimes and partially, other entities we tend to call algorithms.

Nevertheless, this analytical gesture suffers from a certain asymmetry: on the one hand, a small ethnographic report resulting from a PhD thesis, and on the other hand, a whole industry that is constantly growing and innovating. With such limited means, the present investigation could only glimpse the irrigation system of algorithms in their incredible diversity. Worse, by shedding new light on a very limited part of the constituent relationships of algorithms, this inquiry suggested a continent without saying much about it. What about the courses of action involved in getting algorithms out of the laboratories, incorporating them into commercial arrangements, integrating them into software infrastructures, modifying their inner components, maintaining them, improving them, or cursing or loving them? By the very fact of showing that it was possible to bring algorithms back to the ground and consider them products of mundane amendable processes, this investigation probably promised more than it delivered. What value can be attributed to an inquiry that suggests more than asserts?

An Insurgent Document

One can start by stressing the protesting subtext of this investigation. Even if it did not wish to criticize contemporary social studies on algorithms—because they help us to be concerned by our "algorithmic lives" (Mazzotti 2017)—the present inquiry's approach and results nonetheless take a stand against a habit of thought these studies sometimes tend to instill.

This habit, briefly mentioned in the introduction, consists in considering algorithms from an external position and in the light of their effects. I have said it over and over again, this posture is important as it creates political affections. However, by becoming generalized, it also comes up against a limit that takes the form of a looping drama. The argument, initially developed by Ziewitz (2016), is the following: while salutary in many ways, the recent proliferation of studies of the effects of algorithms insidiously tends to make them appear autonomous. Increasingly considered from afar and in terms of the differences they produce, algorithms slowly start to become stand-alone influential entities. This is the first act of the *algorithmic drama*, as Ziewitz calls it: algorithms progressively become, at least within the social science literature, *powerful floating entities*.

Moreover, once the networks allowing them to deploy and persevere are overlooked, algorithms also become more and more mysterious. Indeed, according to this risky standpoint, what can these powerful entities be made of? As the study of the effects of algorithms tends to be privileged to the study of what supports and makes them happen, these entities appear to be made of theoretical, immaterial, and abstract ingredients, loosely referred to as mathematics, code, or a combination of both. Having no grip on what these packages contain, complexity is easily called for help: Whatever the mathematics or the code that form algorithms may refer to, algorithms have to be highly complex entities since they are abstract *and* powerful. How can something be distributed, evanescent, and influential at the same

time? This is the kind of question induced—in hollow—by the multiplication of studies on the effects of algorithms, surreptitiously introducing the
second act of the algorithmic drama: *algorithms become inscrutable*. The end
result is a disempowering loop, for as Ziewitz (2016, 8) wrote, "the opacity
of operations tends to be seen as a new sign of their influence and power."
The algorithmic drama surreptitiously unfolding within the social science
landscape is thus circular: algorithms are powerful because they are inscrutable, because they are powerful, because they are inscrutable ...

The present investigation goes against this trend (which yet remains
important and valuable). Instead of considering algorithms from a distance
and in light of their effects, this book's three case studies—with their theoretical and methodological complements—show that it is in fact possible
to consider algorithms from within the places in which they are concretely
shaped. It is therefore a fundamental, yet fragile, act of resistance *and* organization. It challenges the setup of an algorithmic drama while proposing
ways to renew and sustain this challenge. As it aims to depict algorithms
according to the collective processes that make them happen, this inquiry
is also a constituent impetus that challenges a constituted setup. Again,
there is no innocence.

All the credit, in my opinion, goes to philosopher Antonio Negri for
having detected the double aspect of insurgent acts. In his book *Insurgencies: Constituent Power and the Modern State*, Negri (1999) nicely identifies
a fundamental characteristic of critical gestures: they are always, in fact,
the bearers of articulated visions. It is only from the point of view of the
constituted setup and by virtue of the constitutionalization processes that
were put it in place that insurgent impulses seem disjointed, incomplete,
and utopian. Historically, and philosophically, the opposite is true: beyond
the appearances, the constituted power is quite empty as it mainly falls
back on and recovers the steady innovations of the constituent forces that
are opposed to it. This argument allows Negri to affirm, in turn, that far
from representing marginal and disordered forces to which it is necessary,
at some point, to put an end—in the manner of a Thermidor—constituent
impetuses are topical and coherent and represent the permanent bedrock
of democratic political activities.

Though this book does not endorse all of Negri's claims regarding the
concept of constituent power,[1] it is well in line with Negri's strong proposition that the political, in the sense of politicization processes, cannot avoid

insurgent moves. By suggesting interesting, and surprising, bridges with the pragmatist tradition,[2] Negri (1999, 335) indeed affirms that "the political without constituent power is like an old property, not only languishing but also ruinous, for the workers as well as for its owner." And that is where the political argument of this book lies; it offers an alternative insurgent view on the formation of algorithms in order to feed arguments and suggest renovative modes of organization.

But if this book can be seen as an act of resistance and organization that intends to fuel and lubricate public issues related to algorithms by proposing an alternative account of how they come into existence, why not call it "the constituent of algorithms"? Why did I deliberately choose the term "constitution," seemingly antithetical to the insurgent acts that feed politicization processes? This is where we must also consider this investigation as what it is *materially*: an inscription that circulates more or less. We find here a notion that has accompanied us throughout the book. Thanks to their often durable, mobile, and re-presentable characteristics, inscriptions contribute greatly to the continuous shaping of the collective world. And like any inscription, due to what I have called "Dorothy Smith's law" (cf. introduction), this inscribed volume seeks to establish one reality at the expense of others. Once again, as always, *there is no innocence*: by expressing realities by means of texts, inscriptions also enact these realities. A text, however faithful—and some texts are definitely more faithful than others—is also a wishful accomplishment.

The fixative aspect of this investigation, which comes from its very scriptural form, should not be underestimated. This is even a limit, in my opinion, to Negri's work on constituent power, however interesting and thorough it may be. Although insurrectional impetuses form the driving force of political history—we can keep that—they are nonetheless, very often, scriptural acts that contain a foundational character.[3] The term "constitution" thus appears the most appropriate; if this inquiry participates in the questioning of a constituted setup, it remains constitutive, in its capacity as an inscription, of an affirmation power.

An Impetus to Be Pursued

However, nothing prevents this insurgent document from also being complemented and challenged by other insurgent documents. It is even one of its main ambitions: to inspire a critical dynamic capable of making algorithms ever more graspable. This was the starting point of this investigation, and it is also its end point: to learn more about algorithms by living with them more intimately. And there are certainly many other ways to do just that.

Such alternative paths have been suggested throughout the book in both its theoretical and empirical chapters. Chapter 1, in introducing the methodology of the inquiry, also indicated ways of organizing other inquiries that are grounded in other places and situations. For example, it would be immensely interesting if an ethnographer integrated the team of a start-up trying to design and sell algorithm-related products.[1] With regard to chapter 2, systematic investigations on the work required for the conception, compilation, and aggregation of academic and industrial ground truths would certainly help to link algorithms with more general dynamics related, for example, to the emergence of new forms of on-demand labor. Such an investigative effort could also build analytical bridges between current network technologies that support the commodification of personal data and, for example, blockchain technology which is precisely based on a harsh criticism of this very possibility.[5] In chapter 3, when it came to the progressive setting aside of programming practices from the 1950s onward, more systematic sociohistorical investigations of early electronic computing projects could ignite a fresh new look at "artificial intelligence," a term that, perhaps, has built on other similar invisibilizations of work practices.[6] With regard to chapter 4 and the situated practices of computer programming, conducting further sociological investigations on the organizational and material devices mobilized by programmers in their daily work could contribute to better appreciating this specialized activity that is central to our contemporary societies. Programming practitioners may, in turn, no longer be considered an esoteric community with its own codes but also, and perhaps above all, differentiated groups constantly exploring alternative ways to interact with computers by means of numbered lists of instructions. In chapter 5, although it was about operationalizing a specific understanding of mathematical knowledge, the reader will certainly have noticed the few sources on which my propositions were based. It goes without saying that more sociological analyses of the theoretical work underlying the formation of mathematical statements is, in our increasingly computerized world, more important than ever. Finally, concerning formulating practices, as outlined at the end of chapter 6, analyzing the recent dynamics related to machine learning in light of the practical processes that make them exist could lead to considering the resurrected promises of artificial intelligence through a new lens: What are the costs of this intelligence? How is it artificial? What are its inherent limits? These are urgent topics to be considered at the ground level, not only to fuel controversies but also, perhaps (and always temporarily), to close them.

For now, we are still far from such a generalized sociology of algorithms this book hopes to suggest. We are only at the very beginning of a road that, if we want to democratically integrate the ecology of algorithms into the collective world, is a very long one. With this book, beyond the presented elements that, I hope, have some value in themselves, one can also see an invitation to pursue the investigation of the mundane work underlying the formation and circulation of algorithms—an open-ended and amendable constitution, in short.

Glossary

actant designates any particular human or nonhuman entity. The notion was developed by semiotician Algirdas Julien Greimas before being taken up by Bruno Latour (2005) to expand agency to nonhuman actors and ground his sociological theory, often labeled "actor-network theory."

algorithm is what this book tries to define in an action-oriented way. In view of the inquiry's empirical results, algorithms may be considered, but certainly not reduced to, uncertain products of ground-truthing, programming, and formulating activities.

algorithmic drama refers to the impasse threatening critical studies of algorithms. By mainly considering algorithms from a distance and in terms of their effects, these studies take the risk of being stuck in a dramatic loop: Algorithms are powerful because they are inscrutable, because they are powerful, because they inscrutable, and so on. The term "algorithmic drama" was initially proposed by Malte Ziewitz (2016).

association refers to a connection, or a link, made between at least two actants. An association is an event from which emanates a difference that a text can, sometimes, partially account for.

BRL is the acronym of *Ballistic Research Laboratory*, a now-dismantled center dedicated to ballistics research for the US Army that was located at Aberdeen Proving Ground, Maryland. The BRL played an important role in the history of electronic computing because the ENIAC project was initially launched to accelerate the analysis of ballistic trajectories carried out within the BRL's premises—in collaboration with the Moore School of Electrical Engineering at the University of Pennsylvania.

CCD and **CMOS** are acronyms for *charge-coupled device* and *complementary metal-oxide semiconductor*, respectively. Through the translation of electromagnetic photons into electron charges as well as their amplification and digitalization, these devices enable the production of digital images constituted of discrete square elements called pixels. Organized according to a coordinate system allowing the identification of their locations within a grid, these discrete pixels—to which are typically assigned eight-bit red, green, and blue values in the case of color images—allow computers equipped with dedicated programs to process them. Both CCD and CMOS are central parts of digital cameras. Although they are still the subject of many research efforts, they are now industrially produced and supported by many norms and standards.

chain of reference is a notion initially developed by Bruno Latour and Steve Woolgar (1986) to address the construction of scientific facts. Closely linked with the notion of inscription, a chain of reference allows the maintenance of constants, thus sometimes providing access to that which is distant. Making chains of reference visible, for example, by describing scientific instrumentations in laboratories allows appreciation of the materiality required to produce certified information about remote entities.

cognition is an equivocal term, etymologically linked with the notion of knowledge as it derives from the Latin verb *cognōscere* (get to know). To deflate this notion, which has become hegemonic largely for political reasons, this inquiry—in the wake of the work of Simon Penny (2017)—prefers to attribute to it the more general process of *making sense*.

cognitivism is a specific way to consider cognition. For contingent historical reasons, the general process of making sense has progressively been affiliated with the process of gaining knowledge about remote entities without taking into account the instrumentation enabling this gain. The metaphysical division between a knowing subject and a known object is a direct consequence of this nonconsideration of the material infrastructure involved in the production of knowledge. This, in turn, has forced cognitivism to amalgamate knowledge and reality, thus making the *adaequatio rei et intellectus* the unique, though nonrealistic, yardstick of valid statements and behaviors.

collective world is the immanent process of what is happening. It is close to Wittgenstein's definition of the world as "everything that is the case" (Wittgenstein 1922). The adjective "collective" seeks to underlie the multiplicity of entities involved in this generative process.

Command Window is a space within the Matlab *integrated development environment* (IDE) that allows programmers to see the results of their programming actions on their computer terminal.

composition is the focus of this inquiry; that in which it is trying, at its own level, to participate. Close to compromise, composition expresses a desire for commonality without ignoring the creative readjustments such a desire constantly requires. Composition is an alternative to modernity in that its desire for universality is based on comparative anthropology, thus avoiding—at least potentially—the traps of ethnocentrism.

computationalism is a type of cognitivist metaphysics for which perceptual inputs take the shape of nervous pulses processed by mental models that, in turn, output a different numerical value to the nervous system. According to computationalism, agency is considered the output of both perception and cognition processes and takes the form of bodily movements instructed by nervous pulses. This conception of cognition is closely related to the *computational metaphor of the mind* that establishes an identity relationship between the human mind and (programmed) computers.

constitution refers to both a process and a document. The notion is here preferred to the more traditional one of *construction* because it preserves a fundamental tension of sociological ventures: to describe and contest. The term "constitution" reminds us that a reality comes into being to the detriment of another.

course of action is an accountable sequence of gestures, looks, speeches, movements, and interactions between human and nonhuman actants whose articulations sometimes end up producing *something* (a piece of steel, a plank, a court decision, an algorithm, etc.). Following the seminal work of Jacques Theureau, courses of action are the building blocks of this inquiry. The notion is closely linked to that of *activity* that, in this book, is understood as a set of intertwining courses of actions sharing common finalities. The three parts of this book are all adventurous attempts to present activities taking part in the formation of algorithms; hence their respective gerund titles: ground-truth*ing*, programm*ing*, formulat*ing*.

CSF is the acronym of *Computer Science Faculty*. It is the department to which the Lab belongs. The CSF is part of what I call, for reasons of anonymity, the *European technical institute* (ETI).

digital signal is, in its technical understanding, represented by n number of dimensions depending on the independent variables used to describe the signal. A sampled digital sound is, for example, typically described as a one-dimensional signal whose dependent variables—amplitudes—vary according to time (t); a digital image is typically described as a two-dimensional signal whose dependent variables—intensities— vary according to two axes (x, y) while audiovisual content will be described as a three-dimensional signal with independent variables (x, y, t).

Editor is a space within the Matlab *integrated development environment* (IDE) allowing a programmer to inscribe characters capable of triggering—with the help of an interpreter—electric pulses to compute digital data in desired ways. It is part of the large family of *source-code editors* that can be stand-alone applications or functionalities built into larger software environments.

EDVAC is the acronym of *Electronic Discrete Variable Automatic Computer*. This classified project was launched in August 1944 as the direct continuation of the ENIAC project at the Moore School of Electrical Engineering. The EDVAC played an important role in the history of electronic computing because it was the subject of an influential report written by John von Neumann in 1945. This unfinished report, entitled *First Draft of a Report on the EDVAC*, laid the foundations for what would later be called the *von Neumann architecture*.

ENIAC is the acronym of *Electronic Numerical Integrator and Computer*. This classified project was launched in April 1943 under the direction of John Mauchly and John Presper Eckert at the Moore School of Electrical Engineering. It initially aimed to accelerate the production of firing tables required for long-distance weapons by solving large iterative equations at electronic speed. Although innovative in many ways, the limitations of ENIAC prompted Mauchly, Eckert, and later von Neumann to launch another electronic computing project: the EDVAC.

flat laboratory is a figure of style aiming to address the physical locations in which mathematicians work to produce certified statements. Compared with, for example, laboratories of molecular biology or high-energy physics, the instrumentation of mathematical laboratories tends to take up less space. It is important here not to confuse flatness with the mathematical concept of dimensionality often used to capture and qualify the experience of flatness (or bulkiness). According to the point of view adopted in this book, dimensionality should be considered a product of the relative flatness of mathematical laboratories' equipment.

formula is a mathematical operation expressed in a generic scriptural form. The practical process of enrolling a formula to establish antecedence and posteriority among sets of data is here called *formulating*.

ground truth is an artifact that typically takes the shape of a digital database. Its main function is to relate sets of input-data—images, text, audio—to sets of output-targets—labeled images, labeled text, labeled audio. As ground truths institute problems that not-yet-designed algorithms will have to solve, they also establish their veracity. As this book indicates, many ground truths do not preexist and thus need to be constructed. The collective processes leading to the design and shaping of ground truths heavily impact the nature of the algorithms they help constitute, evaluate, and compare.

image processing is a subfield of computer science that aims to develop and publish computerized methods of calculation capable of processing CDD- and CMOS-derived pixels in meaningful ways. Because digital images can be described as two-dimensional signals whose dependent variables—intensities—vary according to two axes (x, y), image processing is also sometimes called "two-dimensional signal processing." When it focuses on recognition tasks, it is generally called "image recognition."

inscription is a special category of actant that is *durable* (it lives on beyond the here and now of its instantiation), *mobile* (it can move from one place to another without being too much altered), and *re-presentable* (it can—together with suitable infrastructures—carry, transport, and display properties that are not only its own). Due to these capacities, inscriptions greatly participate in shaping the collective world.

INT is the abbreviation for *interpreter*, a complex computer program that translates inscriptions written in high-level programming language into an abstract syntax tree before establishing communication with the computer's hardware. Whenever an interpreter cannot complete its translation, the high-level program cannot perform fully.

Lab stands for the computer science academic *laboratory* that is the field site of the present ethnographic inquiry. The Lab specializes in digital image processing, and its members—PhD students, postdocs, invited researchers, professors—spend a significant amount of their time trying to shape new algorithms and publish them in peer-reviewed journals and conferences.

laboratory study is an STS-inspired genre of ethnographic work that consists in accounting for the mundane work of scientists and technologists. Borrowing from anthropology, it implies staying within an academic or industrial laboratory for a relatively long period of time, collaborating with its members, becoming somewhat competent, and taking a lot of notes on what is going on. At some point, eventually, it also implies leaving the laboratory—at least temporarily—to further compile and analyze the data before submitting, finally, a research report on the scrutinized activity.

machine learning is not only a class of statistical methods but also, and perhaps above all, a lived experience consisting of automating parts of formulating activities. However, this algorithmic delegation for algorithmic design relies on increasing, and often invisibilized, ground-truthing and programming efforts.

mathematics is, in this book, considered integral part of scientific activity. It thus typically consists of producing certified facts about objects shaped or discovered with the help of instruments and devices within (flat) laboratories.

Matlab is a privately held mathematical software for numerical computing built around its own interpreted high-level programming language. Because of its agility in designing problems of linear algebra, Matlab is widely used for research and industrial purposes in computer science, electrical engineering, and economics. Yet as Matlab works mainly with an interpreted programming language, its programs have to be translated by an *interpreter* (INT) before interacting with the hardware. This interpretative step makes it less efficient for processing heavy matrices than, for example, programs directly written in compiled languages such as C or C++.

model is a term that is close to an algorithm. In this book, the distinction between an algorithm and a model can only be retrospective: If what is called a "model" derives from, at least, ground-truthing, programming, and formulating activities, it is considered an algorithm.

problematization is, in this book, the collective process of establishing the terms of a problem. Building on *Science and Technology Studies*, analyzing problematization implies describing the way questions are framed, organized, and progressively transformed into issues for which solutions can be proposed.

process thought is an ontological position supported by a wide and heterogeneous body of philosophical works that share similar sensibilities toward associations—sometimes also called relations. For process thinkers, what things are is what they become in association to other entities, the association itself being part of the process. The emphasis is put on the "how" rather than the "what": instead of asking what *is* something, process thinkers would rather ask *how* something *becomes*. This ontology is about continuous performances instead of binary states.

PROG specifically refers, in this book, to a Matlab computer program aiming to create matrices whose pixel-values correspond to the number of rectangles drawn by human crowdworkers on pixels of digital images.

program is a document whose structure and content, when adequately articulated, makes computers compute data. The practical process of writing a computer program is called *programming*.

re-presentation is the presentation of something *again*. Inscriptions are common re-presentations in that they display properties of other entities over. Re-presentations, in this book, should not be confused with *representations* (without the hyphen), a term that refers to the solution found by cognitivist authors to overcome the distinction between extended things (*res extensa*) and thinking things (*res cogitans*).

saliency detection is a subfield of image processing that aims to detect what attracts people's attention within digital images. Because the topic of these detection efforts is extremely equivocal, saliency detection is a field of research that shows dynamics that may go unnoticed in more traditional subfields such as facial or object recognition.

scenario refers to a narrative operating a triple shifting out toward another place, another time, and other actants while having a hold on its enunciator. As performative narrative resources, scenarios are of crucial importance for programming activities because they institute horizons on which programmers can hold—while being held by them—and establish, in turn, the boundaries of computer programming episodes.

Science and Technology Studies (STS) are a subfield of social science and sociology that aims to document the co-construction of science, technology, and the collective world. What loosely connects the practitioners of this heterogeneous research community is the conviction that science is not just the expression of a logical empiricism, that knowledge of the world does not preexist, and that scientific and technological truths are dependent on collective arrangements, instrumentations, and dynamics.

script commonly refers to a small computer program. Many interlinking scripts and programs calling on each other typically form a *software*. The notion should not be confused with Madeleine Akrich's (1989) "scripts" that, in this book, are close to the notion of *scenario*.

sociology is, in this book, the activity of describing associations (*socius*) by means of specialized texts (*logos*). It aims to help understand what is going on in the collective world and better compose with the heterogeneous entities that populate/shape it. In this book, sociology is differentiated from *social science* that is considered the scientific study of an a priori postulated aggregate, generally called the social (or society).

technical detour is a furtive and difficult-to-record experience that takes the form of a zigzag: Thanks to unpredictable detours, a priori distant entities become the missing pieces in the realization of a project. Technical detours—as conceptualized by Bruno Latour (2013)—involve a form of delegation to newly enrolled entities. They also imply forgetting their brief passages once the new composition has been established.

translation is a work by which actants modify, move, reduce, transform, and articulate other actants to align them with their concerns. This is a specific type of association that produces differences that can, with an appropriate methodology, be reflected in a text. The notion was initially developed by Michel Serres (1974) before being taken up by Madeleine Akrich, Michel Callon, and Bruno Latour to ground their *sociologie de la traduction*, which I call sociology here.

trial is a testing event whose outcome has a strong impact on the becoming of an actant. If the trial is overcome, the actant may manage to associate with other actants, with this new association becoming, in turn, more resistant. If the trial is not overcome, the actant will lose some of its properties, sometimes to point of disappearing.

visibility/invisibility are relative states of work practices. These variable states are products of visibilization, or invisibilization, processes. If complete invisibility of work practices is not desirable, complete visibility is not either. In this book, I have chosen public controversies as indicators of negative invisibilities, suggesting in turn the launching of visibilization processes by means of, for example, sociological inquiries.

Notes

Introduction

1. Process thought refers to a wide and heterogeneous body of philosophical works that share similar sensibilities toward associations, sometimes also called relations (Barad 2007; Butler 2006; Dewey [1927] 2016; James [1912] 2003; Latour 1993b, 2013; Mol 2002; Pickering 1995; Serres 1983; Whitehead [1929] 1978). For process thinkers, as Introna put it (2016, 23), "relations do not connect (causally or otherwise) pre-existing entities (or actors), rather, relations enact entities in the flow of becoming." What things are is what they become in association to other entities, the association itself being part of the process. The emphasis is then put on the "how" rather than the "what": instead of asking what *is* something, process thinkers would rather ask how something *becomes*. This ontology is then about continuous performances instead of binary states. The present volume embraces this ontology of becoming.

2. At the end of the book, a glossary briefly defines technical terms used for this investigation (e.g., actant, collective world, constitution, course of action).

3. This unconventional conception of the social has been initially developed and popularized by Madeleine Akrich, Michel Callon, and Bruno Latour at the Centre de Sociologie de l'Innovation (Akrich, Callon, and Latour 2006; Callon 1986). It is important to note that even though this theoretical standpoint has somewhat made its way through academic research, it remains shared among a minority of scholars.

4. As pointed out by Latour (2005, 5–6), the Latin root *socius* that denotes a companion—an associate—fits well with the conception of the social as what emanates from the association among heterogeneous entities.

5. What connects the practitioners of the heterogeneous research community of *Science and Technology Studies* is the conviction that science is not just the expression of a logical empiricism; that knowledge of the world does not preexist; and that scientific and technological truths are dependent on collective arrangements, instrumentations, and dynamics (Dear and Jasanoff 2010; Jasanoff 2012). For a comprehensive introduction to STS, see Felt et al. (2016).

6. It is important to note that this lowering of capacity to act does not concern the sociology of attachments that precisely tries to document the appearance of delighted objects, as developed by Antoine Hennion (2015, 2017). At the end of chapter 5, I will discuss the important notion of attachment.

7. The notion of "composition"—at least as proposed by Latour (2010a)—is, in my view, an elegant alternative to the widely used notion of "governance." Both nonetheless share some characteristics. First, both notions suppose heterogeneous elements put together—collectives of humans, machines, objects, companies, and institutions trying to collaborate and persevere on the same boat. Second, they share the desire of a common world while accepting the irreducibility of its parts: for both notions, the irreducible entities that constitute the world would rather live in a quite informed community aware of different and competitive interests than in a distrustful and whimsical wasteland. Both composition and governance thus share the same basic topic of inquiry: how to step-by-step transform heterogeneous *collectives* into heterogeneous *common worlds*? Third, they both agree that traditional centralized decisional powers can no longer achieve the constitution of common worlds; to the verticality of orders and injunctions, composition and governance prefer the horizontality of compromises and negotiations. Yet they nonetheless differ on one crucial point: if governance still carries the hope of a smooth—yet heterogeneous—*cosmos*, composition promotes the need of a laborious and constantly readjusted *kakosmos* (Latour 2010a, 487). In other words, if control is still an option for governance, composition is committed to the always surprising "made to do" (Latour 1999b). It is this emphasis on the constant need for creative readjustments that makes me prefer the notion of "composition" over "governance."

8. The next two paragraphs derive from Jaton (2019, 319–320).

9. The single term "algorithm" became increasingly common in the Anglo-American critical literature from the 2000s onward. It would be interesting to learn more about the ways by which the term "algorithm" has come to take over other alternative terms (such as "software," "code," or "software-algorithm") that were also synonymously used in the past, especially in the 1990s.

10. In Jaton and Vinck (submitted), we closely consider the specific dynamic of the recent politicization of algorithms.

11. This controversy has been thoroughly analyzed in Baya-Laffite, Beaude, and Garrigues (2018).

12. As we will see in the empirical chapters of this book, it is not clear whether we should talk about computer scientists or engineers. But as the academic field of computer science is now well established, I choose to use the generic term "computer scientist" to refer to those who work every day to design surprising new algorithms.

13. For thorough discussions on this topic, see Denis (2018, 83–95).

14. Does it mean that "objective knowledge" is impossible? As we will see in chapters 4, 5, and 6, drawing such a conclusion is untenable: despite the irremediable limits of the inscriptions on which scientific practices heavily rely, these practices nonetheless manage to produce certified objective knowledge.

15. In their 2004 paper, Law and Urry build upon an argument initially developed by Haraway (1992, 1997).

16. This partly explains some hostile reactions of scientists regarding STS works on the "construction of scientific facts." On this topic, see Latour (2013, 151–178).

17. For recent examples, see Cardon (2015) and Mackenzie (2017).

18. In chapter 5, I will discuss at greater length the crucial importance of scientific literature for the formation of certified knowledge.

19. The term "infra-ordinary," as opposed to "extra-ordinary," was originally proposed by Pérec (1989). The term was later taken up in Francophone sociology, notably by Lefebvre (2013).

20. See, for example, Bishop (2007), Cormen et al. (2009), Sedgewick and Wayne (2011), Skiena (2008), and Wirth (1976). I will discuss some of these manuals in chapter 1.

21. However, it is crucial to remain alert to the performative aspects of manuals and classes. This topic is well studied in the sociology of finance; see, for example, MacKenzie, Muniesa, and Siu (2007) and Muniesa (2015).

22. This also often concerns social scientists interviewing renowned computer scientists (e.g., Seibel 2009; Biancuzzi and Warden 2009). As these investigations mainly focus on well-respected figures of computer science whose projects have largely succeeded, their results tend to be retrospective, summarized narratives occluding uncertainties and fragilities. On some limitations of biographic interviews, see Bourdieu (1986). On the problematic habit of reducing ethnography to interviews, see Ingold (2014).

23. For a presentation of some of the reasons why scholars started to inquire within scientific laboratory, see Doing (2008), Lynch (2014), and Pestre (2004).

24. On some of the problematic, yet fascinating, dynamics of this rapprochement between computer science and the humanities (literature, history, linguistics, etc.) that gave rise to digital humanities, see Gold (2012), Jaton and Vinck (2016), and Vinck (2016).

25. Among the rare attempts to document computer science work are Bechmann and Bowker (2019), Button and Sharrock (1995), Grosman and Reigeluth (2019), Henriksen and Bechmann (2020), and Mackenzie and Monk (2004). I will come back to some of these studies in the empirical chapters of the book.

26. After a thorough review of the contemporary critical studies of algorithms, Ziewitz (2016) warned that they could be about to reach a problematic impasse. Roughly put, the argument goes as follows: by mainly considering algorithms from a distance and in terms of their effects, critical studies are taking the risk of being stuck in a dramatic loop, constantly rehashing that algorithms are powerful because they are inscrutable, because they are powerful, because they inscrutable, and so on. The present volume can be considered an attempt at somewhat preventing such a drama from taking hold. In the conclusion, when I clarify the political aspect of this inquiry, I come back to this notion of algorithmic drama.

27. Theureau's work is unique in many ways. Building on the French ergonomics tradition (Ombredane and Faverge 1955) and critical readings of Newell and Simon's (1972) cognitive behaviorism as well as Varela's notion of "enactive cognition" (discussed in chapter 3), he has gradually proposed a simple yet effective definition of a course of action as an "observable activity of an agent in a defined state, actively engaged in a physically and socially defined environment and belonging to a defined culture" (Theureau 2003, 59). His analyses of courses of action involved in traffic management (Theureau and Filippi 2000), nuclear reactor control (Theureau et al. 2001), and musical composition (Donin and Theureau 2007) has led him to propose the notion of "courses-of-action centered design" for ergonomic studies.

28. At the beginning of chapter 4, I will briefly consider the problem of "representativeness."

Chapter 1

1. The general issue subtending my research has not fundamentally changed since the date at which I was awarded the research grant.

2. One of the particularities of the CSF was its international focus. During the official events I attended, deans regularly put forward the CSF's capacity to attract foreign students and researchers. This was especially true in the case of the Lab where I was the only "indigenous" scientific collaborator for nearly a year. The lingua franca was in line with this international environment; even though the Lab was located in a French-speaking region, most interactions, presentations, and documents were in English.

3. The history of the development of the charge-coupled device has been documented, though quite partially, in Seitz and Einspruch (1998, 212–228) and Gertner (2013, 250–265).

4. For an accessible introduction to CCDs and image sensors, see Allen and Triantaphillidou (2011, 155–173).

5. CMOS is a more recent variant of CCD where each pixel contains a photodetector *and* an amplifier. This feature currently allows significant size and power reduction

of image sensors. This is one of the reasons why CMOSs now equip most portable devices such as smartphones and compact cameras.

6. It is commonly assumed that the term pixel, as a contraction of "picture element," first appeared in a 1969 paper from Caltech's Jet Propulsion Lab (Leighton et al. 1969). The story is more intricate than that as the term was regularly used in emergent image-processing communities thoughout the 1960s. For a brief history of the term pixel, see Lyon (2006).

7. A digital signal is represented by n number of dimensions depending on the independent variables used to describe the signal. A sampled digital sound is, for example, typically described as a one-dimensional signal whose dependent variables—amplitudes—vary according to time (t); a digital image is typically described as a two-dimensional signal whose dependent variables—intensities—vary according to two axes (x, y), whereas audio-visual content will be described as a three-dimensional signal with independent variables (x, y, t). For an accessible introduction to digital signal processing, see Vetterli, Kovacevic, and Goyal (2014).

8. It was not the only research focus of the Lab. Several researchers also worked on CCD/CMOS architectures and sensors.

9. It is important to note that for digital image processing and recognition to become a major subfield of computer science, digital images first had to become stable entities capable of being processed by computer programs—a long-standing research and development endeavor. Along with the development, standardization, and industrial production of image sensors such as CCDs and, later, CMOSs, theoretical works on data compression—such as those of O'Neal Jr. (1966) on differential pulse code modulation; Ahmed, Natarajan, and Rao (1974) on cosine transform; or Gray (1984) on vector quantization—have first been necessary. The later enrollment of these works for the definition of the now-widespread International Organization for Standardization norm JPEG, approved in 1993, was another decisive step: from that moment, telecommunication providers, software developers, and hardware manufacturers could rely on and coordinate around one single photographic coding technique for digitally compressed representations of still images (Hudson et al. 2017). During the late 1990s, the growing distribution of microcomputers, their gradual increase in terms of processing power, and the development and maintenance of web technologies and standards have also greatly contributed to establishing digital image processing as a mainstream field of study. The current popularity of image processing for research, industry, and defense is thus to be linked with the progressive advent of multimedia communication devices and the blackboxing of their fundamental components operating now as standard technological infrastructure.

10. According to Japan-based industry association Camera & Imaging Products Association (to which, among others, Canon, Nikon, Sony, and Olympus belong), sales of digital cameras have dropped from 62.9 million in 2010 to fewer than

24.25 million in 2017 (Statista 2019). However, according to estimates generated by InfoTrends and Bitkom, the number of pictures taken worldwide increased from 660 billion to 1,200 billion over the same period (Richter 2017). This discrepancy is due, among other things, to the increasing sophistication of smartphone cameras as well as the popularity and sharing functionalities of social-media sites such as Instagram and Facebook (Cakebread 2017).

11. For example, Google, Amazon, Apple, Microsoft, and IBM all propose application programming interface products for image recognition (respectively, Cloud Vision, Amazon Rekognition, Apple Vision, Microsoft Computer Vision, and Watson Visual Recognition).

12. According to 2011 documents obtained by Edward Snowden, the National Security Agency intercepted millions of images per day throughout the year 2010 to develop computerized tracking methods for suspected terrorists (Risen and Poitras 2014). Chinese authorities also heavily invest in facial recognition for security and control purposes (Mozur 2018).

13. See, for example, *International Journal of Computer Vision, IEEE Transactions on Pattern Analysis and Machine Intelligence, IEEE Transactions on Image Processing*, or *Pattern Recognition*.

14. See, for example, IEEE Conference on Computer Vision and Pattern Recognition, European Conference on Computer Vision, IEEE International Conference on Computer Vision, or IEEE International Conference on Image Processing.

15. Giving an example of the close relationships between academic and industrial worlds regarding image-processing algorithms, Jordan Fisher—chief executive officer of Standard Cognition, a start-up that specializes in image recognition for autonomous checkout—says in a recent *TechCrunch* article (Constine 2019): "It's the wild west—applying cutting-edge, state-of-the-art machine learning research that's hot off the press. We read papers then implement it weeks after it's published, putting the ideas out into the wild and making them production-worthy."

16. In 2016 and 2017, papers from Apple and Microsoft research teams won the best-paper award of the IEEE Conference on Computer Vision and Pattern Recognition, the most prestigious conference in image processing and recognition. Moreover, in 2018, Google launched *Distill Research Journal*, its own academic journal aiming at promoting machine learning in the field of image and video recognition.

17. This is for example the case in Knuth (1997a) where the author starts by recalling that "algorithm" is a late transformation of the term "algorism" that itself derives from the name of famous Persian mathematician Abū 'Abd Allāh Muhammad ibn Mūsa al-Khwārizmi—literally, "Father of Abdullah, Mohammed, son of Moses, native of Khwārizm," Khwārizm referring in this case to a region south of the Aral Sea (Zemanek 1981). Knuth then specifies that from its initial acceptation

as the process of doing arithmetic with Arabic numerals, the term algorism gradually became corrupted: "as explained by the *Oxford English Dictionary*, the word 'passed through many pseudo-etymological perversions, including a recent *algorithm*, in which it is learnedly confused' with the Greek root of the word *arithmetic*" (Knuth 1997a, 2).

18. See, for example, the (very) temporary definition of algorithms by Knuth (1997, 4): "The modern meaning for algorithm is quite similar to that of *recipe, process, method, technique, procedure, routine, rigmarole.*"

19. See, for example, Sedgewick and Wade's (2011, 3) definition of algorithms as "methods for solving problems that are suited for computer implementation."

20. See also Cormen et al.'s (2009, 5) definition: "A well-defined computational procedure that takes some value, or set of values, as *input* and produces some value, or set of values, as *output* [being] thus a sequence of computational steps that transform the input into the output."

21. See also Dasgupta, Papadimitriou, and Vazirani's (2006, 12) phrasing: "Whenever we have an algorithm, there are three questions we always ask about it: 1. Is it correct? 2. How much time does it take, as a function of n? 3. And can we do better?" And also Skiena (2008, 4): "There are three desirable properties for a good algorithm. We seek algorithms that are correct and efficient, while being easy to implement."

Chapter 2

1. This chapter expands Jaton (2017). I thank Geoffrey Bowker, Roderic Crooks, and John Seberger for fruitful discussions about some of its topics.

2. Excerpts in quotes are literal transcriptions from audio recordings, slightly reworked for reading comfort. Excerpts not in quotes are retranscriptions from written notes taken on the fly.

3. In chapter 3, I critically discuss the computational metaphor of the mind on which many cognitive studies rely.

4. Studies on attention had already been engaged before the 1970s, notably through the seminal work of Neisser (1967) who suggested the existence of a pre-attentive stage in the human visual processing system.

5. Another important neurobiological model of selective attention method was proposed by Wolfe, Cave, and Franzel (1989). This model of selective attention method later inspired competing low-level feature computational models (e.g., Tsotsos 1989; Tsotsos et al. 1995).

6. The class of algorithms that calculates on low-level features quickly became interesting for the development of autonomous vehicles for which real-time image

processing was sought (Baluja and Pomerleau 1997; Grimson 1986; Mackworth and Freuder 1985).

7. Different high-level detection algorithms can nonetheless be assembled as modules in one same program that could, for example, detect faces and cars and dogs, and so on.

8. At that time, only two saliency-detection algorithms were published, in Itti, Koch, and Niebur (1998) and Ma and Zhang (2003). But the ground truths used for the design and evaluation of these algorithms were similar to those used in laboratory cognitive science. The images of these ground truths were, for example, sets of dots disrupted by a vertical dash. As a consequence, if these first two saliency-detection algorithms could, of course, process natural images, no evaluations of their performances on such images could be conducted.

9. Ground truths assembled by computer science laboratories are generally made available online in the name of reproducible research (Vandewalle, Kovacevic, and Vetterli 2009). The counterpart to this free access is the proper citation of the papers in which these ground truths were first presented.

10. An API, in its broadest sense, is a set of communication protocols that act as an interface among several computer programs. If APIs can take many different forms (e.g., hardware devices, web applications, operating systems), their main function is to stabilize and blackbox elements so that other elements can be built on top of them.

11. For a condensed history of contingent work, see Gray and Suri (2019, 48–63). On what crowdsourcing does to contemporary capitalism, see also Casilli (2019).

12. As Gray and Suri (2019, 55–56) put it: "Following a largely untested management theory, a wave of corporations in the 1980s cut anything that could be defined as 'non-essential business operations'—from cleaning offices to debugging software programs—in order to impress stockholders with their true value, defined in terms of 'return on investment' (in industry lingo, ROI) and 'core competencies.' … Stockholders rewarded those corporations that were willing to use outsourcing to slash costs and reduce full-time-employee ranks."

13. It is important to note, however, that on-demand work is not necessarily alienating. As Gray and Suri (2019, 117) noted: "[on-demand work] can be transformed into something more substantive and fulfilling, when the right mixture of workers' needs and market demands are properly aligned and matched. It can rapidly transmogrify into ghost work when left unchecked or hidden behind software rather than recognized as a rapidly growing world of global employment." Concrete ways to make crowdsourcing more sustainable have been proposed by the National Domestic Workers Alliance and their "Good Work Code" quality label. On this topic, see Scheiber (2016).

14. However, this shared unawareness toward the underlying processes of crowd-sourcing may be valued and maintained for identity reasons, for as Irani (2015, 58) noted: "The transformation of workers into a computational service ... serves not only employers' labor needs and financial interests but also their desire to maintain preferred identities; that is, rather than understanding themselves as managers of information factories, employers can continue to see themselves as much-celebrated programmers, entrepreneurs, and innovators."

15. Matlab is a privately held mathematical software for numerical computing built around its own interpreted high-level programming language. Because of its agility to design problems of linear algebra—all integers being considered scalars—Matlab is widely used for research and industrial purposes in computer science, electrical engineering, and economics. Yet, as Matlab works mainly with an interpreted programming language—just like the language Python that is now Matlab's main competitor for applied research purposes—its programs have to be translated into machine-readable binary code by an *interpreter* in order to make the hardware effectively compute data. This complex interpretative step makes it less efficient for processing heavy matrices than, for example, programs directly written in compiled languages such as C or C++. For a brief history of Matlab, see Haigh (2008).

16. In chapter 6, we will more thoroughly consider the relationship between ground-truthing and formulating activities.

17. The services of the crowdsourcing company costed the Lab around US$950.

18. The numerical features extracted from the training set were related, among others, to "2D Gaussian function," "spatial compactness," "contrast-based filtering," "high-dimensional Gaussian filters," and "element uniqueness." In chapter 6, using the case of the "2D Gaussian function," I will deal with these *formulating practices*.

19. This can be read as a mild critique of the recent, growing, and important literature on algorithm biases. Authors such as Obermeyer et al. (2019), Srivastava and Rossi (2018), and Yapo and Weiss (2018), among others, show that the results of many algorithms are indeed biased by the preconceptions of those who built them. Though this statement is, I believe, completely correct—algorithms derive from problematization practices influenced by habits of thought and action—it also runs the risk of confusing premises with consequences: biases are not the consequences of algorithms but, perhaps, are one of the things that make them come into existence. Certain biases expressed and materialized by ground truths can and, in my opinion, should be considered harmful, unjust, and wrong; racial and gender biases have, for example, to be challenged and disputed. However, the outcome of these disputes may well be other biases expressed in other potentially less harmful, unjust, and incorrect ground truths. As far as algorithms are concerned, one bias calls for another; hence the importance of asserting their existence and making them visible in order to, eventually, align them with values one wishes algorithms to promote.

20. Edwards (2013) uses the term "data image" instead of "ground truth." But I assume that both are somewhat equivalent and refer to digital repositories organized around data whose values vary according to independent variables (that yet need to be defined).

21. At the end of chapter 6, I will come back to the topic of machine learning and its contemporary labeling as "artificial intelligence."

22. This discussion has been reconstructed from notes in Logbook 3, May–October 2014.

23. However, it is interesting to note that BJ blames the reviewers of important conferences in image processing. According to him, the reviewers tend to privilege papers that make "classical improvement" over those that solve—and thus define—new problems. At any rate, there was obviously a problem in the framing of the Group's paper as the reviewers were not convinced by its line of argument. As a consequence, the algorithm could not circulate within academic and industrial communities and its existence remained, for a while, circumscribed to the Lab's servers.

II

1. In computer science and engineering, it is indeed well admitted that computer programming practices are difficult to conduct and their results very uncertain. On this well-documented topic, see Knuth (2002), Rosenberg (2008), and in a more literary way, Ullman (2012a, 2012b).

Chapter 3

1. My point of departure is arbitrary in the sense that I could have started somewhere else, at a different time. Indeed, as Lévy (1995) showed, the premises of what will be called "von Neumann architecture of electronic computers" can be found not only in Alan Turing's 1937 paper but also in the development of the office-machine industry during the 1920s, but also in the mechanic-mathematical works of Charles Babbage during the second half of the nineteenth century, but also in eighteenth century's looms programmed with punched cards, and so on, at least until Leibniz's work on binary arithmetic and Pascal's calculating machine. The history of the computer is fuzzy. As it only appears "after a cascade of diversions and reinterpretations of heterogeneous materials and devices" (Lévy 1995, 636), it is extremely difficult—in fact, almost impossible—to propose any unentangled filiation. Fortunately, this section does not aim to provide any history of the computer: It "just" tries to provide elements that, in my view, participated in the formation of one specific and influential document: von Neumann's report on the EDVAC.

2. For a more precise account of the design of firing tables in the United States during World War II, see Haigh, Priestley, and Rope (2016, 20–23) and Polachek (1997).

3. More than their effective computing capabilities—they required up to several days to be set up (Haigh, Priestley, and Rope 2016, 26) and their results were often less accurate than those provided by hand calculations (Polachek 1997, 25–27)—an important characteristic of differential analyzers was their capacities to attract computing experts around them. For example, by 1940, MIT, the University of Pennsylvania, and the University of Manchester, England—three important institutions for the future development of electronic computing—all possessed a differential analyzer (Campbell-Kelly et al. 2013, 45–50; Owens 1986). On the role of differential analyzers in early US-based computing research, see also Akera (2008, 38–45).

4. The assembling of the numerous factors affecting the projectiles started at the test range in Aberdeen where the velocities of the newly designed shells were measured (Haigh, Priestley, and Rope 2016, 20).

5. Although the differential equations defining the calculation of shells' trajectories are mathematically quite simple, solving them can be very complicated as one needs to model air resistance varying in a nonlinear manner. As Haigh, Priestley, and Rope (2016, 23) put it: "Unlike a calculus teacher, who selects only equations that respond to elegant methods, the mathematicians at the BRL couldn't ignore wind resistance or assign a different problem. Like most differential equations formulated by scientists and engineers, ballistic equations require messier techniques of numerical approximation."

6. Interesting to note that delay-line storage is originally linked to radar technology. More precisely, one problem of the radar technology in 1942 was that cathode-ray tube displays showed moving *and* stationary objects. Consequently, radar screens translated the positions of planes, buildings, or forests in one same messy picture extremely difficult to read. MIT's radiation laboratory subcontracted the development of a moving target indicator (MTI) to the Moore School in order to develop a system that could filter radar signals according to their changing positions. This was the beginning of delay-line storage technology at the Moore School, that at first had nothing to do with computing (Akera 2008, 84–86; Campbell-Kelly et al. 2013, 69–74). Radar technology also significantly helped the design of British highly confidential Colossus computer in 1943–1944 (Lévy 1995, 646).

7. By 1942, in order to speed up the resolution of ballistic differential equations, only a limited range of factors tended to be considered by human computers at the BRL. By simplifying the equations, more firing tables could be produced and distributed, but the drawback was that their precision tended to decrease (Polachek 1997). Of course, on the war front, once soldiers realized that the first volley was not adequately defined, they could still slightly modify the parameters of the long-distance weapon to increase its precision. Yet—and this is the crucial point—*in between* the first volley and the subsequent ones, the opposite side had enough time to take cover, hence making the overall long-distance shooting enterprise less effective. The

nerve of war was precisely the first long-distance volleys that, when accurate, could lead to many casualties. By extension, then, the nerve of war was also, to a certain extent, the ability to include more factors in the differential equations whose solutions were printed out in firing table booklets (Haigh, Priestley, and Rope 2016, 25).

8. Created in 1940, the National Defense Research Committee (NDRC) united the research laboratories of the US Navy and the Department of War with hundreds of US universities' laboratory. The NDRC initially had an important budget to fund applied research projects that could provide significant advantages on future battlefields. It also operated as an advisory organization as in the case of the ENIAC that was considered nearly infeasible due to the important amount of unreliable vacuum tubes it would require. On this topic, see Campbell-Kelly et al. (2013, 70–72).

9. The history of this contract could be the topic of a whole book. For a nice presentation of its most important moments, see Haigh, Priestley, and Rope (2016, 17–33).

10. Based on a proposal by Howard Aiken, the Harvard Mark 1 was developed by IBM for Harvard University between 1937 and late 1943. Though computationally slow, even for the standards of the time, it was an important computing system as it expressed an early convergence of scientific calculation and office-machine technologies. For a more in-depth history of the Harvard Mark 1, see Cohen (1999).

11. Though its shape varied significantly throughout its existence, the ENIAC was fundamentally a network of different units (accumulators, multipliers, and function tables). Each unit had built-in dials and switches. If adequately configured, these dials and switches could define one single operation; for example, "clear the values of the accumulator," "transmit a number to multiplier number 3," "receive a number," and so on. To start processing an operation, each configuration of dials and switches had to be triggered by a "program line" wired directly to the specific unit. All these "program lines" formed a network of wires connecting all the units for one specific series of operations. But as soon as another series of operations was required, the network of wires had to be rearranged in order to fit the new configurations of dials and switches. For more elements about the setup of ENIAC, see Haigh, Priestley, and Rope 2016 (35–57).

12. Von Neumann tried to hire Alan Turing as a postdoctoral assistant at Princeton. Turing refused as he wanted to return to England (MacRae 1999, 187–202).

13. The Manhattan Project was, of course, highly confidential and this prevented von Neumann from specifying his computational needs with the ENIAC team.

14. As suggested by Akera (2008, 119–120) and Swade (2011), and further demonstrated by Haigh, Priestley, and Rope (2014; 2016, 231–257), the notion of "stored program" is a historical artifact: "the 'stored program concept' was never proposed as a specific feature in the agreed source, the *First Draft*, and was only retroactively adopted to pick out certain features of the EDVAC design" (Haigh, Priestley, and Rope 2016, 256).

15. Shortly after the distribution of von Neumann's *First Draft*, Eckert and Mauchly distributed a much longer—and far less famous—counter-report entitled *Automatic High-Speed Computing: A Progress Report on the EDVAC* (Eckert and Mauchly 1945) in which they put the emphasis on the idealized aspect the *First Draft*. The stakes were indeed high for Eckert and Mauchly: if the idealized depiction of the EDVAC by von Neumann was considered a realistic description of the engineering project, no patent could ever be extracted from it. And this is exactly what happened. In 1947, the Ordnance Department's lawyers decided that the *First Draft* was the first publication on the project EDVAC, hence canceling the patents submitted by Eckert and Mauchly in early 1946 (Haigh, Priestley, and Rope 2016, 136–152).

16. This consideration of programming as an applicative and routine activity can also be found in the more comprehensive reports von Neumann coauthored in 1946 and 1947 with Arthur W. Burks and Herman H. Goldstine at Princeton Institute for Advanced Study (Burks, Goldstine, and von Neumann 1946; Goldstine and von Neumann 1947). In these reports, and especially in the 1947 report entitled *Planning and Coding of Problems for an Electronic Computing Instrument*, the implementation of instruction sequences for scientific electronic calculations is carefully considered. But while the logico-mathematical planning of problems to be solved is presented as complex and "dynamic," the further translation of this planning is mainly considered trivial and "static" (Goldstine and von Neumann 1947, 20). Programming is presented, in great detail, as a linear process that is problematic during its initial planning phase but casual during its implementation phase. What the report does not specify—but this was not its purpose—is that errors in the modeling and planning phases become manifest in the implementation phase (as it was often the case when the ENIAC was put in action), making empirical programming processes more whirlwind than linear.

17. In 1955, to alleviate the operating costs of the IBM 701 and the soon-to-be-released IBM 704, several of IBM's customers—among them Paul Armer of the RAND Corporation, Lee Amaya of Lockheed Aircraft, and Frank Wagner of North American Aviation—launched a cooperative association they named "Share." This customer association, and the many others that followed, greatly participated in the early circulation of basic suites of programs. On this topic, see Akera (2001; 2008, 249–274).

18. For a fine-grained historical account of this real-time computing project named "Whirlwind" that was initially designed as a universal aircraft simulator, see Akera (2008, 184–220).

19. For more thorough accounts of the SAGE project, see Redmond and Smith (1980, 2000), Jacobs (1986), Edwards (1996, 75–112), and Campbell-Kelly et al. (2013, 143–166).

20. According to Pugh (1995), this contract gave IBM a significant advantage on the early computer market.

21. In a nutshell, Thurstone Primary Mental Abilities (PMA) test was proposed in 1936 by Louis Leon Thurstone, by then the first president of the Psychometric Society. Originally intended for children, the test sought to measure intelligence differentials using seven factors: word fluency, verbal comprehension, spatial visualization, number facility, associative memory, reasoning, and perceptual speed. For a brief history of the PMA test and psychometrics, see Jones and Thissen (2007).

22. One important insight of the EDSAC project was to use the new concept of program to initialize the system and make it translate further programs from non-binary instructions into binary strings of zeros and ones. David Wheeler, one of Maurice Wilkes' PhD students, wrote in 1949 such very first program he called "Initial Orders" (Richards 2005). This type of program whose function was to transform other programs into binary (the only code cathode-ray tubes, magnetic core, or microprocessors can interact with) were soon called "assemblers" and cast to linguistic terms such as "translation" and "language" (Nofre, Priestley, and Alberts 2014). During the 1950s, as multiple manufacturers invested in the electronic computer market, many different assemblers were designed, thereby creating important problems of compatibility: as (almost) every new computer organized the accumulator and multiplier registers slightly differently, a new assembler was generally required. The problem lay in the one-to-one relationship between an assembler and its hardware. Since an assembler had one instruction for one hardware operation, every modification in the operational organization of the hardware required a new assembler. Yet—and this was the crucial insight of Grace Hopper and then John Backus from IBM (Campbell-Kelly et al. 2014, 167–188)—if, instead of a program with a one-to-one relationship with the hardware, one could provide a more complex program that would transform lines of code into another program with somehow equivalent machine-instructions, one may be able to stabilize computer programming languages since any substantial modification of the hardware could be integrated within the "transformer" program that lay in between the programmer's code and the hardware. This is the fundamental idea of *compilers*, programs that take as input a program written in so-called high-level computer language and outputs *another* program—often called "executable"—whose content can interact with specific hardware. In the late 1950s, besides their greater readability, a tremendous advantage of the first high-level computer programming languages such as FORTRAN or COBOL over assembly language lay in their compilers whose constant maintenance could compensate and "absorb" the frequent modifications of the hardware. For example, if two different computers both had a FORTRAN compiler—a crucial and costly condition—the same FORTRAN program could be run on both computers despite their different internal organizations.

23. Between 1964 and 1967, IBM invested heavily in the development of an operating system for its computer System 360. The impressive backlogs, bugs, and overheads of this colossal software project made Frederick Brooks—its former manager—call it "a multi-million-dollar mistake" (Brooks 1975).

24. In 1968, an article by cofounder of Informatics General Corporation Werner Frank popularized the idea that the cost of software production will outpace the cost of computer hardware in the near future (Frank 1968). Though speculative in many respects, this claim was fairly reused and embellished by commentators until the 1980s. Though Frank himself later acknowledged that he unintentionally generated a myth (Frank 1983), this story "reinforced a popular perception that programmer productivity was lagging, especially compared to the phenomenal advances in computer hardware" (Abbate 2012, 93).

25. The topic of "logical statement performances" is recurrent in behavioral studies of computer programming, especially during the 1970s. This has to do with a controversy initiated by Edsger Dijkstra over the GOTO statement as allowed by high-level computer programming languages such as BASIC or early versions of FORTRAN (Dijkstra 1968). According to Dijkstra, these branch statements that create "jumps" inside a program make the localization of errors extremely tedious and should thus be avoided. He then proposed "structured programming," a methodology that consists in subdividing programs in shorter "modules" for more efficient maintenance (Dijkstra 1972). Behavioral studies of computer programming in the 1970s typically tried to evaluate the asserted benefits of this methodology.

26. To prove his second incompleteness theorem, Gödel first had to show that any syntaxic proposition could be expressed as a number. Turing's 1937 demonstration highly relied on this seminal insight. On the links between Gödel's incompleteness theorem and Turing's propositions regarding the *Entscheidungsproblem*, see Dupuy (1994, 22–30).

27. Neural networks, particularly those defined as "deep" and "convolutional," have recently been the focus of much attention. However, it is important to note that the notion of neural networks as initially proposed by McCulloch and Pitts (who preferred to use the notion of "networks of neurons") in their 1943 paper, and later taken up by von Neumann in his 1945 report, is very different from its current acceptance. As Cardon, Cointet, and Mazières (2018) have shown, McCulloch and Pitts's neural networks that were initially logical activation functions were worked on by Donald O. Hebb (1949) who associated them with the idea of learning, which was itself reworked by, among others, Frank Rosenblatt (1958, 1962) and his notion of Perceptron. The progressive probabilization of the inference rules suggested by Marvin Minsky (Minsky and Papert 1970), the works on the back-propagation algorithm (Werbos 1974; LeCun 1985; Rumelhart, Hinton, and Williams 1986) and on Boltzmann machines (Hinton, Sejnowski, and Ackley 1984) then actively participated in the association of the notions of "convolution" (LeCun et al. 1989) and, more recently, "depth" (Krizhevsky, Sutskever, and Hinton 2012). The term "neural network" may have survived this translation process but it now refers to very different world-enacting procedures. At the end of chapter 6, I will consider this topic related to machine learning and artificial intelligence.

28. The division between "extended things" and "thinking things" derives, to a large extent, from Cartesian dualism. For thorough discussions of Descartes's aporia, see the work of Damasio (2005).

29. As we saw in chapter 2, saliency detection in image processing is directly confronted with this issue. Hence the need to carefully frame and constrict the saliency problem with appropriate ground truths.

30. One may trace these critics back to the Greek Sophists (Cassin 2014). James (1909) and Merleau-Ponty (2013) are also important opposition figures. In developmental psychology, the "social development theory" proposed by Vygotsky (1978) is also a fierce critic of cognitivism.

Chapter 4

1. To conduct this project, I had to become competent in Python, PHP, JavaScript, and Matlab programming languages.

2. It is important to note that this line-by-line translation is what is experienced by the programmer. In the trajectory of INT and most other interpreters, the numbered list of written symbols is translated into an abstract syntax tree that does not always conserve the line-by-line representation of the Editor.

3. It is difficult to know exactly how INT managed to deal with these three values at T1. It may by default consider that only the first two values of image-size—width and height—generally matter.

4. In the Matlab programming language, every statement that is not conditional and that does not end with an semicolon is, by default, printed by the interpreter in the Command Window. This is different from many other high-level programming languages for which printing operations should be specified by an instruction (typically, the instruction "print").

5. In chapter 5, where I will consider the formation of mathematical knowledge, I will more thoroughly examine the shaping of scientific facts as proposed by STS.

6. This may be a limitation of *Software Studies*, as for example presented in Fuller (2008) and in the journal *Computational Culture*. By considering completed code, these studies tend to overlook the practical operations that led to the completion of the code. Of course, this glance remains important as it allows us to consider the performative effects of software-related cultural products, something my action-oriented method is not quite able to do.

7. The successive operations required to assemble chains of reference in the case of program-testing are well documented, though in a literary way, by Ullman (2012b).

8. It is interesting to note that DF's alignment practices would have been greatly facilitated by the next version of Matlab. Indeed, the 2017 version of Matlab's

interpreter automatically recognizes this type of dimension error during matrix incrementation processes and directly indicates the related breakpoint, the line at which the problem occurred (in our case, at line 9).

9. Donald Knuth, one of the most prominent programming theorists, stressed the importance of program intelligibility by proposing the notion of *literate programming*: a computer programming method that primarily focuses on the task of explaining programs to fellow programmers rather than "just" instructing computers.

10. To my knowledge, there are only three exceptions: Vinck (1991), Latour (2006), and Latour (2010b).

11. This discussion has been reconstructed from notes in Logbook 8, November 2015–March 2016.

12. Some STS authors use the term "script" to define these particular narratives that engage those who enunciate them (Akrich 1989; Latour 2013). If I use the term "scenario," it is mainly for sake of clarity as "script" is often used by computer scientists and programmers—and myself in this book—to describe small programs such as PROG.

Chapter 5

1. Here, my style of presentation and use of scenes are greatly inspired by Latour (1987).

2. I am following here Rosental's (2003) book.

3. I am following here the work of MacKenzie (1999).

4. This is taken from Logbook 1, October 2013–February 2014.

5. With their distinction between *apodeixis* (rigorous demonstration) and *epideixis* (rhetorical maneuvering), Platonists philosophers may have initiated such grand narratives (Cassin 2014; Latour 1999). According to Leo Corry (1997), this way of presenting mathematics culminated with Bourbaki's structuralist conception of mathematical truth. On this topic, see also Lefebvre (2001, 56–68). For a philosophical exploration of grand narratives, see the classic book by Lyotard (1984).

6. Yet "likes" and "retweets" that support claims published on Facebook or Twitter may, sometimes, work as significant external allies. On this topic, see Ringelhan, Wollersheim, and Welpe (2015).

7. Before the 1878 foundation of the *American Journal of Mathematics* (AJM), there was no stable academic facility for the publication of mathematical research in the United States (Kent 2008). The situation in England was a bit different: built on the ashes of the *Cambridge and Dublin Mathematical Journal*, the *Quarterly Journal of Pure and Applied Mathematics* (QJPAM) published its first issue in 1855 (Crilly 2004). Yet for both Kempe's and Heawood's papers, the editorial boards of their journals—as

indicated on their front matters—were rather small compared with today's standards: five members for *AJM* in 1879 (J. J. Sylvester, W. E. Story, S. Newcomb, H. A. Newton, H. A. Rowland) and four members for *QJPAM* in 1890 (N. M. Ferrers, A. Cayley, J. W. L. Glaisher, A. R. Forsyth).

8. According to the document in American Association for Artificial Intelligence (1993).

9. See, for example, the *Journal of Informetrics*.

10. In a nutshell, Kempe circumscribed the problem to maps drawn on a plane that contain at least one region called "country" with fewer than six neighbors. He could then limit himself to five cases, countries from one to up to five neighbors. Proving that "four colorability" is preserved for countries with three neighbors was, obviously, not a problem. Yet in order to prove it for countries with four neighbors, Kempe used an argument known as the "Kempe chains" (MacKenzie 1999, 19–20). This argument stipulates that for a country X with four neighbor countries A, B, C, D, two opposite neighbor countries, say A and C, are either joined by a continuous chain of, say, red and green countries, or they are not. If they are joined by such a red-green chain, A can be colored red and C can be colored green. But as we are dealing with a map drawn on a plane, the two other opposite neighbor countries of X—B and C—*cannot* be joined by a continuous chain of blue and yellow countries (one way or another, this chain is indeed interrupted by a green or red country). As a consequence, these two opposite neighbor countries can be colored blue and X can be colored yellow. Four colorability is thus preserved for countries with four neighbors. Kempe thought that this method also worked for countries with five neighbors. But Heawood's figure shows a case of failure of this method where E's red-green region (vertically cross-hatched in figure 5.1) intersects B's yellow-red region (horizontally cross-hatched), thus forcing both countries to be colored red. Consequently, X has to be colored differently than red, blue, yellow, and green. In such a case, four colorability is *not* preserved.

11. On this topic, see the work of Lefebvre (2001).

12. For rhetorical habits in the life sciences, see Latour and Woolgar (1986, 119–148) and Knorr-Cetina (1981, 94–130). For a thorough comparison among scientific disciplines—excluding mathematics—see Penrose and Katz (2010).

13. Despite the efforts made by Serres (1995, 2002).

14. There was, of course, no scientific institution at that time; experimental protocols, peer witnessing, and, later, academic papers are products of the seventeenth century (Shapin and Shaffer 1989). Yet, as Netz (2003, 271–312) showed, theorems written on wax tablets and parchments did circulate among a restricted audience of (very!) skeptical readers.

15. This is at least Netz's (2003, 271–304) hypothesis, supported by the work of Lloyd (1990, 2005). As Latour summarized it: "It is precisely *because* the public life in

Greece was so invasive, so polemical, so inconclusive, that the invention, by 'highly specialized networks of autodidacts', of *another way* to bring an endless discussion to a close took such a tantalizing aspect" (Latour 2008, 449).

16. So surprising that this careful and highly specialized method of conviction mastered by a peripheral community of autodidacts who took great care to stick to forms was soon "borrowed" by Plato and extended to content in order to, among other things, silence the Sophists. This is at least the argument made by Cassin (2014), Latour (1999b, 216–235), and Netz (2004, 275–282).

17. Aristotle seems to be one of the first to compile geometrical texts and systematize their logical arguments (Bobzien 2002). During late antiquity, commentators such as Eutocius annotated many geometrical works and compiled their main results to facilitate their systematic comparisons (Netz 1998). According to Netz (2004), these collections of standardized geometrical compilations further helped Islamic mathematicians such as al-Kwarizmi and Khayyam to constitute the algebraic language.

18. During the late nineteenth century's so-called crisis of foundations in mathematics, the formalist school—headed by David Hibert—tried to establish the foundations of mathematics on logical principles (Corry 1997). This led to famous failures such as Russell and Whitehead's three volumes of *Principia Mathematica* (Whitehead and Russell 1910, 1911, 1913). Thanks to the philological work of Netz, we now better understand why such an endeavor has failed: it was the very practice of mathematics—lettered diagrams carefully indexed to small Greek sentences—that led to the formulation of the rules of logic and not the other way round.

19. Except, to a certain extent, Lefebvre (2001) and Mialet (2012). It seems then that Latour's remark remains true: few scholars have had the courage to do a careful anthropological study of mathematics (Latour 1987, 246).

20. This is taken from Latour (1987, chapter 2) and Wade (1981, chapter 13).

21. This is taken from Pickering and Stephanides (1992) and Hankins (1980, 280–312).

22. Very schematically, peptides are chemical elements made of chains of amino acids. They are known for interacting intimately with hormones. As there are many different amino acids (twenty for the case of humans), there exists—potentially—billions of different peptides made of combinations of two to fifty amino-acids. It is important to note that in 1972, at the time of Guillemin's experiment, peptides could already be assembled—and probed—within well-equipped laboratories.

23. At the time of Hamilton, the standard algebraic notation for a complex number—so-called absurd quantities such as square roots of negative numbers—was $x + iy$, where $i^2 = -1$ and x and y are real numbers. These advances in early complex algebra were problematic to geometers: if positive real numbers could be considered measurable quantities, negative real numbers and their square roots were difficult to represent as shapes on a plane. A way to overcome this impasse was to consider x and y as

coordinates of the end point of a segment terminating at the origin. Therefore, "the x-axis of the plane measured the real component of a given complex number represented as such a line segment, and the y axis the imaginary part, the part multiplied by i in the algebraic expression" (Pickering and Stephanides 1992, 145). With this visualization of complex numbers, algebraic geometers such as Hamilton could relate complex geometrical operations on segments and complex algebraic operations on equations. A bridge between geometry and complex algebra was thus built. Yet geometry is not confined to planes: if a two-dimensional segment $[0, x+iy]$ can represent a complex number, there is a priori no reason why a three-dimensional segment $[0, x+iy+jz]$ could not represent another complex number. Characterizing the behavior of such a segment was the stated goal of Hamilton's experiment.

24. Hamilton's inquiry into the relationships between complex number theory and geometry was not a purely exploratory endeavor. As Pickering and Stephanides noted, "the hope was to construct an algebraic replica of transformations of line segments in three-dimensional space and this to develop a new and possibly useful algebraic system appropriate to calculations in three-dimensional geometry" (Pickering and Stephanides 1992, 146).

25. Contrary to Hamilton, ancient Greek geometers could only refer to their lettered diagrams with short but still cumbersome Greek sentences (Netz 2003, 127–167). Along with Greek geometers' emphasis on differentiation, the absence of a condensed language such as algebra—that precisely required compiled collections of geometrical works in order to be constituted (Netz 1998)—may have participated in limiting the scope of ancient Greek geometrical propositions (Netz 2004, 11–54).

26. Regarding these instruments, it is worth mentioning that here we retrieve what we were discussing about in the last section: all of them—except, perhaps, noncommutative algebra—are blackboxed polished facts that were, initially, written claims. Rat pituitary cell cultures, algebraic notations, radioimmunoassays, coordinate spaces and even Pythagoras's theorem all had to overcome trials in order to gain conviction strength and become established, certified facts.

27. This topological characteristic of mathematical laboratories may be a reason why they have rarely been sites for ethnographic inquiries (Latour 2008, 444).

28. Of course, as we saw in chapter 4, such inscriptions are meaningless without the whole series of inscriptions previously required to produce them. It is only by aligning the "final" inscriptions to former ones, thus creating a chain of reference, that Guillemin can produce information about his peptide (Latour 2013, chapter 3).

29. Here we retrieve something we already encountered in chapters 3 and 4: the "cognitive" practice of aligning inscriptions. Just as DF in front of his computer terminal, Brazeau, Guillemin, and Hamilton never stop grasping inscriptions they acquire from experiments. These inscriptions can, in turn, be considered takes suggesting further actions.

30. Again, this is taken from Latour (1987, chapter 2) and Wade (1981, chapter 13).

31. Again, this is taken from Pickering and Stephanides (1992) and Hankins (1980, 280–312).

32. Brazeau and Guillemin published their results in *Science* (Brazeau et al. 1973). After having presented his results at the Royal Irish Academy in November 1843, Hamilton published a paper on quaternions in *The London, Edinburg and Dublin Philosophical Magazine and Journal of Science* (Hamilton 1844). An important thing to note about quaternions is that after Hamilton named them that way, he still had to define the complex quantities k^2, ik, kj, and i^2 in order to complete his system. According to a letter Hamilton wrote in 1865, the solution to this problem—the well-known $i^2 = j^2 = k^2 = ijk = -1$—appeared to him as he was walking along the Royal Canal in Dublin. If this moment was indubitably important, it would be erroneous to call it "the discovery of quaternions" (Buchman 2009). As shown by Pickering and Stephanides (1992), quaternions were already defined as objects before the attribution of values to the imaginary quantities' products. In fact, when compared with the experimental work required to define the problem of these products' values, what happened on Dublin's Royal Canal appears relatively minor.

33. This is the recurrent problem of biographies of important mathematicians; as they tend to use nature to explain great achievements, they often ignore the many instruments and inscriptions that were needed to shape the "discovered" objects. Biographies of great mathematicians are thus often—yet not always (see the amazing comic strip *Logicomix* [Doxiàdis et al. 2010])—unrealistic stories of solitary geniuses chosen by nature.

34. Accepting the dual aspect of nature—the *consequence* of settled controversies as well as the *retrospective cause* of noncontroversial facts—provides a fresh new look at the classical opposition between Platonism and Intuitionism in the philosophy of mathematics. It seems indeed that the oddity of both Platonism—for which mathematical objects come from the outer world of ideas—and Intuitionism—for which mathematical objects come from the inner world of human consciousness—comes from their shared starting point: they both consider certified noncontroversial mathematical facts. Yet as soon as one accounts for controversies in mathematics—that is, mathematics *in the making*—nature from above (the outer-world of ideas) or nature from below (the inner-world of human consciousness) cannot be considered resources anymore as both are precisely what is at stake during the controversies. It is interesting to note, however, that both antagonist unempirical conceptions of the origin of mathematics led to important performative disagreements about the practice of mathematics, notably through the acceptance, or refusal, of the law of excluded middle. On this fascinating topic, see Rotman (2006) and Corry (1997).

35. According to Netz (2004, 181–186), the constant search for differentiation and originality in ancient mathematical texts had the effect of multiplying individual

proofs of similar problems stated differently. In short, Greek geometers were not interested in systems; they were interested in authentic proofs with a specific "aura" (Netz 2004, 58–63).

36. Netz suggests that the polemical dynamics of ancient mathematical texts prevented Greek mathematicians from normalizing their works, demonstrations, and problems. As he noted: "The strategy we have seen so far—of the Greek mathematician trying to isolate his work from its context—is seen now as both prudent and effective. It is prudent because it is a way of protecting the work, in advance, from being dragged into inter-textual polemics over which you do not have control. And it is effective because it makes your work shine, as if beyond polemic. When Greek mathematicians set out the ground for their text, by an explicit introduction or, implicitly, by the mathematical statement of the problem, what they aim to do is to wipe the slate clean: to make the new proposition appear, as far as possible, as a *sui generis* event—the first *genuine* solution of the problem at hand" (Netz 2004, 62–63).

37. To a certain extent, as we will shall see in chapter 6, mathematical software such as Wolfram Mathematica and Matlab can be considered repositories of polished, compiled, and standardized mathematical certified knowledge.

38. Very schematically, a neuron cell is made of three parts. There is first the "dendrite": the structure that allows a neuron to receive an electro-chemical signal. There is then the "cell body": the spherical part of the neuron that contains the nucleus of the cell and reacts to the signal. There is finally the "axon": the extended cell membrane that sends information to other dendrites.

39. It is important to note that the inevitable losses that go along with reduction processes can be used to criticize the products of these reductions. This is exactly what I did in chapter 3 when I was dealing with the computational metaphor of the mind. I used what some reductions did not take into account in order to criticize the product of these reductions.

Chapter 6

1. BJ's face-detection algorithm computes the size of a face as the ratio of the area of the face-detection rectangle to the size of the image; hence the very small size-values of faces in figure 6.3.

2. Remember that this comparison exercise was the main reason why the Group's paper on the algorithm was initially rejected by the committee of the image-processing conference (see chapter 2).

3. It is important to note that this spreadsheet form required not so trivial Matlab parsing scripts written by the Group. The construction of a ground-truth database thus also sometimes requires computer programming practices as described in chapter 4.

4. Napier initiated the theory of logarithms mainly to facilitate manual numerical calculations, notably in astronomy. On this topic, see the old but enjoyable work by Cajori (1913).

5. This discussion was reconstructed from notes in Logbook 2, February 2014– May 2014.

6. With lower-level programming languages such as C or C++, it might be trickier to transform this scenario into a completed program.

7. If it is not time consuming to approximate square roots of positive real numbers, it is more complicated to get precise results. Nowadays, computers start by expressing the positive real number in floating point notation $m * 2^e$ where m is a number between 1 and 2 and e is its exponent (MacKenzie 1993). Thanks to this initial translation, computer languages can then use the Newton-Raphson iteration method to calculate the *reciprocal* of square root before finally multiplying this result with the initial real number to get the final answer. Calculating k-means of five clusters is also not that trivial. It can be summarized by a list of six operations: (1) place five arbitrary random centroids within the given dataset; (2) compute the distances of every point of the dataset from all centroids; (3) assign every point of the dataset to its nearest centroid; (4) compute the center of gravity of every centroid-assigned group of points; (5) assign each centroid to the position of the center of gravity of its group; and (6) reiterate the operation until no centroid changes its assignment anymore.

8. Remember that INT stands for the Matlab interpreter that translates instructions written in the Editor into machine code, the only language that can make processors trigger electric pulses.

9. Information retrieved from Matlab Central Community Forum (MATLAB Answers 2017)

10. This discussion has been reconstructed from notes in Logbook 3, February– May 2014.

11. This discussion has been reconstructed from notes in Logbook 3, February– May 2014.

12. Fei-Fei Li is now a professor at Stanford University. Between 2017 and 2018, she was chief scientist at Google Cloud.

13. Image classification in digital image processing consists of categorizing the content of images into predefined labels. For an accessible introduction to image classification, see Kamavisdar, Saluja, and Agrawal (2013).

14. The beginnings of the ImageNet ground truth project were difficult. As Gershgorn noted it: "Li's first idea was to hire undergraduate students for $10 an hour to manually find images and add them to the dataset. But back-of-the-napkin math

quickly made Li realize that at the undergrads' rate of collecting images it would take 90 years to complete. After the undergrad task force was disbanded, Li and the team went back to the drawing board. What if computer-vision algorithms could pick the photos from the internet, and humans would then just curate the images? But after a few months of tinkering with algorithms, the team came to the conclusion that this technique wasn't sustainable either—future algorithms would be constricted to only judging what algorithms were capable of recognizing at the time the dataset was compiled. Undergrads were time-consuming, algorithms were flawed, and the team didn't have money—Li said the project failed to win any of the federal grants she applied for, receiving comments on proposals that it was shameful Princeton would research this topic, and that the only strength of proposal was that Li was a woman" (Gershgorn 2017).

15. To minimize crowdworkers' labeling errors, Fei-Fei Li and her team asked different workers to label the same image—one label being considered a vote, the majority of votes "winning" the labeling task. However, depending on the complexity of the labeling task—categories such as "Burmese cat" being difficult to accurately identify—Fei-Fei Li and her team have varied the levels of consensus required. To determine these content-related required levels of consensus, they have developed an algorithm whose functioning is, however, not detailed in the paper (Deng et al. 2009, 252).

16. Once assembled, the ImageNet dataset and ground truth did not generate immediate interest among the image recognition community. Far from it: the first publication of the project in the 2009 Computer Vision and Pattern Recognition (Deng et al. 2009) was taken from a poster stuck in a corner of the Fontainebleau Resort at Miami Beach (Gershgorn 2017).

17. In a nutshell, ILSVRC challenges, in the wake of PASCAL VOC challenges, consist of two related components: (1) a publicly available ground truth and (2) an annual competition whose results are discussed during dedicated workshops. As Russakovsky et al. summarized it: "The publically released dataset contains a set of manually annotated training images. A set of test images is also released, with the manual annotations withheld. Participants train their algorithms using the training images and then automatically annotate the test images. These predicted annotations are submitted to the evaluation server. Results of the evaluation are revealed at the end of the competition period and authors are invited to share insights at the workshop held at the International Conference on Computer Vision (ICCV) or European Conference on Computer Vision (ECCV) in alternate years" (Russakovsky et al. 2015, 211).

18. AlexNet, as the algorithm presented in Krizhevsky, Sutskever, and Hinton (2012) ended up being called, has brought back to the forefront of image processing the convolutional neural network learning techniques developed by Joshua Bengio, Geoffrey Hinton, and Yann LeCun since the 1980s. Today, convolutional neural networks for text, image, and video processing are ubiquitous, empowering products

distributed by large tech companies such as Google, Facebook, or Microsoft. Moreover, Bengio, Hinton, and LeCun received the Turing Prize Award in 2018, generally considered the highest distinction in computer science.

19. These criticisms were summarized by Marvin Minsky, the head of the MIT Artificial Intelligence Research Group, and Seymour Papert in their book *Perceptrons: An Introduction to Computational Geometry* (1969).

20. Boltzmann machines are expansions of spin glass-inspired neural networks. By including a stochastic decision rule, Ackley, Hinton, and Sejnokwski (1985) could make a neural network reach an appreciable learning equilibrium. As Domingos explained, "the probability of finding the network in a particular state was given by the well-known Boltzmann distribution from thermodynamics, so they called their network a Boltzmann machine" (Domingos 2015, 103).

21. As noted in Cardon, Cointet, and Mazières (2018), there is a debate regarding the anteriority of backprop algorithm: "This method has been formulated and used many times before the publication of [Rumelhart Hinton, and Williams 1986]'s article, notably by Linnainmaa in 1970, Werbos in 1974 and LeCun in 1985" (Cardon, Cointet, and Mazières 2018, 198; my translation).

22. This second marginalization of connectionists during the 1990s can be related to the spread of Support Vector Machines (SVMs), audacious learning techniques that are very effective on small ground truths. Moreover, while SVMs manage to find, during the learning of the loss function, the global error minimum, convolutional neural networks can only find local minimums (a limit that will prove to be less problematic with the advent of large ground truths, such as ImageNet, and the increase in the computing power of computers). On this specialized topic, see Domingos (2015, 107–111) and Cardon, Cointet, and Mazières (2018, 200–202).

Conclusion

1. Though, like Negri, this book is drawn to the idea of contributing to founding a philosophy capable of going beyond modernity understood as "the definition and development of a totalizing thought that assumes human and collective creativity in order to insert them into the instrumental rationality of the capitalist mode of production" (Negri 1999, 323).

2. Curiously, even though Negri explicitly positions himself as an opponent of the Anglo-American liberal tradition, his conclusions regarding the dual aspect of insurrectional acts are quite aligned with propositions made by American pragmatist writers such as Walter Lippmann and John Dewey. Indeed, whereas for these two authors, the political can only be expressed by means of issues that redefine our whole living together (Dewey [1927] 2016; Lippmann [1925] 1993; Marres 2005), for Negri, the political, as Michael Hardt notes, "is defined by the forces that challenge

the stability of the constituted order ... and the constituent processes that invent alternative forms of social organization. ... The political exists only where innovation and constituent processes are at play" (Hardt 1999, ix).

3. This, I believe, is a potential way of somewhat reconciling Negri—at least, his writings—with the great German legal tradition that he is also explicitly opposed to. If Negri is certainly right to refuse the exteriority of constituent power vis-à-vis constituted power, thus emptying legal constitutions of any power of political innovation, he is probably wrong to dismiss Georg Jellinek's and Hans Kelsen's propositions as to the scriptural, and therefore ontological, weight of constituent texts. On this tension between *Sollen* (what ought to be) and *Sein* (what is) within constitutive processes, see Negri (1999, 5–35) as well as Jellinek ([1914] 2016) and Kelsen (1991).

4. This is the topic of Anne Henriksen's and Cornelius Heimstädt's PhD theses (currently being conducted at Aarhus University and Mines ParisTech, respectively), as well as Nick Seaver's forthcoming book (Seaver forthcoming).

5. The moral economy of blockchain technology is the topic of Clément Gasull's PhD thesis, currently being conducted at Mines ParisTech.

6. This is part of Vassileios Gallanos's PhD thesis, currently being conducted at the University of Edinburgh.

References

Abbate, Janet. 2012. *Recoding Gender: Women's Changing Participation in Computing*. Cambridge, MA: MIT Press.

Achanta, Radhakrishna, Sheila Hemami, Francisco Estrada, and Sabine Susstrunk. 2009. "Frequency-Tuned Salient Region Detection." In *IEEE Conference on Computer Vision and Pattern Recognition*, Miami, FL, June, 1597–1604. New York: IEEE.

Ackley, David H., Geoffrey E. Hinton, and Terrence J. Sejnowski. 1985. "A Learning Algorithm for Boltzmann Machines." *Cognitive Science* 9, no. 1: 147–169.

Adelson, Beth. 1981. "Problem Solving and the Development of Abstract Categories in Programming Languages." *Memory & Cognition* 9, no. 4: 422–433.

Ahmed, Faheem, Luiz F. Capretz, Salah Bouktif, and Piers Campbell. 2012. "Soft Skills Requirements in Software Development Jobs: A Cross-Cultural Empirical Study." *Journal of Systems and Information Technology* 14: 58–81.

Ahmed, Faheem, Luiz F. Capretz, and Piers Campbell. 2012. "Evaluating the Demand for Soft Skills in Software Development." *IT Professional* 14, no. 1: 44–49.

Ahmed, Nassir U., T. Natarajan, and K.R. Rao. 1974. "Discrete Cosine Transform." *IEEE Transactions on Computers* 23, no. 1: 90–93.

Akera, Atsushi. 2001. "Voluntarism and the Fruits of Collaboration: The IBM User Group, Share." *Technology and Culture* 42, no. 4: 710–736.

Akera, Atsushi. 2008. *Calculating a Natural World: Scientists, Engineers, and Computers during the Rise of U.S. Cold War Research*. Cambridge, MA: MIT Press.

Akrich, Madeleine. 1989. "La construction d'un système socio-technique: Esquisse pour une anthropologie des techniques." *Anthropologie et Sociétés* 13, no. 2: 31–54.

Akrich, Madeleine, Michel Callon, and Bruno Latour. 2006. *Sociologie de la traduction: Textes fondateurs*. Paris: Presses de l'École des Mines.

Albrecht, Sandra L. 1982. "Industrial Home Work in the United States: Historical Dimensions and Contemporary Perspective." *Economic and Industrial Democracy* 3, no. 4: 413–430.

Allen, Elizabeth, and Sophie Triantaphillidou, eds. 2011. *The Manual of Photography.* 10th ed. Burlington, MA: Focal Press.

Alpaydin, Ethem. 2010. *Introduction to Machine Learning.* 2nd ed. Cambridge, MA: MIT Press.

Alpaydin, Ethem. 2016. *Machine Learning: The New AI.* Cambridge, MA: MIT Press.

Alpert, Sharon, Meirav Galun, Achi Brandt, and Ronen Basri R. 2007. "Image Segmentation by Probabilistic Bottom-Up Aggregation and Cue Integration." In *2007 IEEE Conference on Computer Vision and Pattern Recognition.* New York: IEEE.*2007.* DOI: https//:10.1109/CVPR.2007.383017.

American Association for Artificial Intelligence. 1993. "Organization of the American Association for Artificial Intelligence." The Eleventh National Conference on Artificial Intelligence (AAAI-93), July 11–15, Washington, DC. http://www.aaai.org /Conferences/AAAI/1993/aaai93committee.pdf (last accessed March 2017).

Ananny, Mike, and Kate Crawford. 2018. "Seeing without Knowing: Limitations of the Transparency Ideal and Its Application to Algorithmic Accountability." *New Media & Society* 20, no.3: 973–989.

Anderson, Christopher W. 2011. "Deliberative, Agonistic, and Algorithmic Audiences: Journalism's Vision of Its Public in an Age of Audience Transparency." *International Journal of Communication* 5: 550–566.

Anderson, Drew. 2017. "GLAAD and HRC Call on Stanford University & Responsible Media to Debunk Dangerous & Flawed Report Claiming to Identify LGBTQ People through Facial Recognition Technology." *GLAAD.org*, September 8. https://www .glaad.org/blog/glaad-and-hrc-call-stanford-university-responsible-media-debunk -dangerous-flawed-report (last accessed February 2018).

Anderson, John R. 1983. *The Architecture of Cognition.* Cambridge, MA: Harvard University Press.

Angwin, Julia, Jeff Larson, Surya Mattu, and Lauren Kirchner. 2016. "Machine Bias: There's Software Used across the Counter to Predict Future Criminals. And It's Biased against Blacks." *ProPublica*, May 23. https://www.propublica.org/article/machine-bias -risk-assessments-in-criminal-sentencing.

Antognazza, Maria R. 2011. *Leibniz: An Intellectual Biography.* Reprint. Cambridge: Cambridge University Press.

Ashby, Ross W. 1952. *Design for a Brain.* New York: Wiley.

Aspray, William. 1990. *John von Neumann and the Origins of Modern Computing.* Cambridge, MA: MIT Press.

Aspray, William, and Philip Kitcher, eds. 1988. *History and Philosophy of Modern Mathematics*. Minneapolis: University of Minnesota Press.

Austin, John L. 1975. *How to Do Things with Words*. 2nd ed. Cambridge, MA: Harvard University Press.

Badinter, Elisabeth. 1981. *Mother Love: Myth and Reality*. New York: Macmillan.

Baluja, Shumeet, and Dean A. Pomerleau. 1997. "Expectation-Based Selective Attention for Visual Monitoring and Control of a Robot Vehicle." *Robotics and Autonomous Systems* 22: 329–344.

Barad, Karen. 2007. *Meeting the Universe Halfway: Quantum Physics and the Entanglement of Matter and Meaning*. Durham, NC: Duke University Press.

Bardi, Jason S. 2007. *The Calculus Wars: Newton, Leibniz, and the Greatest Mathematical Clash of All Time*. New York: Basic Books.

Barfield, Woodrow. 1986. "Expert-Novice Differences for Software: Implications for Problem-Solving and Knowledge Acquisition." *Behaviour & Information Technology* 5, no. 1: 15–29.

Barocas, Solon, and Andrew D. Selbst. 2016. "Big Data's Disparate Impact." *California Law Review* 104: 671–732.

Barrett, Justin L. 2007. "Cognitive Science of Religion: What Is It and Why Is It?" *Religion Compass* 1, no. 6: 768–786.

Baya-Laffite, Nicolas, Boris Beaude, and Jérémie Garrigues. 2018. "Le Deep Learning au service de la prédication de l'orientation sexuelle dans l'espace public: Déconstruction d'une alerte ambigüe." *Réseaux* 211, no. 211: 137–172.

Bechmann, Anja, and Geoffrey C. Bowker. 2019. "Unsupervised by Any Other Name: Hidden Layers of Knowledge Production in Artificial Intelligence on Social Media." *Big Data & Society* 6, no. 1. https://doi.org/10.1177/2053951718819569.

Beer, David. 2009. "Power through the Algorithm? Participatory Web Cultures and the Technological Unconscious." *New Media & Society* 11, no. 6: 985–1002.

Bengio, Yoshua. 2009. "Learning Deep Architectures for AI." *Foundations and Trends in Machine Learning* 2, no. 1: 1–127.

Bengio, Yoshua, Réjean Ducharme, Pascal Vincent, and Christian Jauvin. 2003 "A Neural Probabilistic Language Model." *Journal of Machine Learning Research* 3: 1137–1155.

Bensaude-Vincent, Bernadette. 1995. "Mendeleyev: The Story of a Discovery." In *A History of Scientific Thought: Elements of a History of Science*, edited by Michel Serres, 556–582. Oxford: Blackwell.

Berg, Nate. 2014. "Predicting Crime, LAPD-Style." *Guardian*, June 25. https://www.theguardian.com/cities/2014/jun/25/predicting-crime-lapd-los-angeles-police-data-analysis-algorithm-minority-report.

Berggren, John L. 1986. *Episodes in the Mathematics of Medieval Islam*. Berlin: Springer.

Bhattacharyya, Siddhartha, Hrishikesh Bhaumik, Anirban Mukherjee, and Sourav De. 2018. *Machine Learning for Big Data Analysis*. Berlin: Walter de Gruyter.

Biancuzzi, Federico, and Shane Warden. 2009. *Masterminds of Programming: Conversations with the Creators of Major Programming Languages*. Sebastopol, CA: O'Reilly.

Birch, Kean, and Fabian Muniesa, eds. 2020. *Assetization: Turning Things into Assets in Technoscientific Capitalism*. Cambridge, MA: MIT Press.

Bishop, Chistopher M. 2007. *Pattern Recognition and Machine Learning*. New York: Springer.

Blaiwes, Arthur S. 1974. "Formats for Presenting Procedural Instructions." *Journal of Applied Psychology* 59, no. 6: 683–686.

Bloom, Alan M. 1980. "Advances in the Use of Programmer Aptitude Tests." In *Advances in Computer Programming Management*, edited by Thomas A. Rullo, Vol. 1: 31–60. Philadelphia: Hayden, 1980.

Bloor, David. 1981. "The Strengths of the Strong Programme." *Philosophy of the Social Sciences* 11, no. 2: 199–213.

Bobzien, Susanne. 2002. "The Development of Modus Ponens in Antiquity: From Aristotle to the 2nd Century AD." *Phronesis* 47, no. 4: 359–394.

Boltanski, Luc, and Laurent Thévenot. 2006. *On Justification: Economies of Worth*. Princeton, NJ: Princeton University Press.

Bonaccorsi, Andrea, and Cristina Rossi. 2006. "Comparing Motivations of Individual Programmers and Firms to Take Part in the Open Source Movement: From Community to Business." *Knowledge, Technology & Policy* 18, no. 4: 40–64.

Borji, Ali. 2012. "Boosting Bottom-up and Top-down Visual Features for Saliency Estimation." In *2012 IEEE Conference on Computer Vision and Pattern Recognition*, Providence, RI, June, 438–445. New York: IEEE.

Bostrom, Nick. 2017. "Strategic Implications of Openness in AI Development." *Global Policy* 8, no. 2: 135–48.

Bottazzini, Umberto. 1986. *The Higher Calculus: A History of Real and Complex Analysis from Euler to Weierstrass*. Berlin: Springer.

Bourdieu, Pierre. 1986. "L'illusion biographique." *Actes de la recherche en sciences sociales* 62, no. 1: 69–72.

Bowker, Geoffrey C. 1993. "How to Be Universal: Some Cybernetic Strategies, 1943--70." *Social Studies of Science* 23, no. 1: 107–127.

Boyer, Carl B. 1959. *The History of the Calculus and Its Conceptual Development*. New York: Dover Publications.

Bozdag, Engin. 2013. "Bias in Algorithmic Filtering and Personalization." *Ethics and Information Technology* 15, no. 3: 209–227.

Brazeau, Paul, Wylie Vale, Roger Burgus, Nicholas Ling, Madalyn Butcher, Jean Rivier, and Roger Guillemin. 1973. "Hypothalamic Polypeptide That Inhibits the Secretion of Immunoreactive Pituitary Growth Hormone." *Science* 179, no. 4068: 77–79.

Brockell, Gillian. 2018. "Dear Tech Companies, I Don't Want to See Pregnancy Ads after My Child Was Stillborn." *Washington Post*, December 12.

Brooke, J. B., and K. D. Duncan. 1980a. "An Experimental Study of Flowcharts as an Aid to Identification of Procedural Faults." *Ergonomics* 23, no. 4: 387–399.

Brooke, J. B., and K. D. Duncan. 1980b. "Experimental Studies of Flowchart Use at Different Stages of Program Debugging." *Ergonomics* 23, no. 11: 1057–1091.

Brooks, Frederick. 1975. *The Mythical Man-Month: Essays on Software Engineering*. Reading, MA: Addison-Wesley Professional.

Brooks, John. 1976. *Telephone: The First Hundred Years*. New York: Harper & Row.

Brooks, Ruven. 1977. "Towards a Theory of the Cognitive Processes in Computer Programming." *International Journal of Man-Machine Studies* 9, no. 6: 737–751.

Brooks, Ruven. 1980. "Studying Programmer Behavior Experimentally: The Problems of Proper Methodology." *Communications of the ACM* 23, no. 4: 207–213.

Bucher, Taina. 2012. "Want to Be on the Top? Algorithmic Power and the Threat of Invisibility on Facebook." *New Media & Society* 14, no. 7: 1164–1180.

Buchman, Amy. 2009 "A Brief History of Quaternions and the Theory of Holomorphic Functions of Quaternionic Variables." Paper, November. https://ui.adsabs.harvard.edu/abs/2011arXiv1111.6088B.

Burks, Alice R., and Arthur W. Burks. 1989. *The First Electronic Computer: The Atanasoff Story*. Ann Arbor, MI: University of Michigan Press.

Burks, Arthur W., Herman H. Goldstine, and John von Neumann. 1946. *Preliminary Discussion of the Logical Design of an Electronic Computer Instrument*. Princeton, NJ: Institute for Advanced Study.

Burrell, Jenna. 2016. "How the Machine 'Thinks': Understanding Opacity in Machine Learning Algorithms." *Big Data & Society* 3, no. 1: 1–12.

Butler, Judith. 2006. *Gender Trouble: Feminism and the Subversion of Identity*. New York and London: Routledge.

Button, Graham, and Wes Sharrock. 1995. "The Mundane Work of Writing and Reading Computer Programs." In *Situated Order: Studies in the Social Organization of Talk and Embodied Activities*, edited by Paul T. Have and George Psathas, 231–258. Washington, DC: University Press of America.

Cajori, Florian. 1913. "History of the Exponential and Logarithmic Concepts." *The American Mathematical Monthly* 20, no. 1: 5–14.

Cakebread, Caroline. 2017. "People Will Take 1.2 Trillion Digital Photos This Year—Thanks to Smartphones." *Business Insider*, August 31. https://www.businessinsider.fr/us/12-trillion-photos-to-be-taken-in-2017-thanks-to-smartphones-chart-2017-8/.

Callon, Michel. 1986. "Some Elements of a Sociology of Translation: Domestication of the Scallops and the Fishermen of St Brieuc Bay." In *Power, Action and Belief: A New Sociology of Knowledge?* edited by John Law, 196–223. London: Routledge & Kegan Paul.

Callon, Michel. 1999. "Le Réseau Comme Forme Émergente et Comme Modalité de Coordination." In *Réseau et Coodination*, edited by Michel Callon, Patrick Cohendet, Nicolas Curlen, Jean-Michel Dalle, François Eymard-Duvernay, Dominique Foray and Eric Schenk, 13–63. Paris: Economica.

Callon, Michel. 2017. *L'emprise des marchés: Comprendre leur fonctionnement pour pouvoir les changer*. Paris: La Découverte.

Campbell-Kelly, Martin. 2003. *From Airline Reservations to Sonic the Hedgehog: A History of the Software Industry*. Cambridge, MA: MIT Press.

Campbell-Kelly, Martin, William Aspray, Nathan Ensmenger, and Jeffrey R. Yost. 2013. *Computer: A History of the Information Machine*. 3rd ed. Boulder, CO: Westview Press.

Capretz, Fernando L. 2014. "Bringing the Human Factor to Software Engineering." *IEEE Software* 31, no. 2: 104–104.

Card, Stuart K., Thomas P. Moran, and Allen Newell. 1986. *The Psychology of Human-Computer Interaction*. Hillsdale, NJ: Lawrence Erlbaum.

Cardon, Dominique. 2015. *À quoi rêvent les algorithmes. Nos vies à l'heure du Big Data*. Paris: Le Seuil.

Cardon, Dominique, Jean-Philippe Cointet, and Antoine Mazières. 2018. "La revanche des neurones. L'invention des machines inductives et la controverse de l'intelligence artificielle." *Réseaux* 211, no. 5: 173–220.

Carnap, Rudolf. 1937. *The Logical Syntax of Language*. Chicago: Open Court Publishing.

Carroll, John M., John C. Thomas, and Ashok Malhotra. 1980. "Presentation and Representation in Design Problem-Solving." *British Journal of Psychology* 71, no. 1: 143–153.

Casilli, Antonio. 2019. *En attendant les robots: Enquête sur le travail du clic*. Paris: Le Seuil.

Cassin, Barbara. 2014. *Sophistical Practice: Toward a Consistent Relativism*. New York: Fordham University Press.

Cerf, Moran, Paxon E. Frady, and Christof Koch. 2009. "Faces and Text Attract Gaze Independent of the Task: Experimental Data and Computer Model." *Journal of Vision* 9, no. 12: 101–115.

Chang, Kai-Yueh, Tyng-Luh Liu, Hwann-Tzong Chen, and Shang-Hong Lai. 2011. "Fusing Generic Objectness and Visual Saliency for Salient Object Detection." In *2011 IEEE International Conference on Computer Vision*, Barcelona, November. New York: IEEE, pp. 914–921.

Chen, Li-Qun, Xing Xie, Xin Fan, Wei-Ying Ma, Hong-Jiang Zhang, and He-Qin Zhou. 2003. "A Visual Attention Model for Adapting Images on Small Displays." *Multimedia Systems* 9, no. 4: 353–364.

Cheng, Ming-Ming, Guo-Xin Zhang, N. J. Mitra, Xiaolei Huang, and Shi-Min Hu. 2011. "Global Contrast Based Salient Region Detection." In *CVPR 2011: The 24th IEEE Conference on Computer Vision and Pattern Recognition*, 409–416. Washington, DC: IEEE Computer Society.

Clark, Andy. 1998. *Being There: Putting Brain, Body, and World Together Again*. Cambridge, MA: MIT Press.

Clark, Andy, and Chalmers David. 1998. "The Extended Mind." *Analysis* 58, no. 1: 7–19.

Cobb, John B. 2006. *Dieu et le monde*. Paris: Van Dieren.

Cohen, Bernard I. 1999. *Howard Aiken: Portrait of a Computer Pioneer*. Cambridge, MA: MIT Press.

Collins, Charlotte A., Irwin Olsen, Peter S. Zammit, Louise Heslop, Aviva Petrie, Terence A. Partridge, and Jennifer E. Morgan. 2005. "Stem Cell Function, Self-Renewal, and Behavioral Heterogeneity of Cells from the Adult Muscle Satellite Cell Niche." *Cell* 122, no. 2: 289–301.

Collins, Harry M. 1975. "The Seven Sexes: A Study in the Sociology of a Phenomenon, or the Replication of Experiments in Physics." *Sociology* 9, no. 2: 205–224.

Collins, Harry M. 1992. *Changing Order: Replication and Induction in Scientific Practice*. Chicago: University of Chicago Press.

Constine, Josh. 2019. "To Automate Bigger Stores than Amazon, Standard Cognition Buys Explorer.Ai." *TechCrunch* (blog), January 7. https://techcrunch.com/2019/01/07/autonomous-checkout/.

Coombs, M. J., R. Gibson, and J. L. Alty. 1982. "Learning a First Computer Language: Strategies for Making Sense." *International Journal of Man-Machine Studies* 16, no. 4: 449–486.

Corfield, David. 2006. *Towards a Philosophy of Real Mathematics*. Rev. ed. Cambridge: Cambridge University Press.

Cormen, Thomas H, Charles E. Leiserson, Ronald L. Rivest, and Clifford Stein. 2009. *Introduction to Algorithms*. 3rd ed. Cambridge, MA: MIT Press.

Corry, Leo. 1997. "The Origins of Eternal Truth in Modern Mathematics: Hilbert to Bourbaki and Beyond." *Science in Context* 10, no. 2: 253–296.

Crawford, Kate, and Ryan Calo. 2016. "There Is a Blind Spot in AI Research." *Nature* 538, no. 7625: 311–313.

Crevier, Daniel. 1993. *AI: The Tumultuous History of the Search for Artificial Intelligence*. New York: Basic Books.

Crilly, Tony. 2004. "The Cambridge Mathematical Journal and Its Descendants: The Linchpin of a Research Community in the Early and Mid-Victorian Age." *Historia Mathematica* 31, no. 4: 455–497.

Crooks, Roderic N. 2019. "Times Thirty: Access, Maintenance, and Justice." *Science, Technology, & Human Values* 44, no. 1: 118–142.

Cruz, Shirley, Fabio da Silva, and Luiz Capretz. 2015. "Forty Years of Research on Personality in Software Engineering: A Mapping Study." *Computers in Human Behavior* 46: 94–113.

Curtis, Bill. 1981. "Substantiating Programmer Variability." *Proceedings of the IEEE* 69, no. 7: 846.

Curtis, Bill. 1988. "Five Paradigms in the Psychology of Programming." In *Handbook of Human-Computer Interaction*, edited by Martin Helander, 87–105. Amsterdam: Elsevier North-Holland.

Curtis, Bill, Sylvia B. Sheppard, Elizabeth Kruesi-Bailey, John Bailey, and Deborah A. Boehm-Davis. 1989. "Experimental Evaluation of Software Documentation Formats." *Journal of Systems and Software* 9, no. 2: 167–207.

Daganzo, Carlos F. 1995. "The Cell Transmission Model, Part II: Network Traffic." *Transportation Research Part B: Methodological* 29, no. 2: 79–93.

Daganzo, Carlos F. 2002. "A Behavioral Theory of Multi-Lane Traffic Flow. Part I: Long Homogeneous Freeway Sections." *Transportation Research Part B: Methodological* 36, no. 2: 131–158.

Damasio, Anthony. 2005. *Descartes' Error: Emotion, Reason, and the Human Brain*. Reprint. London: Penguin Books.

Dasgupta, Sanjoy, Christos Papadimitriou, and Umesh Vazirani. 2006. *Algorithms.* 1st ed. Boston: McGraw-Hill Education.

Dauben, Joseph W. 1990. *Georg Cantor: His Mathematics and Philosophy of the Infinite.* Reprint ed. Princeton, NJ: Princeton University Press.

Dear, Peter. 1987. "Jesuit Mathematical Science and the Reconstitution of Experience in the Early Seventeenth Century." *Studies in History and Philosophy of Science Part A* 18, no. 2: 133–175.

Dear, Peter, and Sheila Jasanoff. 2010. "Dismantling Boundaries in Science and Technology Studies." *Isis* 101, no. 4: 759–774.

Dekowska, Monika, Michał Kuniecki, and Piotr Jaśkowski. 2008. "Facing Facts: Neuronal Mechanisms of Face Perception." *Acta Neurobiologiae Experimentalis* 68, no. 2: 229–252.

de la Bellacasa, Maria P. 2011 "Matters of Care in Technoscience: Assembling Neglected Things." *Social Studies of Science* 41, no. 1: 85–106.

Deleuze, Gilles. 1989. "Qu'est-ce qu'un dispositif?" In *Michel Foucault philosophe: rencontre international Paris 9, 10, 11, janvier 1988.* Paris: Seuil.

Deleuze, Gilles. 1992. *Fold: Leibniz and the Baroque.* Minneapolis: University of Minnesota Press.

Deleuze, Gilles. 1995. *Difference and Repetition.* New York: Columbia University Press.

Demazière, Didier, François Horn, and Marc Zune. 2007. "The Functioning of a Free Software Community: Entanglement of Three Regulation Modes—Control, Autonomous and Ditributed." *Science Studies* 20, no. 2: 34–54.

Denelesky, Garland Y., and Michael G. McKee. 1974. "Prediction of Computer Programmer Training and Job Performance Using the Aabp Test1." *Personnel Psychology* 27, no. 1: 129–137.

Deng, Jia, Alexander C. Berg, Kai Li, and Li Fei-Fei. 2010. "What Does Classifying More Than 10,000 Image Categories Tell Us?" In *Computer Vision—ECCV 2010*, edited by Kostas Daniilidis, Petros Maragos, and Nikos Paragios, 71–84. Berlin: Springer.

Deng, Jia, Alexander C. Berg, and Li Fei-Fei. 2011a. "Hierarchical Semantic Indexing for Large Scale Image Retrieval." In *CVPR 2011: The 24th IEEE Conference on Computer Vision and Pattern Recognitio*, 785–792. Washington, DC: IEEE Computer Society.

Deng, Jia, Wei Dong, Richard Socher, Li-Jia Li, Kai Li, and Li Fei Fei. 2009. "ImageNet: A Large-Scale Hierarchical Image Database." In *2009 IEEE Conference on Computer Vision and Pattern Recognition*, Miami, FL, June, 248–255. New York: IEEE.

Deng, Jia, Sanjeev Satheesh, Alexander C. Berg, and Li Fei-Fei. 2011b. "Fast and Balanced: Efficient Label Tree Learning for Large Scale Object Recognition." In *Advances in Neural Information Processing Systems 24*, edited by J. Shawe-Taylor, R. S. Zemel, P. L. Bartlett, F. Pereira, and K. Q. Weinberger, 567–575. Red Hook, NY: Curran Associates.

Deng, Jia, Olga Russakovsky, Jonathan Krause, Michael S. Bernstein, Alex Berg, and Li Fei-Fei. 2014. "Scalable Multi-Label Annotation." In *Proceedings of the SIGCHI Conference on Human Factors in Computing Systems*, 3099–3102. New York: ACM.

Denis, Jérôme. 2018. *Le travail invisible des données: Éléments pour une sociologie des infrastructures scripturales*. Paris: Presses de l'École des Mines.

Denis, Jérôme, and David Pontille. 2015. "Material Ordering and the Care of Things." *Science, Technology, & Human Values* 40, no. 3: 338–367.

Dennett, Daniel C. 1984. "Cognitive Wheels: The Frame Problem of AI." In *Minds, Machines and Evolution*, edited by Christopher Hookway, 129–150. Cambridge: Cambridge University Press.

Dennis, Michael A. 1989. "Graphic Understanding: Instruments and Interpretation in Robert Hooke's Micrographia." *Science in Context* 3, no. 2: 309–364.

Desrosières, Alain. 2010. *The Politics of Large Numbers: A History of Statistical Reasoning*. Translated by Camille Naish. New ed. Cambridge, MA: Harvard University Press.

Dewey, John. (1927) 2016. *The Public and Its Problems*. Athens, OH: Ohio University Press.

Diakopoulos, Nicholas. 2014. "Algorithmic Accountability." *Digital Journalism* 3, no. 3: 398–415.

Dijkstra, Edsger W. 1968. "Letters to the Editor: Go to Statement Considered Harmful." *Communications of the ACM* 11, no. 3: 147–148.

Dijkstra, Edsger W. 1972. "Notes on Structured Programming." In *Structured Programming*, edited by Ole-Johan Dahl, Edsger W. Dijkstra, and Charles A. R. Hoare, 1–82. London: Academic Press.

Di Paolo, Ezequiel A. 2005. "Autopoiesis, Adaptivity, Teleology, Agency." *Phenomenology and the Cognitive Sciences* 4, no. 4: 429–452.

Doganova, Liliana. 2012 *Valoriser la science. Les partenariats des start-up technologiques*. Paris: Presses de l'École des mines.

Doing, Park. 2008. "Give Me a Laboratory and I Will Raise a Discipline: The Past, Present, and Future Politics of Laboratory Studies." In *The Handbook of Science and Technology Studies*. 3rd ed, edited by Edward J. Hackett, Olga Amsterdamska, Michael Lynch, and Judy Wajcman, 279–295. Cambridge, MA: MIT Press.

Domingos, Pedro. 2015. *The Master Algorithm: How the Quest for the Ultimate Learning Machine Will Remake Our World*. New York: Basic Books.

Domínguez Rubio, Fernando. 2014. "Preserving the Unpreservable: Docile and Unruly Objects at MoMA." *Theory and Society* 43, no. 6: 617–645.

Domínguez Rubio, Fernando. 2016. "On the Discrepancy between Objects and Things: An Ecological Approach." *Journal of Material Culture* 21, no. 1: 59–86.

Donin, Nicolas, and Jacques Theureau. 2007. "Theoretical and Methodological Issues Related to Long Term Creative Cognition: The Case of Musical Composition." *Cognition, Technology & Work* 9: 233–251.

Doxiàdis, Apóstolos K., Christos Papadimitriou, Alecos Papadatos, and Annie Di Donna. 2010. *Logicomix*. Paris: Vuibert.

Draper, Stephen W. 1992. "Critical Notice. Activity Theory: The New Direction for HCI?" *International Journal of Man-Machine Studies* 37, no. 6: 812–821.

Dreyfus, Hubert L. 1992. *What Computers Still Can't Do: A Critique of Artificial Reason*. Rev. ed. Cambridge, MA: MIT Press.

Dreyfus, Hubert L. 1998. "The Current Relevance of Merleau-Ponty's Phenomenology of Embodiment." *Electronic Journal of Analytic Philosophy* 4: 15–34.

Dunsmore, H. E., and J. D. Gannon. 1979. "Data Referencing: An Empirical Investigation." *Computer* 12, no. 12: 50–59.

Dupuy, Jean-Pierre. 1994. *Aux origines des sciences cognitives*. Paris: La Découverte.

Eason, Robert G., Russell M. Harter, and C. T. White. 1969. "Effects of Attention and Arousal on Visually Evoked Cortical Potentials and Reaction Time in Man." *Physiology & Behavior* 4, no. 3: 283–289.

Eckert, John P., and John W. Mauchly. 1945. *Automatic High Speed Computing: A Progress Report on the EDVAC*. Philadelphia: University of Pennsylvania, September 30.

Edge, David O. 1976 "Quantitative Measures of Communication in Sciences." In *International Symposium on Quantitative Measures in the History of Science*, Berkeley, CA, September.

Edwards, Paul N. 1996. *The Closed World: Computers and the Politics of Discourse in Cold War America*. Cambridge, MA: MIT Press.

Edwards, Paul N. 2013. *A Vast Machine: Computer Models, Climate Data, and the Politics of Global Warming*. Cambridge, MA: MIT Press.

Elazary, Lior, and Laurent Itti. 2008. "Interesting Objects Are Visually Salient." *Journal of Vision* 8, no. 3: 1–15.

Elkan, Charles. 1993. "The Paradoxical Success of Fuzzy Logic." In *Proceedings of the Eleventh National Conference on Artificial Intelligence*, 698–703. Palo Alto, CA: Association for the Advancement of Artificial Intelligence.

Elkan, Charles, H. R. Berenji, B. Chandrasekaran, C. J. S. de Silva, Y. Attikiouzel, D. Dubois, H. Prade, P. Smets, C. Freksa, O. N. Garcia, G. J. Klir, Bo Yuan, E. H. Mamdani, F. J. Pelletier, E. H. Ruspini, B. Turksen, N. Vadlee, M. M. Jamshidi, Pel-Zhuang Wang, Sie-Keng Tan, S. Tan, R. R. Yager, and L. A. Zadeh. 1994. "The Paradoxical Success of Fuzzy Logic." *IEEE Expert* 9, no. 4: 3–49.

Elliott, Margaret S., and Walt Scacchi. 2008. "Mobilization of Software Developers: The Free Software Movement." *Information Technology & People* 21, no. 1: 4–33.

Ensmenger, Nathan L. 2010. *The Computer Boys Take Over: Computers, Programmers, and the Politics of Technical Expertise*. Cambridge, MA: MIT Press.

Espeland, Wendy Nelson, and Michael Sauder. 2016. *Engines of Anxiety: Academic Rankings, Reputation, and Accountability*. New York: Russell Sage Foundation.

Estellés-Arolas, Enrique, and Fernando González-Ladrón-de-Guevara. 2012. "Towards an Integrated Crowdsourcing Definition." *Journal of Information Science* 38, no. 2: 189–200.

Everest, Mary B. 2007. *Philosophy and Fun of Algebra*. New York: Read Books.

Ewald, William. 2007. *From Kant to Hilbert*. Volume 1: *A Source Book in the Foundations of Mathematics*. Reprint ed. Oxford: Oxford University Press.

Fellbaum, Christiane, ed. 1998. *WordNet: An Electronic Lexical Database*. Cambridge, MA: A Bradford Book.

Felt, Ulrike, Raymond Fouché, Clark A. Miller, and Laurel Smith-Doerr. 2016. *The Handbook of Science and Technology Studies*. 4th ed. Cambridge, MA: MIT Press.

Ferreirós, José. 2007. *Labyrinth of Thought—A History of Set Theory and Its Role*. Berlin: Springer.

Ferreirós, José. 2008. "The Crisis in the Foundations of Mathematics." In *Princeton Companion to Mathematical Proof*, edited by Timothy Gowers, 142–156. Princeton, NJ: Princeton University Press.

Finlay, Steven. 2017. *Artificial Intelligence and Machine Learning for Business: A No-Nonsense Guide to Data Driven Technologies*. 2nd ed. London: Relativistic.

Fisher, Jennifer. 2007. *On the Philosophy of Logic*. Belmont, CA: Wadsworth.

Flor, Nick V., and Edwin L. Hutchins. 1991. "Analyzing Ditributed Cognition in Software Teams: A Case Study of Team Programming During Perfective Software Maintenance." In *Empirical Studies of Programmers: Fourth Workshop*, edited by Jurgen Koenemann-Belliveau, Thomas Moher, and Scott P. Robertson, 36–62. Norwood, NJ: Ablex Publishing.

Fodor, Jerry A. 1975. *The Language of Thought*. Cambridge, MA: Harvard University Press.

Fodor, Jerry A. 1987. *Psychosemantics: The Problem of Meaning in the Philosophy of Mind*. Cambridge, MA: MIT Press.

Forsythe, Diana E. 2002. *Studying Those Who Study Us: An Anthropologist in the World of Artificial Intelligence*. Stanford, CA: Stanford University Press.

Frank, Werner L. 1968. "Software for Terminal Oriented Systems." *Datamation* 1968 (June): 30–36.

Frank, Werner L. 1983. "The History of Myth No. 1." *Datamation*, May 1983: 252–263.

Fujimura, Joan H. 1987. "Constructing 'Do-Able' Problems in Cancer Research: Articulating Alignment." *Social Studies of Science* 17, no. 2: 257–293.

Fuller, Matthew, ed. 2008. *Software Studies: A Lexicon*. Cambridge, MA: MIT Press.

Gallagher, Shaun. 2005. *How the Body Shapes the Mind*. Oxford: Clarendon Press.

Gandy, Oscar H. 2002. "Data Mining and Surveillance in the Post-9.11 Environment." In *The Intensification of Surveillance Crime, Terrorism and Warfare in the Information Age*, edited by Kristie Ball and Frank Webster, 113–137. London: Pluto Press.

Gannon, John D. 1976. "An Experiment for the Evaluation of Language Features." *International Journal of Man-Machine Studies* 8: 61–73.

Garfinkel, Harold. 1981. "The Work of a Discovering Science Constructed with Materials from the Optically Discovered Pulsar." *Philosophy of the Social Sciences* 11, no. 2: 131.

Gershgorn, Dave. 2017. "The Data That Transformed AI Research—and Possibly the World." *Quartz*, July 26. https://qz.com/1034972/the-data-that-changed-the-direction-of-ai-research-and-possibly-the-world/.

Gertner, Jon. 2013. *The Idea Factory: Bell Labs and the Great Age of American Innovation*. New York: Penguin.

Gibson, James J. 1986. *The Ecological Approach to Visual Perception*. London: Lawrence Erlbaum Associates.

Gibson, James. 2014. *The Ecological Approach to Visual Perception*. Classic ed. London: Psychology Press.

Gillepsie. Tarleton. 2014. "The Relevance of Algorithms." In *Media Technologies: Essays on Communication, Materiality, and Society*, edited by Tarleton Gillepsie, Pablo Boczkowski, and Kirsten Foot, 167–194. Cambridge, MA: MIT Press.

Gitelman, Lisa. 2014. *Paper Knowledge: Toward a Media History of Documents*. Durham, NC: Duke University Press Books.

Gödel, Kurt. 1931. "Über Formal Unentscheidbare Sätze der Principia Mathematica und Verwandter Systeme I." *Monatshefte für Mathematik und Physik* 38, no. 1: 173–198.

Goferman, Stas, Lihi Zelnik-Manor, and Ayellet Tal. 2012. "Context-Aware Saliency Detection." *IEEE Transactions on Pattern Analysis and Machine Intelligence* 34, no. 10: 1915–1926.

Gold, Matthew K., ed. 2012. *Debates in the Digital Humanities*. Minneapolis: University of Minnesota Press.

Goldstine, Herman H. (1972) 1980. *The Computer from Pascal to von Neumann*. Princeton, NJ: Princeton University Press.

Goldstine, Herman H., and John von Neumann. 1947. *Planning and Coding of Problems for an Electronic Computing Instrument: Report on the Mathematical and Logical Aspects of an Electronic Computing Instrument*. Princeton NJ: Institute for Advanced Study.

Good, Andrew. 2017. "An Algorithm Helps Protect Mars Curiosity's Wheels." *National Aeronautic and Space Administration*, June 29. https://www.nasa.gov/feature /jpl/an-algorithm-helps-protect-mars-curiositys-wheels (last accessed October 2017).

Gooday, Graeme. 1990. "Precision Measurement and the Genesis of Physics Teaching Laboratories in Victorian Britain." *The British Journal for the History of Science* 23, no. 1: 25–51.

Gooding, David, Trevor Pinch, and Simon Schaffer, eds. 1989. *The Uses of Experiment: Studies in the Natural Sciences*. Cambridge: Cambridge University Press.

Goody, Jack. 1977. *The Domestication of the Savage Mind*. Cambridge: Cambridge University Press.

Gray, Mary L., and Siddharth Suri. 2019. *Ghost Work: How to Stop Silicon Valley from Building a New Global Underclass*. Boston: Houghton Mifflin Harcourt.

Gray, Robert. 1984. "Vector Quantization." *IEEE ASSP Magazine* 1, no. 2: 4–29.

Green, Thomas R. G. 1977. "Conditional Program Statements and Their Comprehensibility to Professional Programmers." *Journal of Occupational Psychology* 50, no. 2: 93–109.

Green, Thomas R. G. 1980. "Programming as a Cognitive Activity." In *Human Interaction with Computers*, edited by Harold T. Smith and Thomas R. G. Green, 277–320. London: Academic Press.

Greimas, Algirdas J. 1983. *Structural Semantics: An Attempt at a Method*. Lincoln: University of Nebraska Press.

Grier, David A. 2005. *When Computers Were Human*. Princeton, NJ: Princeton University Press.

Grimson, W. L. Eric. 1986. "The Combinatorics of Local Constraints in Model-Based Recognition and Localization from Sparse Data." *Journal of the ACM* 33, no. 4: 658–686.

Grimson, Eric, and Tomas Lozano-Perez. 1983. "Model-Based Recognition and Localization from Sparse Range or Tactile Data." *The International Journal of Robotics Research* 3, no. 3: 3–35.

Grosman, Jérémy, and Tyler Reigeluth. 2019. "Perspectives on Algorithmic Normativities: Engineers, Objects, Activities." *Big Data & Society* 6, no. 2: 2053951719858742.

Guo, Hongwei. 2011. "A Simple Algorithm for Fitting a Gaussian Function." *IEEE Signal Processing Magazine* 28, no. 5: 134–137.

Gurvitz, Yossi. 2017. "When Kafka Met Orwell: Arrest by Algorithm." *Mondoweiss*, July 3. https://mondoweiss.net/2017/07/orwell-arrest-algorithm/.

Hacking, Ian. 1983. *Representing and Intervening: Introductory Topics in the Philosophy of Natural Science*. Cambridge: Cambridge University Press.

Hacking, Ian. 2014. *Why Is There Philosophy of Mathematics at All?* Cambridge: Cambridge University Press.

Hagen, Nathan, and Eustace L. Dereniak. 2008. "Gaussian Profile Estimation in Two Dimensions." *Applied Optics* 47, no. 36: 6842–6851.

Haigh, Thomas. 2008. "Cleve Moler: Mathematical Software Pioneer and Creator of Matlab." *IEEE Annals of the History of Computing* 30, no. 1: 87–91.

Haigh, Thomas. 2011. "Charles W. Bachman: Database Software Pioneer." *IEEE Annals of the History of Computing* 33, no. 4: 70–80.

Haigh, Thomas, Mark Priestley, and Crispin Rope. 2014. "Los Alamos Bets on ENIAC: Nuclear Monte Carlo Simulations, 1947–1948." *IEEE Annals of the History of Computing* 36, no. 3: 42–63.

Haigh, Thomas, Mark Priestley, and Crispin Rope. 2016. *ENIAC in Action: Making and Remaking the Modern Computer*. Cambridge, MA: MIT Press.

Hallinan, Blake, and Ted Striphas. 2014. "Recommended for You: The Netflix Prize and the Production of Algorithmic Culture." *New Media & Society* 18, no. 1: 117–137.

Hamilton, William R. 1844. "On Quaternions; Or on a New System of Imaginaries in Algebra." *The London, Edinburg and Dublin Philosophical Magazine and Journal of Science* 25: 1–13.

Hankins, Thomas L. 1980. *Sir William Rowan Hamilton*. Baltimore: Johns Hopkins University Press.

Haraway, Donna. 1992. "The Promises of Monsters: A Regenerative Politics for Inappropriate/d Others." In *Cultural Studies*, edited by Lawrence Grossberg, Carry Nelson, and Paula A. Treichler, 295–337. New York: Routledge.

Haraway, Donna. 1997. *Modest_Witness@Second_Millenium: Female_Man©_Meets_OncomouseTM: Feminism and Technoscience*. New York: Routledge.

Hardt, Michael. 1999. "Foreword: Three Keys to Understanding Constituent Power." In Antonio Negri, *Insurgencies. Constituent Power and the Modern State*, vii–xiii. Minneapolis: University of Minnesota Press.

Hars, Alexander, and Shaosong Ou. 2001. "Working for Free? Motivations of Participating in Open Source Projects." *International Journal of Electric Commerce* 6, no. 3: 25–39.

Haugeland, John. 1989. *Artificial Intelligence: The Very Idea*. Reprint ed. Cambridge, MA: Bradford Book.

Haugeland John. 2000. *Having Thought: Essays in the Metaphysics of Mind*. New ed. Cambridge, MA: Harvard University Press.

Hayles, Katherine N. 1999. *How We Became Posthuman: Virtual Bodies in Cybernetics, Literature, and Informatics*. Chicago: University of Chicago Press.

He, Kaiming, Xiangyu Zhang, Shaoqing Ren, and Jian Sun. 2016. "Deep Residual Learning for Image Recognition." In *2016 IEEE Conference on Computer Vision and Pattern Recognition*, Las Vegas, NV, June–July, 770–778. New York: IEEE.

Heath, Thomas. 1981a. *A History of Greek Mathematics, Volume I: From Thales to Euclid*. Revised ed. New York: Dover Publications.

Heath, Thomas. 1981b. *A History of Greek Mathematics, Volume II: From Aristarchus to Diophantus*. Revised ed. New York: Dover Publications.

Heawood, Percy J. 1890. "Map-Colour Theorem." *Quarterly Journal of Mathematics* 24: 332–339.

Hebb, Donald O. 1949. *The Organization of Behaviour: A Neuropsychological Theory*. New York: Wiley.

Heinke, Dietmar, and Glyn W. Humphreys. 2004. "Computational Models of Visual Selective Attention: A Review." In *Connectionist Models in Cognitive Psychology*, edited by George Houghton, 273–312. London: Psychology Press.

Hennion, Antoine. 2015. *The Passion for Music: A Sociology of Mediation*. Farnham: Ashgate Publishing.

Hennion, Antoine. 2017. "Attachments, You Say? How a Concept Collectively Emerges in One Research Group." *Journal of Cultural Economy* 10, no. 1: 112–121.

Henriksen, Anne, and Anja Bechmann. 2020. "Building Truths in AI: Making Predictive Algorithms Doable in Healthcare." *Information, Communication & Society* 23, no. 6: 802–816.

Hesseling, Dennis E. 2004. *Gnomes in the Fog: The Reception of Brouwer's Intuitionism in the 1920s*. Basel: Birkhäuser.

Hine, Christine. 2008. *Systematics as Cyberscience: Computers, Change, and Continuity in Science*. Cambridge, MA: MIT Press.

Hinton, Geoffrey E., Terrence J. Sejnowski, and David H. Ackley. 1984. *Boltzmann Machines: Constraints Satisfaction Networks That Learn*. Technical Report No. CMU-CS-84-119. Pittsburgh, PA: Carnegie-Mellon University.

Hjelmås, Erik, and Boon K. Low. 2001. "Face Detection: A Survey." *Computer Vision and Image Understanding* 83, no. 3: 236–274.

Hoffman, Donna L., and Thomas P. Novak. 1998. "Bridging the Racial Divide on the Internet." *Science* 280, no. 5362: 390–391.

Hollan, James, Edwin Hutchins, and David Kirsh. 2000. "Distributed Cognition: Toward a New Foundation for Human-Computer Interaction Research." *ACM Transactions on Computer-Human Interaction* 7, no. 2: 174–196.

Hopfield, John J. 1982. "Neural Networks and Physical Systems with Emergent Collective Computational Abilities." *Proceedings of the National Academy of Sciences* 79, no. 8: 2554–2558.

Howe, Jeff. 2006. "The Rise of Crowdsourcing." *Wired*, June 1. https://www.wired.com/2006/06/crowds/.

Hudson, Graham, Alain Léger, Birger Niss, and István Sebestyén. 2017. "JPEG at 25: Still Going Strong." *IEEE MultiMedia* 24, no. 2: 96–103.

Hughes, Thomas Parke. 1983. *Networks of Power: Electrification in Western Society, 1880–1930*. Baltimore: Johns Hopkins University Press.

Hurley, Susan L. 2002. *Consciousness in Action*. Cambridge, MA: Harvard University Press.

Husserl, Edmund. 2012. *Philosophy of Arithmetic: Psychological and Logical Investigations with Supplementary Texts from 1887–1901*. Berlin: Springer Science & Business Media.

Hutchins, Edwin. 1995. *Cognition in the Wild*. Cambridge, MA: MIT Press.

Iacoboni, Marco. 2001. "Playing Tennis with the Cerebellum." *Nature Neuroscience* 4, no. 6: 555–556.

Ingold, Tim. 2014. "That's Enough about Ethnography!" *HAU: Journal of Ethnographic Theory* 4, no. 1: 383–395.

Introna, Lucas D. 2016. "Algorithms, Governance, and Governmentality: On Governing Academic Writing." *Science Technology Human Values* 41, no. 1: 17–49.

Introna, Lucas D., and Helen Nissenbaum. 2000. "Shaping the Web: Why the Politics of Search Engines Matters." *The Information Society* 16, no. 3: 169–185.

Introna, Lucas D., and David Wood. 2002. "Picturing Algorithmic Surveillance: The Politics of Facial Recognition Systems." *Surveillance & Society* 2, no. 2–3: 177–198.

Irani, Lilly. 2015. "Difference and Dependence among Digital Workers: The Case of Amazon Mechanical Turk." *South Atlantic Quaterly* 114, no. 1: 225–234.

Isaac, Mike. 2016. "Facebook, in Cross Hairs after Election, Is Said to Question Its Influence." *New York Times*, November 12. https://www.nytimes.com/2016/11/14 /technology/facebook-is-said-to-question-its-influence-in-election.html.

Isaac, Mike, and Sydney Ember. 2016. "Shocker! Facebook Changes Its Algorithm to Avoid 'Clickbait.'" *New York Times*, August 4. https://www.nytimes.com/2016/08/05 /technology/facebook-moves-to-push-clickbait-lower-in-the-news-feed.html.

Itti, Laurent. 2000. "Models of Bottom-Up and Top-Down Visual Attention." PhD diss., California Institute of Technology.

Itti, Laurent, and Christof Koch. 2001. "Computational Modelling of Visual Attention." *Nature Reviews Neuroscience* 2, no. 3: 194–203.

Itti, Laurent, Christof Koch, and Jochen Braun. 2000. "Revisiting Spatial Vision: Toward a Unifying Model." *Journal of the Optical Society of America: A, Optics, Image Science, and Vision* 17, no. 11: 1899–1917.

Itti, Laurent, Christof Koch, and Ernst Niebur. 1998. "A Model of Saliency-Based Visual Attention for Rapid Scene Analysis." *IEEE Transactions on Pattern Analysis and Machine Intelligence* 20, no. 11: 1254–1259. https://doi.org/10.1109/34.730558.

Jacobs, John F. 1986. *The SAGE Air Defense Systems: A Personal History*. Bedford, MA: MITRE Corporation.

James, William. 1909. *A Pluralistic Universe: Hibbert Lectures to Manchester College on the Present Situation in Philosophy*. London: Longmans, Green.

James, William. (1912) 2003. *Essays in Radical Empiricism*. Mineola, NY: Dover Publications.

Jasanoff, Sheila. 2012. "Genealogies of STS." *Social Studies of Science* 42, no. 3: 435–441.

Jaton, Florian. 2017. "We Get the Algorithms of Our Ground Truths: Designing Referential Databases in Digital Image Processing." *Social Studies of Science* 47, no. 6: 811–840.

Jaton, Florian. 2019. "Pardonnez cette platitude: de l'intérêt des ethnographies de laboratoire pour l'étude des processus algorithmiques." *Zilsel* 5: 315–339.

Jaton, Florian, and Dominique Vinck. 2016. "Unfolding Frictions in Database Projects." *Revue d'anthropologie des connaissances* 10, no. 4: a–m.

Jaton, Florian, and Dominique Vinck. Forthcoming. "Politicizing Algorithms by Other Means: Toward Inquiries for Affective Dissension." *Perspectives on Science.*

Jeffries, Robin, Althea A. Turner, Peter G. Polson, and Michael E. Atwood. 1981. "The Processes Involved in Designing Software." In *Cognitive Skills and Their Acquisition*, edited by John R. Anderson, 255–283. Hillsdale, NJ: Lawrence Erlbaum.

Jellinek, Georg. (1914) 2016. *Allgemeine Staatslehre Und Politik: Vorlesungsmitschrift Von Max Ernst Mayer Aus Dem Sommersemester*, edited by Andreas Funke and Sascha Ziemann. Tübingen: Mohr Siebrek Ek.

Jennions, Michael D., and Anders Pape Møller. 2003. "A Survey of the Statistical Power of Research in Behavioral Ecology and Animal Behavior." *Behavioral Ecology* 14, no. 3: 438–445.

Jet Propulsion Laboratory (JPL). 2015. "NASA Facts: Mars Exploration Rover." NASA Facts, JPL 400-1537. https://www.jpl.nasa.gov/news/fact_sheets/mars-science -laboratory.pdf (last accessed October 2017).

Jiang, Bowen, Lihe Zhang, Huchuan Lu, Chuan Yang, and Ming-Hsuan Yang. 2013. "Saliency Detection via Absorbing Markov Chain." In *2013 IEEE International Conference on Computer Vision*, Sydney, Australia, December, 1665–1672. New York: IEEE.

Jones, Lyle V., and David Thissen. 2006. "A History and Overview of Psychometrics." In *Handbook of Statistics*, edited by C. R. Rao and S. Sinharay, 1–27. Amsterdam: Elsevier.

Jones, Matthew L. 2018. "How We Became Instrumentalists (Again): Data Positivism since World War II." *Historical Studies in the Natural Sciences* 48, no. 5: 673–684.

Jordan, Michael I., and Tom M. Mitchell. 2015. "Machine Learning: Trends, Perspectives, and Prospects." *Science* 349, no. 6245 (July 17): 255–260. https://doi.org/10 .1126/science.aaa8415.

Judd, Tilke, Frédo Durand, and Antonio Torralba. A Benchmark of Computational Models of Saliency to Predict Human Fixations. Report No. MIT-CSAIL-TR-2012-001. Cambridge, MA: MIT. http://dspace.mit.edu/handle/1721.1/68590 (last accessed January 2017).

Kamavisdar, Pooja, Sonam Saluja, and Sonu Agrawal. 2013. "A Survey on Image Classification Approaches and Techniques." *International Journal of Advanced Research in Computer and Communication Engineering* 2, no. 1: 1005–1009.

Kammann, Richard. 1975. "The Comprehensibility of Printed Instructions and the Flowchart Alternative." *Human Factors: The Journal of the Human Factors and Ergonomics Society* 17, no. 2: 183–191.

Karthikeyan, Shanmugavadivel, Vignesh Jagadeesh, and B. S. Manjunath. 2013. "Learning Top Down Scene Context for Visual Attention Modelling in Natural

Images." In *2013 IEEE International Conference on Image Processing*, Melbourne, Victoria, Australia, September, 211–215. New York: IEEE.

Kelsen, Hans. 1991. *General Theory of Norms*. Oxford: Clarendon Press.

Kempe, Alfred B. 1879. "On the Geographical Problem of the Four Colours." *American Journal of Mathematics* 2, no. 3: 193–200.

Kent, Deborah. 2008. "The Mathematical Miscellany and The Cambridge Miscellany of Mathematics: Closely Connected Attempts to Introduce Research-Level Mathematics in America, 1836–1843." *Historia Mathematica* 35, no. 2: 102–122.

Klein, Philip N. 2013. *Coding the Matrix: Linear Algebra through Applications to Computer Science*. 1st ed. London: Newtonian Press.

Kline, Morris. 1990a. *Mathematical Thought from Ancient to Modern Times*, Volume 1. New ed. New York: Oxford University Press.

Kline, Morris. 1990b. *Mathematical Thought from Ancient to Modern Times*. Volume 2. New ed. New York: Oxford University Press.

Kline, Morris. 1990c. *Mathematical Thought from Ancient to Modern Times,* Volume. 3. New ed. New York: Oxford University Press.

Kling, Rob, ed. 1996. *Computerization and Controversy: Value Conflicts and Social Choices*. 2nd ed. San Diego, CA: Morgan Kaufmann.

Knorr-Cetina, Karin D. 1981. *The Manufacture of Knowledge: An Essay on the Constructivist and Contextual Nature of Science*. New York: Pergamon Press.

Knorr-Cetina, Karin D. 1999. *Epistemic Cultures: How the Sciences Make Knowledge*. Cambridge, MA: Harvard University Press.

Knorr-Cetina, Karin D., and Michael J. Mulkay. 1983. *Science Observed: Perspectives on the Social Study of Science*. London: Sage Publications.

Knuth, Donald E. 1992. *Literate Programming*. Stanford, CA: Center for the Study of Language and Information.

Knuth, Donald E. 1997a. *The Art of Computer Programming*. Volume 1: *Fundamental Algorithms*. 3rd ed. Reading, MA: Addison-Wesley Professional.

Knuth, Donald E. 1997b. *The Art of Computer Programming*. Volume 2: *Seminumerical Algorithms*. 3rd ed. Reading, MA: Addison-Wesley Professional.

Knuth, Donald E. 1998. *The Art of Computer Programming*. Volume 3: *Sorting and Searching*. 2nd ed. Reading, MA: Addison-Wesley Professional.

Knuth, Donald E. 2002. "All Questions Answered." *Notices of the AMS* 49, no. 3: 318–324.

Knuth, Donald E. 2011. *The Art of Computer Programming*. Volume 4A: *Combinatorial Algorithms, Part 1*. 1st ed. Upper Saddle River, NJ: Addison-Wesley Professional.

Koblitz, Neal. 2012. *A Course in Number Theory and Cryptography*. Berlin: Springer Science & Business Media.

Koch, Christof, and Shimon Ullman. 1985. "Shifts in Selective Visual Attention: Towards the Underlying Neural Circuitry." *Human Neurobiology* 4, no. 4: 219–227.

Kraemer, Felicitas, Kees van Overveld, and Martin Peterson. 2010. "Is There an Ethics of Algorithms?" *Ethics and Information Technology* 13, no. 3: 251–260.

Krizhevsky, Alex, Ilya Sutskever, and Geoffrey E. Hinton. 2012. "ImageNet Classification with Deep Convolutional Neural Networks." In *Proceedings of the 25th International Conference on Neural Information Processing Systems*, Stateline, NV, September, 1097–1105. Red Hook, NY: Curran Associates.

Kushner, Scott. 2013. "The Freelance Translation Machine: Algorithmic Culture and the Invisible Industry." *New Media & Society* 15, no. 8: 1241–1258.

Lakatos, Imre. 1976. *Proofs and Refutations: The Logic of Mathematical Discovery*. Cambridge: Cambridge University Press.

Landini, Francesca, and Giancarlo Navach. 2017. "Nutella Maker Fights Back on Palm Oil after Cancer Risk Study." *Reuters*, January 11. https://www.reuters.com/article/us-italy-ferrero-nutella-insight-idUSKBN14V0MK.

Lapowsky, Issie. 2016. "Here's How Facebook Actually Won Trump the Presidency." *Wired*. November 15. https://www.wired.com/2016/11/facebook-won-trump-election-not-just-fake-news/.

Lappin, Joseph S., and William R. Uttal. 1976. "Does Prior Knowledge Facilitate the Detection of Visual Targets in Random Noise?" *Perception & Psychophysics* 20, no. 5: 367–374.

Latour, Bruno. 1987. *Science in Action: How to Follow Scientists and Engineers through Society*. Cambridge, MA: Harvard University Press.

Latour, Bruno. 1992 "Where Are the Missing Masses? The Sociology of a Few Mundane Artifacts." In *Shaping Technology/Building Society: Studies in Sociotechnical Change*, edited by Wiebe E. Bijker and John Law, 225–258. Cambridge, MA: MIT Press.

Latour, Bruno. 1993a. *The Pasteurization of France*. Cambridge, MA: Harvard University Press.

Latour, Bruno. 1993b. *We Have Never Been Modern*. Cambridge, MA: Harvard University Press.

Latour, Bruno. 1996. "Sur les pratiques des théoriciens." In *Savoirs théoriques et savoirs pratiques*, edited by Jean-Marie Barbier, 131–146. Paris: PUF.

Latour, Bruno. 1999a. *Pandora's Hope: Essays on the Reality of Science Studies*. Cambridge, MA: Harvard University Press.

Latour, Bruno. 1999b. "Factures/Fractures: From the Concept of Network to the Concept of Attachment." *RES: Anthropology and Aesthetics* 36: 20–31.

Latour, Bruno. 2005. *Reassembling the Social: An Introduction to Actor-Network-Theory.* Oxford: Oxford University Press.

Latour, Bruno. 2006. *Petites leçons de sociologie des sciences.* Paris: La Découverte.

Latour, Bruno. 2008. "Review Essay: The Netz-Works of Greek Deductions." *Social Studies of Science* 38, no. 3: 441–459.

Latour, Bruno. 2010a. "An Attempt at a 'Compositionist Manifesto.'" *New Literary History* 41, no. 3: 471–490.

Latour, Bruno. 2010b. *Cogitamus: Six lettres sur les humanités scientifiques.* Paris: La Découverte.

Latour, Bruno. 2013. *An Inquiry into Modes of Existence: An Anthropology of the Moderns.* Translated by C. Porter. Cambridge, MA: Harvard University Press.

Latour, Bruno, Philippe Mauguin, and Geneviève Teil. 1992. "A Note on Socio-Technical Graphs." *Social Studies of Science* 22, no. 1: 33–57.

Latour, Bruno, and Steve Woolgar. 1986. *Laboratory Life: The Construction of Scientific Facts.* 2nd ed. Princeton, NJ: Princeton University Press.

Law, John, and John Urry. 2004. "Enacting the social." *Economy and Society* 33, no. 3: 390–410.

Lawrence, Steve, and C. Lee Giles. 1999. "Accessibility of Information on the Web." *Nature* 400, no. 6740: 107.

Lea, Tess, and Paul Pholeros. 2010. "This Is Not a Pipe: The Treacheries of Indigenous Housing." *Public Culture* 22, no. 1: 187–209.

Leadem, Rose. 2017. "Nutella's New Jars Are Designed by an Algorithm." *Entrepreneur*, June 5. https://www.entrepreneur.com/article/295350.

LeCun, Yann, Yoshua Bengio, and Geoffrey Hinton. 2015. "Deep Learning." *Nature* 521: 436–444.

LeCun, Yves. 1985. "A Learning Scheme for Asymmetric Threshold Networks." In *Proceedings of Cognitiva 85*, 599–604. Paris, France.

LeCun, Yves, B. Boser, J. S. Denker, D. Henderson, R. E. Howard, W. Hubbard, and L. D. Jackel. 1989. "Backpropagation Applied to Handwritten Zip Code Recognition." *Neural Computation* 1, no. 4: 541–551.

Lécuyer, Christophe, David C. Brock, and Jay Last. 2010. *Makers of the Microchip: A Documentary History of Fairchild Semiconductor.* Cambridge, MA: MIT Press.

Leese, Matthias. 2014 "The New Profiling: Algorithms, Black Boxes, and the Railure of Anti-Discriminatory Safeguards in the European Union." *Security Dialogue* 45, no. 5: 494–511.

Lefebvre, Muriel. 2001. "Écritures et Espace de Médiation: Étude Anthropologique Des Pratiques Graphiques Dans Une Communauté de Mathématiciens." PhD diss., Université de Strasbourg, Strasbourg, France.

Lefebvre, Muriel. 2013. "L'infra-ordinaire de la recherche: Écritures scientifiques personnelles, archives et mémoire de la recherche." *Sciences de la société*, no. 89: 3–17.

Lehr, David, and Paul Ohm. 2017. "Playing with the Data: What Legal Scholars Should Learn about Machine Learning." *U.C. Davis Law Review* 51: 653–717.

Leighton, Robert B., Norman H. Horowitz, Bruce C. Murray, Robert P. Sharp, Alan G. Herriman, Andrew T. Young, Bradford A. Smith, Merton E. Davies, and Conway B. Leovy. 1969. "Mariner 6 Television Pictures: First Report." *Science* 165, no. 3894: 685–690.

Lenglet, Marc. 2011. "Conflicting Codes and Codings: How Algorithmic Trading Is Reshaping Financial Regulation." *Theory, Culture & Society* 28, no. 6: 44–66.

Lépinay, Vincent A. 2011. *Codes of Finance: Engineering Derivatives in a Global Bank.* Princeton, NJ: Princeton University Press.

Lerner, Gerda. 1986. *The Creation of Patriarchy.* Oxford: Oxford University Press.

Lerner, Josh, and Jean Tirole. 2002. "Some Simple Economics of Open Source." *Journal of Industrial Economics* 50, no. 2: 197–234.

Lettvin, Jerome L. 1989. "Introduction." In *Collected Works of Warren McCulloch*, edited by Rook McCulloch, 7–20. Salinas, CA: Intersystems.

Levin, Sam. 2017. "New AI can guess whether you're gay or straight from a photograph." *Guardian*, September 8. https://www.theguardian.com/technology/2017/sep/07/new-artificial-intelligence-can-tell-whether-youre-gay-or-straight-from-a-photograph.

Lévy, Pierre. 1995. "The Invention of the Computer." In *A History of Scientific Thought. Elements of a History of Science*, edited by Michel Serres, 636–663. Oxford: Blackwell.

Lewis, Seth C, and Oscar Westlund. 2014. "Big Data and Journalism: Epistemology, Expertise, Economics, and Ethics." *Digital Journalism* 3, no. 3: 447–466.

Light, Jennifer S. 1999. "When Computers Were Women." *Technology and Culture* 40, no. 3: 455–483.

Lippmann, Walter. (1925) 1993. *The Phantom Public.* Reprint ed. New Brunswick, NJ: Transaction Publishers.

Lippmann, Walter. 1982. *The Essential Lippmann: A Political Philosophy for Liberal Democracy*. Cambridge, MA: Harvard University Press.

Liptak, Adam. 2017. "Sent to Prison by a Software Program's Secret Algorithms." *New York Times*, May 1. https://www.nytimes.com/2017/05/01/us/politics/sent-to-prison-by-a-software-programs-secret-algorithms.html.

Little, Anthony C., Benedict C. Jones, and Lisa M. DeBruine. 2011. "The Many Faces of Research on Face Perception." *Philosophical Transactions of the Royal Society B: Biological Sciences* 366, no. 1571: 1634–1637.

Liu, Tie, Jian Sun, Nan-Ning Zheng, Xiaoou Tang, and Heung-Yeung Shum. 2007. "Learning to Detect a Salient Object." In *Proceedings of the 2007 IEEE Conference on Computer Vision and Pattern Recognition*, Minneapolis, MN, June, 1–8. New York: IEEE.

Lloyd, Geoffrey E. R. 1990. *Demystifying Mentalities*. Cambridge: Cambridge University Press.

Lloyd, Geoffrey E. R. 2005. *The Delusions of Invulnerability: Wisdom and Morality in Ancient Greece, China and Today*. London: Duckworth.

Lorber, Judith, and Susan A. Farrell, eds. 1991. *The Social Construction of Gender*. Newbury Park, CA: Sage Publications.

Lowe, David G. 1987. "Three-Dimensional Object Recognition from Single Two-Dimensional Images." *Artificial Intelligence*. 31, no. 3: 355–395.

Lowe, David G. 1999. "Object Recognition from Local Scale-Invariant Features." In *Proceedings of the International Conference on Computer Vision, Kerkyra, Corfu, Greece, September 20–25, 1999*, 1150–1157. Washington, DC: IEEE Computer Society.

Lucas, H. C., and R. B. Kaplan. 1976. "A Structured Programming Experiment." *The Computer Journal* 19, no. 2: 136–138.

Lynch, Michael. 1985. *Art and Artifact in Laboratory Science: A Study of Shop Work and Shop Talk in a Research Laboratory*. London: Routledge Kegan & Paul.

Lynch, Michael. 2014. "From Normative to Descriptive and Back: Science and Technology Studies and the Practice Turn." In *Science after the Practice Turn in the Philosophy, History, and Social Studies of Science*, edited by Léna Soler, Sjoerd Zwart, Michael Lynch, and Vincent Israel-Jost, 93–113. London: Routledge.

Lyon, Richard F. 2006. "A Brief History of 'Pixel.'" In *Proceedings of SPIE Digital Photography II*, edited by Nitin Sampat, Jeffrey M. Dicarlo and Russel A. Martin, 1–15. Bellingham, WA: SPIE Press.

Lyotard, Jean-François. 1984. *The Postmodern Condition: A Report on Knowledge*. Minneapolis: University of Minnesota Press.

Ma, Yu-Fei, and Hong-Jiang Zhang. 2003. "Contrast-Based Image Attention Analysis by Using Fuzzy Growing." In *Proceedings of the Eleventh ACM International Conference on Multimedia*, Berkeley, CA, November, 374–381. New York: ACM.

Mackenzie, Adrian. 2017. *Machine Learners: Archaeology of a Data Practice*. Cambridge, MA: MIT Press.

Mackenzie, Adrian, and Simon Monk. 2004. "From Cards to Code: How Extreme Programming Re-Embodies Programming as a Collective Practice." *Computer Supported Cooperative Work* 13, no. 1: 91–117.

MacKenzie, Donald. 1993. "Negotiating Arithmetic, Constructing Proof: The Sociology of Mathematics and Information Technology." *Social Studies of Science* 23, no. 1: 37–65.

MacKenzie, Donald. 1999. "Slaying the Kraken: The Sociohistory of a Mathematical Proof." *Social Studies of Science* 29, no. 1: 7–60.

MacKenzie, Donald. 2000. "A Worm in the Bud? Computers, Systems, and the Safety-Case Problem." In *Systems, Experts, and Computers: The Systems Approach in Management and Engineering, World War II and After*, edited by Agatha C. Hughes and Thomas P. Hughes, 161–190. Cambridge, MA: MIT Press.

MacKenzie, Donald. 2004. *Mechanizing Proof: Computing, Risk, and Trust*. Cambridge, MA: MIT Press.

MacKenzie, Donald. 2006. "Computers and the Sociology of Mathematical Proof." In *18 Unconventional Essays on the Nature of Mathematics*, edited by Reuben Hersh, 128–146. New York: Springer Science & Business Media.

MacKenzie, Donald. 2014. "A Sociology of Algorithms: High-Frequency Trading and the Shaping of Markets." Working paper, University of Edinburgh. http://www.sps.ed.ac.uk/__data/assets/pdf_file/0004/156298/Algorithms25.pdf (last accessed March 2017).

MacKenzie, Donald, Fabian Muniesa, and Lucia Siu, eds. 2007. *Do Economists Make Markets? On the Performativity of Economics*. Princeton, NJ: Princeton University Press.

Mackworth, Alan K., and Eugene C. Freuder. 1985. "The Complexity of Some Polynomial Network Consistency Algorithms for Constraint Satisfaction Problems." *Artificial Intelligence*. 25, no. 1: 65–74.

MacRae, Norman. 1999. *John Von Neumann: The Scientific Genius Who Pioneered the Modern Computer, Game Theory, Nuclear Deterrence, and Much More*. 2nd ed. Providence, RI: American Mathematical Society.

Mahdawi, Arwa. 2018. "To a man with an algorithm all things look like an advertising opportunity." *Guardian*, December 15. https://www.theguardian.com/commentisfree/2018/dec/15/week-in-patriarchy-facebook-parenting-advertising.

Malafouris, Lambros. 2004. "The Cognitive Basis of Material Engagement: Where Brain, Body and Culture Conflate." In *Rethinking Materiality: The Engagement of Mind with the Material World*, edited by Elizabeth DeMarrais, Chris Gosden, and Colin Renfrew, 53–62. Cambridge: McDonald Institute for Archeological Research.

Mancosu, Paolo, ed. 1997. *From Brouwer to Hilbert: The Debate on the Foundations of Mathematics in the 1920s*. Oxford: Oxford University Press.

Markoff, John. 2012. "For Web Images, Creating New Technology to Seek and Find." *New York Times*, November 19. https://www.nytimes.com/2012/11/20/science/for -web-images-creating-new-technology-to-seek-and-find.html.

Marres, Noortje. 2005. "Issues Spark a Public into Being: A Key but Often Forgotten Point of the Lippmann-Dewey Debate." In *Making Things Public*, edited by Bruno Latour and Peter Weibel, 208–217. Cambridge, MA: MIT Press.

MATLAB Answers. 2017. "How Does Matlab's Gaussian Fit Function Select Peak Centers?" MathWorks.com. https://www.mathworks.com/matlabcentral/answers/342610 -how-does-matlab-s-gaussian-fit-function-select-peak-centers (last accessed March 2018).

Mauchly, John W. (1942) 1982. "The Use of High Speed Vacuum Tube Devices for Calculating." In *The Origins of Digital Computers*, edited by Brian Randell, 355–358. Berlin: Springer.

Mayer, Richard E. 1976. "Comprehension as Affected by Structure of Problem Representation." *Memory & Cognition* 4, no. 3: 249–255.

Mazzotti, Massimo. 2017. "Algorithmic Life." *Los Angeles Review of Books*, January 20. https://lareviewofbooks.org/article/algorithmic-life/.

McCulloch, Warren S., and Walter Pitts. (1943) 1990. "A Logical Calculus of the Ideas Immanent in Nervous Activity." *The Bulletin of Mathematical Biophysics* 5, no. 4: 115–133.

McGee, Kyle. 2015. *Latour and the Passage of Law*. Edinburgh: Edinburgh University Press.

McKeithen, Katherine B., Judith S. Reitman, Henry H. Rueter, and Stephen C. Hirtle. 1981. "Knowledge Organization and Skill Differences in Computer Programmers." *Cognitive Psychology* 13, no. 3: 307–325.

Merleau-Ponty, Maurice. 2013. *Phenomenology of Perception*. Abingdon: Routledge, 2013.

Mialet, Hélène. 2012. *Hawking Incorporated: Stephen Hawking and the Anthropology of the Knowing Subject*. Chicago: University of Chicago Press.

Michalski, Ryszard S., Jaime G. Carbonell, and Tom M. Mitchell. 2014. *Machine Learning: An Artificial Intelligence Approach*. Amsterdam: Elsevier.

Minsky, Marvin, and Seymour A. Papert. 1969. *Perceptrons: An Introduction to Computational Geometry*. Cambridge, MA: MIT Press.

Minsky, Marvin, and Seymour A. Papert. 1970. "Proposal to ARPA for Research on Artificial Intelligence at MIT, 1970–1971." Artificial Intelligence Lab Publication, memo no. 185, MIT.

Mirowski, Philip. 2002. *Machine Dreams: Economics Becomes a Cyborg Science*. Cambridge: Cambridge University Press.

Mody, Cyrus C. 2017. *The Long Arm of Moore's Law: Microelectronics and American Science*. Cambridge, MA: MIT Press.

Moher, Thomas, and Michael G. Schneider. 1981. "Methods for Improving Controlled Experimentation in Software Engineering." In *Proceedings of the 5th International Conference on Software Engineering*, San Diego, CA, March, 224–233. New York: IEEE.

Mol, Annemarie. 2002. *The Body Multiple: Ontology in Medical Practice*. Durham, NC: Duke University Press.

Montfort, Nick, Patsy Baudoin, John Bell, Ian Bogost, Jeremy Douglass, Mark C. Marino, Michael Mateas, Casey Reas, Mark Sample, and Noah Vawter. 2013. *10 PRINT CHR$(205.5+RND(1));: GOTO 10*. Bellingham, WA: MIT Press.

Mosseri, Adam. 2017. "Showing More Informative Links in News Feed." *Facebook's Newsroom*, June 30. https://about.fb.com/news/2017/06/news-feed-fyi-showing-more-informative-links-in-news-feed/ (last accessed October 2017).

Movahedi, Vida, and James H. Elder. 2010. "Design and Perceptual Validation of Performance Measures for Salient Object Segmentation." In *2010 IEEE Computer Society Conference on Computer Vision and Pattern Recognition Workshops*, San Francisco, CA, June, 49–56. New York: IEEE.

Mozur, Paul. 2018. "Inside China's Dystopian Dreams: A.I., Shame and Lots of Cameras." *New York Times*, July 8. https://www.nytimes.com/2018/07/08/business/china-surveillance-technology.html.

Müller, Vincent C, ed. 2015. *Risks of Artificial Intelligence*. Boca Raton, FL: Chapman and Hall.

Muniesa, Fabian. 2011a. "Is a Stock Exchange a Computer Solution? Explicitness, Algorithms and the Arizona Stock Exchange." *International Journal of Actor-Network Theory and Technological Innovation* 3, no. 1: 1–15.

Muniesa, Fabian. 2011b. "A Flank Movement in the Understanding of Valuation." *Sociological Review* 59, no. 2: 24–38.

Muniesa, Fabian. 2015. *The Provoked Economy: Economic Reality and the Performative Turn*. London: Routledge.

Muniesa, Fabian, Yvan Millo, and Michel Callon. 2007. "An Introduction to Market Devices." In *Market Devices*, edited by Michel Callon, Yuval Millo, and Fabian Muniesa, 1–12. London: Blackwell.

Myers, Glenford J., Corey Sandler, and Tom Badgett. 2011. *The Art of Software Testing*. 3rd ed. Hoboken, NJ: Wiley.

Nagi, J., F. Ducatelle, G. A. Di Caro, D. Cireşan, U. Meier, A. Giusti, F. Nagi, J. Schmidhuber, and L. M. Gambardella. 2011. "Max-Pooling Convolutional Neural Networks for Vision-Based Hand Gesture Recognition." In *2011 IEEE International Conference on Signal and Image Processing Applications*, Kuala Lumpur, November, 342–347. New York: IEEE.

Nathan, Tobie, and Nathalie Zajde. 2012. *Psychothérapie démocratique*. Paris: Odile Jacob.

Naur, Peter, and Brian Randell. 1969. *Software Engineering: Report on a Conference Sponsored by the NATO Science Committee, Garmisch, Germany, 7th to 11th October 1968*. Brussels: NATO Scientific Affairs Division.

Negri, Antonio. 1999. *Insurgencies: Constituent Power and the Modern State*. Minneapolis: University of Minnesota Press.

Neisser, Ulric. 1967. *Cognitive Psychology*. Upper Saddle River, NJ: Prentice Hall.

Netz, Reviel. 1998. "Deuteronomic Texts: Late Antiquity and the History of Mathematics." *Revue D'Histoire Des Mathématiques* 4, no. 2: 261–288.

Netz, Reviel. 2003. *The Shaping of Deduction in Greek Mathematics: A Study in Cognitive History*. Cambridge: Cambridge University Press.

Netz, Reviel. 2004. *The Transformation of Mathematics in the Early Mediterranean World: From Problems to Equations*. Cambridge: Cambridge University Press.

Newell, Allen, and Herbert A. Simon. 1972. *Human Problem Solving*. Upper Saddle River, NJ: Prentice-Hall, 1972.

Neyland, Daniel. 2016. "Bearing Account-able Witness to the Ethical Algorithmic System." *Science Technology & Human Values* 41, no. 1: 50–76.

Nissenbaum, Helen. 2004. "Hackers and the Contested Ontology of Cyberspace." *New Media & Society* 6, no. 2: 195–217.

Noble, Safiya Umoja. 2018. *Algorithms of Oppression: How Search Engines Reinforce Racism*. New York: New York University Press.

Noë, Alva. 2004. *Action in Perception*. Cambridge, MA: MIT Press.

Nofre, David, Mark Priestley, and Gerard Alberts. 2014. "When Technology Became Language: The Origins of the Linguistic Conception of Computer Programming, 1950–1960." *Technology and Culture* 55, no. 1: 40–75.

Nudd, Tim. 2017. "Nutella's Unique Product Now Comes in 7 Million Unique Jars." *Adweek*, June 6. https://www.adweek.com/creativity/nutellas-unique-product-now-comes-in-7-million-unique-jars/.

Nye, David E. 1992. *Electrifying America: Social Meanings of a New Technology, 1880–1940*. Cambridge, MA: MIT Press.

Obermeyer, Ziad, Brian Powers, Christine Vogeli, and Sendhil Mullainathan. 2019. "Dissecting Racial Bias in an Algorithm Used to Manage the Health of Populations." *Science* 366, no. 6464: 447–453.

Ombredane, André, and Jean-Marie Faverge. 1955. *L'analyse du travail*. Paris: PUF.

O'Neal Jr., B. 1966. "Predictive Quantizing Systems (Differential Pulse Code Modulation) for the Transmission of Television Signals." *Bell System Technical Journal* 45, no. 5: 689–721.

O'Neil, Cathy. 2016. *Weapons of Math Destruction: How Big Data Increases Inequality and Threatens Democracy*. New York: Crown.

Ormerod, Tom. 1990. "Human Cognition and Programming." In *Psychology of Programming*, edited by J. M. Hoc, T. R. G. Green, R. Samurcay, and D.J. Gilmore, 63–82. London: Academic Press.

O'Shea, Donal. 2008. *The Poincaré Conjecture: In Search of the Shape of the Universe*. New York: Walker Books.

Otsu, Nobuyuki. 1979. "A Threshold Selection Method from Gray-Level Histograms." *IEEE Transactions on Systems, Man and Cybernetics* 9, no. 1: 62–66.

Owens, Larry. 1986. "Vannevar Bush and the Differential Analyzer: The Text and Context of an Early Computer." *Technology and Culture* 27, no. 1: 63–95.

Parker, Charlie. 2018. "It's Watching You. Police Big Brother Surveillance Technology to Spy on Your Social Media in Search for Hate Crime." *The Sun*, December 14. https://www.thesun.co.uk/news/7968627/big-brother-surveillance-technology-spy-social-media-police-search-hate-crime/.

Parrington, Norman, and Marc Roper. 1989. *Understanding Software Testing*. Chichester: John Wiley.

Pasquale, Frank. 2015. *The Black Box Society: The Secret Algorithms That Control Money and Information*. Cambridge, MA: Harvard University Press.

Pennington, Nancy. 1987. "Stimulus Structures and Mental Representations in Expert Comprehension of Computer Programs." *Cognitive Psychology* 19, no. 3: 295–341.

Pennington, Shelley, and Belinda Westover. 1989. "Types of Homework." In *A Hidden Workforce: Homeworkers in England, 1850–1985*, edited by Shelley Pennington and Belinda Westover, 44–65. London: Palgrave Macmillan UK.

Penny, Simon. 2017. *Making Sense: Cognition, Computing, Art, and Embodiment*. Cambridge, MA: MIT Press.

Penrose, Ann M., and Steven B. Katz. 2010. *Writing in the Sciences: Exploring Conventions of Scientific Discourse*. 3rd ed. New York: Longman.

Pérec, Georges. 1989. *L'Infra-ordinaire*. Paris: Seuil.

Pestre, Dominique. 2004. "Thirty Years of Science Studies: Knowledge, Society and the Political." *History and Technology: An International Journal* 20, no. 4: 351–369.

Piccinini, Gualtiero. 2004. "The First Computational Theory of Mind and Brain: A Close Look at McCulloch and Pitts's 'Logical Calculus of Ideas Immanent in Nervous Activity.'" *Synthese* 141, no. 2: 175–215.

Pickering, Andrew. 1995. *The Mangle of Practice: Time, Agency, and Science*. Chicago: University of Chicago Press.

Pickering, Andrew. 2011. *The Cybernetic Brain: Sketches of Another Future*. Chicago: University of Chicago Press.

Pickering, Andrew, and Adam Stephanides. 1992. "Constructing Quaternions: On the Analysis of Conceptual Practice." In *Science as Practice and Culture*, edited by Andrew Pickering, 139–167. Chicago: University of Chicago Press.

Plasek, Aaron. 2018. "On the Cruelty of Really Writing a History of Machine Learning." *IEEE Annals of the History of Computing* 38, no. 4: 6–8.

Polachek, Harry. 1997. "Before the ENIAC." *IEEE Annals of the History of Computing* 19, no. 2: 25–30.

Pu, Ida M. 2005. *Fundamental Data Compression*. Oxford: Butterworth-Heinemann.

Pugh, Emerson W. 1995. *Building IBM: Shaping an Industry and Its Technology*. Cambridge, MA: MIT Press.

Putnam, Hilary. (1961) 1980. "Brains and Behavior." In *Readings in Philosophy of Psychology*, edited by Ned Block, 24–36. Cambridge, MA: Harvard University Press.

Pylyshyn, Zenon W., ed. 1987. *The Robots Dilemma: The Frame Problem in Artificial Intelligence*. Norwood, NJ: Praeger.

Pylyshyn, Zenon W. 1989. "Computing in Cognitive Science." In *Foundations of Cognitive Science*, edited by Michael N. Posner, 63–91. Cambridge, MA: MIT Press.

Ramón y Cajal, Santiago. 1968. *The Structure of Ammon's Horn*. Springfield, IL: C. C. Thomas.

Ratcliffe, Matthew. 2009. "Belonging to the World through the Feeling Body." *Philosophy, Psychiatry, & Psychology* 16, no. 2: 205–211.

Ratcliffe, Matthew. 2010. "The Phenomenology of Mood and the Meaning of Life." In *The Oxford Handbook of Philosophy of Emotion*, edited by Peter Goldie, 349–371. Oxford: Oxford University Press.

Redmond, Kent C., and Thomas M. Smith. 1980. *Project Whirlwind: History of a Pioneer Computer*. 1st ed. Bedford, MA: Digital Press.

Redmond, Kent C., and Thomas M. Smith. 2000. *From Whirlwind to MITRE: The R&D Story of The SAGE Air Defense Computer*. Cambridge, MA: MIT Press.

Rheinberger, Hans-Jörg. 1997. *Toward a History of Epistemic Things: Synthesizing Proteins in the Test Tube*. Stanford, CA: Stanford University Press.

Richards, Martin. 2005. "EDSAC Initial Orders and Squares Program." University of Cambridge Computer Laboratory. http://www.cl.cam.ac.uk/~mr10/edsacposter.pdf (last accessed May 2016).

Richter, Felix. 2017. "Smartphones Cause Photography Boom." *Statista Infographics,* August 31. https://www.statista.com/chart/10913/number-of-photos-taken-worldwide / (last accessed January 2019).

Ringelhan, Stefanie, Jutta Wollersheim, and Isabell M. Welpe. 2015. "I Like, I Cite? Do Facebook Likes Predict the Impact of Scientific Work?" *PLOS ONE* 10, no. 8: e0134389.

Risen, James, and Laura Poitras. 2014. "N.S.A. Collecting Millions of Faces from Web Images." *New York Times*, May 31. https://www.nytimes.com/2014/06/01/us/nsa -collecting-millions-of-faces-from-web-images.html.

Ritter, James. 1995. "Measure for Measure: Mathematics in Egypt and Mesopotamia." In *History of Scientific Thought. Elements of a History of Science*, edited by Michel Serres, 44–72. Oxford: Blackwell.

Roberts, Rachel. 2017. "Online Hate Crime to Be Tackled by New National Police Hub, Home Secretary Says." *Independent*, October 8. https://www.independent.co.uk /news/uk/politics/online-hate-crime-amber-rudd-home-office-national-police-hub -facebook-twitter-trolls-a7988411.html.

Rorty, Richard. 1980. *Philosophy and the Mirror of Nature*. Princeton, NJ: Princeton University Press.

Rosenberg, Scott. 2008. *Dreaming in Code: Two Dozen Programmers, Three Years, 4,732 Bugs, and One Quest for Transcendent Software*. Reprint ed. New York: Three Rivers Press.

Rosenblatt, Frank. 1958. "The Perceptron: A Probabilistic Model for Information Storage and Organization in the Brain." *Psychological Review* 65, no. 6: 386–408.

Rosenblatt, Frank. 1962. *Principles of Neurodynamics: Perceptrons and the Theory of Brain Mechanisms*. New York: Spartan Books.

Rosental, Claude. 2003. *La trame de l'évidence: Sociologie de la demonstration en logique.* Paris: PUF.

Rosental, Claude. 2004. "Fuzzyfying the World: Social Practices of Showing the Properties of Fuzzy Logic." In *Growing Explanations: Historical Perspectives on Recent Science*, edited by Norton M. Wise, 159–178. Durham, NC: Duke University Press.

Rotman, Brian. 1995. "Thinking Dia-Grams: Mathematics, Writing, and Virtual Reality." *The South Atlantic Quarterly* 94, no. 2: 389–415.

Rotman, Brian. 2006. "Towards a Semiotics of Mathematics." In *18 Unconventional Essays on the Nature of Mathematics*, edited by Reuben Hersh, 97–127. New York: Springer Science & Business Media.

Rowan, Thomas C. 1957. "Psychological Tests and Selection of Computer Programmers." *Journal of the ACM* 4, no. 3: 348–353.

Rumelhart, David E., Geoffrey E. Hinton, and Ronald J. Williams. 1986. "Learning Representations by Back-Propagating Errors." *Nature* 323, no. 6088: 533.

Russakovsky, Olga, Jia Deng, Hao Su, Jonathan Krause, Sanjeev Satheesh, Sean Ma, Zhiheng Huang, Andrej Karpathy, Aditya Khosla, Michel Bernstein, Alexander C. Berg, and Li Fei-Fei. 2015. "ImageNet Large Scale Visual Recognition Challenge." *International Journal of Computer Vision* 115, no. 3: 211–252.

Sackman, Harold, W. J. Erikson, and E. E. Grant. 1968. "Exploratory Experimental Studies Comparing Online and Offline Programming Performance." *Communications of the ACM* 11, no. 1: 3–11.

Sandvig, Christian, Hamilton Kevin, Karahalios Karrie, and Cedric Langbort. 2016. "When the Algorithm Itself Is a Racist: Diagnosing Ethical Harm in the Basic Components of Software." *International Journal of Communication* 10: 4972–4990.

Santella, Anthony, Maneesh Agrawala, Doug DeCarlo, David Salesin, and Michael Cohen. 2006. "Gaze-Based Interaction for Semi-Automatic Photo Cropping." In *Proceedings of the SIGCHI Conference on Human Factors in Computing Systems*, Montreal, QC, Canada, April, 771–780. New York: ACM.

Scheiber, Noam. 2016. "Uber Drivers and Others in the Gig Economy Take a Stand." *New York Times*, February 2. https://www.nytimes.com/2016/02/03/business/uber -drivers-and-others-in-the-gig-economy-take-a-stand.html.

Schmidhuber, Jürgen. 2015. "Deep Learning in Neural Networks: An Overview." *Neural Networks* 61: 85–117.

Seaver, Nick. 2013. "Knowing Algorithms." Paper presented at Media in Transition 8, Cambridge, MA. https://static1.squarespace.com/static/55eb004ee4b0518639d59d9b /t/55ece1bfe4b030b2e8302e1e/1441587647177/seaverMiT8.pdf (last accessed April 2017).

Seaver, Nick. Forthcoming. *Computing Taste: The Making of Algorithmic Music Recommendation*. Chicago: University of Chicago Press.

Sedgewick, Robert, and Kevin Wayne. 2011. *Algorithms*. 4th ed. Upper Saddle River, NJ: Addison-Wesley Professional.

Seibel, Peter. 2009. *Coders at Work: Reflections on the Craft of Programming*. New York: Apress.

Seitz, Frederick, and Norman G. Einspruch. 1998. *Electronic Genie: The Tangled History of Silicon*. Urbana: University of Illinois Press.

Serres, Michel. 1974. *Hermès III: La traduction*. Paris: Editions de Minuit.

Serres, Michel. 1983. *Hermes: Literature, Science, Philosophy*. Baltimore: The Johns Hopkins University Press.

Serres, Michel. 1995. "Gnomon: The Beginnings of Geometry in Greece." In *History of Scientific Thought: Elements of a History of Science*, edited by Michel Serres, 77–123. Oxford: Blackwell.

Serres, Michel. 2002. *Origins of Geometry*. Manchester: Clinamen Press Limited.

Sha, Xin W. 2005. "Differential Geometrical Performance and Poiesis." *Configurations* 12, no. 1: 133–160.

Shannon, Claude E. 1948. "A Mathematical Theory of Communication." *Bell System Technical Journal* 27, no. 3: 379–423.

Shapin, Steven, and Simon Schaffer. 1989. *Leviathan and the Air-Pump: Hobbes, Boyle, and the Experimental Life*. Princeton, NJ: Princeton University Press.

Sharkey, Jim. 2017. "New Driving Algorithm Helps Protect Curiosity Rover's Wheels." *Spaceflight Insider*, July 4. https://www.spaceflightinsider.com/space-centers/jet-propulsion-laboratory/new-driving-algorithm-helps-protect-curiosity-rovers-wheels/ (last accessed October 2017).

Shen, Xiaohui, and Ying Wu. 2012. "A Unified Approach to Salient Object Detection via Low Rank Matrix Recovery." In *Proceedings of the 2012 IEEE Conference on Computer Vision and Pattern Recognition*, Providence, RI, June, 853–860. New York: IEEE.

Sheppard, Sylvia B., Bill Curtis, Phil Milliman, and Tom Love. 1979. "Modern Coding Practices and Programmer Performance." *Computer* 12, no. 12: 41–49.

Shiffrin, Richard M., and Gerald T. Gardner. 1972. "Visual Processing Capacity and Attentional Control." *Journal of Experimental Psychology* 93, no. 1: 72–82.

Shneiderman, Ben, and Richard Mayer. 1979. "Syntactic/Semantic Interactions in Programmer Behavior: A Model and Experimental Results." *International Journal of Computer & Information Sciences* 8, no. 3: 219–238.

Shneiderman, Ben, Richard Mayer, Don McKay, and Peter Heller. 1977. "Experimental Investigations of the Utility of Detailed Flowcharts in Programming." *Communications of the ACM* 20, no. 6: 373–381.

Sime, Max E., Andrew T. Arblaster, and Thomas G. Green. 1977. "Reducing Programming Errors in Nested Conditionals by Prescribing a Writing Procedure." *International Journal of Man-Machine Studies* 9, no. 1: 119–126.

Sime, Max E., Thomas G. Green, and D. J. Guest. 1973. "Psychological Evaluation of Two Conditional Constructions Used in Computer Languages." *International Journal of Man-Machine Studies* 5, no. 1: 105–113.

Sime, Max E., Thomas G. Green, and D. J. Guest. 1977. "Scope Marking in Computer Conditionals—A Psychological Evaluation." *International Journal of Man-Machine Studies* 9, no. 1: 107–118.

Simon, Herbert A., and Craig A. Kaplan. 1989. "Foundations of Cognitive Science." In *Foundations of Cognitive Science*, edited by Michael I. Posner, 1–47. Cambridge, MA: MIT Press.

Simondon, Gilbert. 2017. *On the Mode of Existence of Technical Objects*. Minneapolis, MN: Univocal Publishing.

Skiena, Steven S. 2008. *The Algorithm Design Manual*. 2nd ed. London: Springer.

Smith, Andrew. 2018. "Franken-Algorithms: The Deadly Consequences of Unpredictable Code." *Guardian*, August 30. https://www.theguardian.com/technology/2018/aug/29/coding-algorithms-frankenalgos-program-danger.

Smith, Blair R. 1983. "The IBM 701—Marketing and Customer Relations." *IEEE Annals of the History of Computing* 5, no. 2: 170–172.

Smith, Dorothy E. 1974. "The Social Construction of Documentary Reality." *Sociological Inquiry* 44, no. 4: 257–268.

Soloway, Elliot. 1986. "Learning to Program = Learning to Construct Mechanisms and Explanations." *Communications of the ACM* 29, no. 9: 850–858.

Sormani, Philippe. 2014. *Respecifying Lab Ethnography: An Ethnomethodological Study of Experimental Physics*. 1st ed. Farnham, UK: Routledge.

Souriau, Étienne. (1943) 2015. *The Different Modes of Existence*. Translated by E. Beranek and T. Howles. Minneapolis, MN: Univocal Publishing.

Srivastava, Biplav, and Francesca Rossi. 2018. "Towards Composable Bias Rating of AI Services." In *Proceedings of the 2018 AAAI/ACM Conference on AI, Ethics, and Society*, New Orleans, LA, February, 284–289. New York: ACM.

Star, Susan L. 1983. "Simplification in Scientific Work: An Example from Neuroscience Research." *Social Studies of Science* 13, no. 2: 205–228.

Star, Susan L. 1989. *Regions of the Mind: Brain Research and the Quest for Scientific Certainty*. Stanford, CA: Stanford University Press.

Star, Susan L., and Anselm Strauss. 1999. "Layers of Silence, Arenas of Voice: The Ecology of Visible and Invisible Work." *Computer Supported Cooperative Work* 8, no. 1–2: 9–30.

Statista. 2019. "Digital Still Cameras CIPA Company Shipments 1999–2018." *Statista .com*. https://www.statista.com/statistics/264337/cipa-companies-shipments-of-digital -cameras-since-1999/ (last accessed January 2019).

Steiner, Christopher. 2012. *Automate This: How Algorithms Came to Rule Our World*. New York: Penguin.

Stern, Nancy B. 1981. *From ENIAC to UNIVAC: Appraisal of the Eckert-Mauchly Computers*. Bedford, MA: Digital Press.

Strebel, Ignaz, Alain Bovet, and Philippe Sormani, eds. 2018. *Repair Work Ethnographies: Revisiting Breakdown, Relocating Materiality*. Basingstoke: Palgrave Macmillan.

Suchman, Lucy. 1987. *Plans and Situated Actions: The Problem of Human-Machine Communication*. Cambridge: Cambridge University Press.

Suchman, Lucy. 1995. "Making Work Visible." *Communications of the ACM* 38, no. 9: 56–64.

Suchman, Lucy. 2007. *Human-Machine Reconfigurations: Plans and Situated Actions*. 2nd ed. Cambridge: Cambridge University Press.

Suchman, Lucy, Dominik Gerst, and Hannes Krämer. 2019. "'If You Want to Understand the Big Issues, You Need to Understand the Everyday Practices That Constitute Them.' Lucy Suchman in Conversation with Dominik Gerst & Hannes Krämer." *Forum Qualitative Sozialforschung/Forum: Qualitative Social Research* 20, no. 2: Art. 1.

Sutton, John. 2007. "Batting, Habit, and Memory: The Embodied Mind and the Nature of Skill." *Sport in Society* 10, no. 5: 763–786.

Swade, Doron. 2011. "Inventing the User: EDSAC in Context." *The Computer Journal* 54, no. 1: 143–147.

Tarjan, Robert E. 1983. *Data Structures and Network Algorithms*. Philadelphia: SIAM.

Tent, M. B. W. 2006. *The Prince of Mathematics: Carl Friedrich Gauss*. Wellesley, MA: A. K. Peters/CRC Press.

Theureau, Jacques. 2003. "Course-of-Action Analysis and Course-of-Action Centered Design." In *Handbook of Cognitive Task Design*, edited by Erik Hollnagel, 55–81. Hillsdale, NJ: Lawrence Erlbaum.

Theureau, Jacques, and Geneviève Filippi. 2000. "Analysing Cooperative Work in an Urban Traffic Control Room for the Design of a Coordination Support System." In

Workplace Studies, edited by Paul Luff, Jon Hindmarsh, and Christian Heath, 68–81. Cambridge: Cambridge University Press.

Theureau, Jacques, Geneviève Filippi, Geneviève Saliou, and Pierre Vermersch. 2001. "Development of a Methodology for Analysing the Dynamic Collective Organisation of the Reactor Operator's and Supervisor's Courses of Experience While Controlling a Nuclear Reactor in Accidental Situations in Full Scope Simulated Control Rooms." In *CSAPC'01: Proceedings of the Eighth Conference on Cognitive Science Approaches to Process Control*, edited by R. Onken. Munich, September.

Thévenot, Laurent. 1984. "Rules and Implements: Investments in Forms." *Social Science Information* 23, no. 1: 1–45.

Thomas, Walker H. 1953. "Fundamentals of Digital Computer Programming." *Proceedings of the IRE* 41, no. 10: 1245–1249.

Thompson, Evan. 2005. "Sensorimotor Subjectivity and the Enactive Approach to Experience." *Phenomenology and the Cognitive Sciences* 4, no. 4: 407–427.

Thompson, Evan. 2010. *Mind in Life: Biology, Phenomenology, and the Sciences of Mind*. Cambridge, MA: Belknap Press.

Tiles, Mary. 2004. *The Philosophy of Set Theory: An Historical Introduction to Cantor's Paradise*. Mineola, NY: Dover Publications.

Traweek, Sharon. 1992. *Beamtimes and Lifetimes: The World of High Energy Physicists*. Cambridge, MA: Harvard University Press.

Tsotsos, John K. 1988. "A 'Complexity Level' Analysis of Immediate Vision." *International Journal of Computer Vision* 1, no. 4: 303–320.

Tsotsos, John K. 1989. "The Complexity of Perceptual Search Tasks." In *Proceedings of the Eleventh International Joint Conference on Artificial Intelligence*. Volume 2: 1571–1577. San Francisco, CA: Morgan Kaufmann.

Tsotsos, John K. 1990. "Analyzing Vision at the Complexity Level." *Behavioral and Brain Sciences* 13, no. 3: 423–445.

Tsotsos, John K., Scan M. Culhane, Winky Yan Kei Wai, Yuzhong Lai, Neal Davis, and Fernando Nuflo. 1995. "Modeling Visual Attention via Selective Tuning." *Artificial Intelligence* 78, no. 1–2: 507–545.

Turing, Alan M. 1937. "On Computable Numbers, with an Application to the Entscheidungsproblem." *Proceedings of the London Mathematical Society* 42, no. 1: 230–265.

Turing, Alan M. 1950. "Computing Machinery and Intelligence." *Mind* 59, no. 236: 433–460.

Ullman, Ellen. 2012a. *Close to the Machine: Technophilia and Its Discontents*. Reprint ed. New York: Picador.

Ullman, Ellen. 2012b. *The Bug*. New York: Picador.

Vandewalle Patrick, Jelena Kovacevic, and Martin Vetterli. 2009. "Reproducible Research in Signal Processing." *IEEE Signal Processing Magazine* 26, no. 3: 37–47.

Vapnik, Vladimir. 1999. *The Nature of Statistical Learning Theory*. 2nd ed. New York: Springer.

Varela, Francisco J., Evan T. Thompson, and Eleanor Rosch. 1991. *The Embodied Mind: Cognitive Science and Human Experience*. Revised ed. Cambridge, MA: MIT Press.

Vessey, Iris. 1989. "Toward a Theory of Computer Program Bugs: An Empirical Test." *International Journal of Man-Machine Studies* 30, no. 1: 23–46.

Vetterli, Martin, Jelena Kovacevic, and Vivek K. Goyal. 2014. *Foundations of Signal Processing*. Cambridge: Cambridge University Press.

Villani, Cédric. 2016. *Birth of a Theorem: A Mathematical Adventure*. Reprint ed. New York: Farrar, Straus and Giroux.

Vinck, Dominique. 1991. "La Coordination Du Travail Scientifique: Étude de Deux Formes Specifiques: Le Laboratoire et Le Reseau." PhD diss., École Nationale Supérieure des Mines de Paris, Paris, France.

Vinck, Dominique, ed. 2003. *Everyday Engineering: An Ethnography of Design and Innovation*. Cambridge, MA: MIT Press.

Vinck, Dominique. 2011. "Taking Intermediary Objects and Equipping Work into Account in the Study of Engineering Practices." *Engineering Studies* 3, no. 1: 25–44.

Vinck, Dominique. 2016. *Humanités numériques: La culture face aux nouvelles technologies*. Paris: Le Cavalier Bleu.

von Neumann, John. (1945) 1993. "First Draft of a Report on the EDVAC." *IEEE Annals of the History of Computing* 15, no. 4: 27–75.

von Neumann, John. (1958) 2012. *The Computer and the Brain*. 3rd ed. New Haven, CT: Yale University Press.

Vygotsky, Lev S. 1978. *Mind in Society: Development of Higher Psychological Processes*. Cambridge, MA: Harvard University Press.

Wade, Nicholas. 1981. *The Nobel Duel*. 1st ed. Garden City, NY: Doubleday.

Wang, Wei, Yizhou Wang, Qinming Huang, and Wen Gao. 2010. "Measuring Visual Saliency by Site Entropy Rate." In *2010 IEEE Conference on Computer Vision and Pattern Recognition*, San Francisco, CA, June, 2368–2375. New York: IEEE.

Wang, Zheshen, and Baoxin Li. 2008. "A Two-Stage Approach to Saliency Detection in Images." In *Proceedings of the IEEE International Conference on Acoustics, Speech and Signal Processing*, Las Vegas, NV, March–April, 965–968. New York: IEEE.

Ward, Dave, Tom Roberts, and Andy Clark. 2011. "Knowing What We Can Do: Actions, Intentions, and the Construction of Phenomenal Experience." *Synthese* 181, no. 3: 375–394.

Ward, Dave, and Mog Stapleton. 2012. "Es Are Good: Cognition as Enacted, Embodied, Embedded, Affective and Extended." In *Consciousness in Interaction: The Role of the Natural and Social Context in Shaping Consciousness*, edited by Fabio Paglieri, 89–104. Amsterdam: John Benjamins.

Warneken, Felix, and Alexandra G. Rosati. 2015. "Cognitive Capacities for Cooking in Chimpanzees." *Proceeding of the Royal Society. B: Biological Sciences* 282, no. 1809: 20150229.

Warwick, Andrew. 1992. "Cambridge Mathematics and Cavendish Physics: Cunningham, Campbell and Einstein's Relativity 1905–1911 Part I: The Uses of Theory." *Studies in History and Philosophy of Science Part A* 23, no. 4: 625–656.

Warwick, Andrew. 1993. "Cambridge Mathematics and Cavendish Physics: Cunningham, Campbell and Einstein's Relativity 1905–1911 Part II: Comparing Traditions in Cambridge Physics." *Studies in History and Philosophy of Science Part A* 24, no. 1: 1–25.

Watson, John B. 1930. *Behaviorism*. London: Kegan Paul Trench Trubner.

Webster, Guy. 2015. "Curiosity Mars Rover Checking Possible Smoother Route." *Jet Propulsion Laboratory News*, January 2014. https://www.jpl.nasa.gov/news/news.php?release=2014-028 (last accessed October 2017).

Weil, David. 2014. *The Fissured Workplace: Why Work Became So Bad for So Many and What Can Be Done to Improve It*. Cambridge, MA: Harvard University Press.

Weinberg, Gerald M. 1971. *The Psychology of Computer Programming*. Hoboken, NJ: Van Nostrand Reinhold.

Weissman, Larry. 1974. "Psychological Complexity of Computer Programs: An Experimental Methodology." *SIGPLAN Notices* 9, no. 6: 25–36.

Werbos, Paul. 1974. "Beyond Regression: New Tools for Prediction and Analysis in the Behavioral Sciences." PhD diss., Harvard University.

Whitehead, Alfred N. (1929) 1978. *Process and Reality*. Edited by D. R. Griffin and D. W. Sherburne. New York: Free Press.

Whitehead, Alfred N., and Bertrand Russell. 1910. *Principia Mathematica*. Cambridge: Cambridge University Press.

Whitehead, Alfred N., and Bertrand Russell. 1911. *Principia Mathematica*. Volume II. Cambridge: Cambridge University Press.

Whitehead, Alfred N., and Bertrand Russell. 1913. *Principia Mathematica*. Volume III. Cambridge: Cambridge University Press.

Wiedenbeck, Susan. 1985. "Novice/Expert Differences in Programming Skills." *International Journal of Man-Machine Studies* 23, no. 4: 383–390.

Wilkes, Maurice. 1985. *Memoirs of a Computer Pioneer*. Cambridge, MA: MIT Press.

Wirth, Niklaus. 1976. *Algorithms + Data Structures = Programs*. Englewood Cliffs, NJ: Prentice Hall.

Wittgenstein, Ludwig. 1922. *Tractatus Logico-Philosophicus*. London: Kegan Paul Trench Trubner.

Wolfe, Jack M. 1971. "Perspectives on Testing for Programming Aptitude." In *Proceedings of the 1971 Twenty-Sixth Annual Conference*, 268–277. New York: ACM.

Wolfe, Jeremy M., Kyle R. Cave, and Susan L. Franzel. 1989. "Guided Search: An Alternative to the Feature Integration Model for Visual Search." *Journal of Experimental Psychology. Human Perception and Performance* 15, no. 3: 419–433.

Wright, Patricia, and Fraser Reid. 1973. "Written Information: Some Alternatives to Prose for Expressing the Outcomes of Complex Contingencies." *Journal of Applied Psychology* 57, no.2: 160–166.

Yapo, Adrienne, and Joseph Weiss. 2018. "Ethical Implications of Bias in Machine Learning." In *Proceedings of the Fifty-First Hawaii International Conference on System Sciences*, Waikoloa Village, HI, January, 5365–5372. Atlanta, GA: Association for Information Systems.

Yates, Joanne. 1989. *Control through Communication: The Rise of System in American Management*. Baltimore: Johns Hopkings University Press.

Zemanek, H. 1981. "Dixit Algorizmi." In *Algorithms in Modern Mathematics and Computer Science Proceedings: Urgench, Uzbek SSR, September 16–22, 1979*, edited by A. P. Ershov and D. E. Knuth. Berlin: Springer.

Zhao, Qi, and Christof Koch. 2011. "Learning a Saliency Map Using Fixated Locations in Natural Scenes." *Journal of Vision* 11, no. 3: 9.

Zhou, Bolei, Aditya Khosla, Agata Lapedriza, Aude Oliva, and Antonio Torralba. 2016. "Learning Deep Features for Discriminative Localization." In *Proceedings of the 2016 IEEE Conference on Computer Vision and Pattern Recognition*, Las Vegas, NV, June–July, 2921–2929. New York: IEEE.

Ziewitz, Malte. 2016. "Governing Algorithms Myth, Mess, and Methods." *Science Technology & Human Values* 41, no. 1: 3–16.

Zuckerberg, Mark. 2016. "I Want to Share Some Thoughts on Facebook and the Election." *Facebook*, November 12. https://www.facebook.com/zuck/posts/10103253901916271 (last accessed October 2017).

Zunshine, Lisa, ed. 2015. *The Oxford Handbook of Cognitive Literary Studies*. Oxford: Oxford University Press.

Zureik, Elia, and Karen Hindle. 2004. "Governance, Security and Technology: The Case of Biometrics." *Studies in Political Economy* 73, no. 1: 113–137.

Index

Inside Technology Series

Edited by Wiebe E. Bijker, W. Bernard Carlson, and Trevor Pinch

Florian Jaton, *The Constitution of Algorithms: Ground-Truthing, Programming, Formulating*

Kean Birch and Fabian Muniesa, *Assetization: Turning Things into Assets in Technoscientific Capitalism*

David Demortain, *The Science of Bureaucracy: Risk Decision-Making and the US Environmental Protection Agency*

Nancy Campbell, *OD: Naloxone and the Politics of Overdose*

Lukas Engelmann and Christos Lynteris, *Sulphuric Utopias: The History of Maritime Fumigation*

Zara Mirmalek, *Making Time on Mars*

Joeri Bruynincx, *Listening in the Field: Recording and the Science of Birdsong*

Edward Jones-Imhotep, *The Unreliable Nation: Hostile Nature and Technological Failure in the Cold War*

Jennifer L. Lieberman, *Power Lines: Electricity in American Life and Letters, 1882–1952*

Jess Bier, *Mapping Israel, Mapping Palestine: Occupied Landscapes of International Technoscience*

Benoît Godin, *Models of Innovation: The History of an Idea*

Stephen Hilgartner, *Reordering Life: Knowledge and Control in the Genomics Revolution*

Brice Laurent, *Democratic Experiments: Problematizing Nanotechnology and Democracy in Europe and the United States*

Cyrus C. M. Mody, *The Long Arm of Moore's Law: Microelectronics and American Science*

Tiago Saraiva, *Fascist Pigs: Technoscientific Organisms and the History of Fascism*

Teun Zuiderent-Jerak, *Situated Interventions: Sociological Experiments in Healthcare*

Basile Zimmermann, *Technology and Cultural Difference: Electronic Music Devices, Social Networking Sites, and Computer Encodings in Contemporary China*

Andrew J. Nelson, *The Sound of Innovation: Stanford and the Computer Music Revolution*

Sonja D. Schmid, *Producing Power: The Pre-Chernobyl History of the Soviet Nuclear Industry*

Casey O'Donnell, *Developer's Dilemma: The Secret World of Videogame Creators*

Christina Dunbar-Hester, *Low Power to the People: Pirates, Protest, and Politics in FM Radio Activism*

Eden Medina, Ivan da Costa Marques, and Christina Holmes, editors, *Beyond Imported Magic: Essays on Science, Technology, and Society in Latin America*

Anique Hommels, Jessica Mesman, and Wiebe E. Bijker, editors, *Vulnerability in Technological Cultures: New Directions in Research and Governance*

Amit Prasad, *Imperial Technoscience: Transnational Histories of MRI in the United States, Britain, and India*

Charis Thompson, *Good Science: The Ethical Choreography of Stem Cell Research*

Tarleton Gillespie, Pablo J. Boczkowski, and Kirsten A. Foot, editors, *Media Technologies: Essays on Communication, Materiality, and Society*

Catelijne Coopmans, Janet Vertesi, Michael Lynch, and Steve Woolgar, editors, *Representation in Scientific Practice Revisited*

Rebecca Slayton, *Arguments that Count: Physics, Computing, and Missile Defense, 1949–2012*

Stathis Arapostathis and Graeme Gooday, *Patently Contestable: Electrical Technologies and Inventor Identities on Trial in Britain*

Jens Lachmund, *Greening Berlin: The Co-Production of Science, Politics, and Urban Nature*

Chikako Takeshita, *The Global Biopolitics of the IUD: How Science Constructs Contraceptive Users and Women's Bodies*

Cyrus C. M. Mody, *Instrumental Community: Probe Microscopy and the Path to Nanotechnology*

Morana Alač, *Handling Digital Brains: A Laboratory Study of Multimodal Semiotic Interaction in the Age of Computers*

Gabrielle Hecht, editor, *Entangled Geographies: Empire and Technopolitics in the Global Cold War*

Michael E. Gorman, editor, *Trading Zones and Interactional Expertise: Creating New Kinds of Collaboration*

Matthias Gross, *Ignorance and Surprise: Science, Society, and Ecological Design*

Andrew Feenberg, *Between Reason and Experience: Essays in Technology and Modernity*

Wiebe E. Bijker, Roland Bal, and Ruud Hendricks, *The Paradox of Scientific Authority: The Role of Scientific Advice in Democracies*

Park Doing, *Velvet Revolution at the Synchrotron: Biology, Physics, and Change in Science*

Gabrielle Hecht, *The Radiance of France: Nuclear Power and National Identity after World War II*

Richard Rottenburg, *Far-Fetched Facts: A Parable of Development Aid*

Michel Callon, Pierre Lascoumes, and Yannick Barthe, *Acting in an Uncertain World: An Essay on Technical Democracy*

Ruth Oldenziel and Karin Zachmann, editors, *Cold War Kitchen: Americanization, Technology, and European Users*

Deborah G. Johnson and Jameson W. Wetmore, editors, *Technology and Society: Building Our Sociotechnical Future*

Trevor Pinch and Richard Swedberg, editors, *Living in a Material World: Economic Sociology Meets Science and Technology Studies*

Christopher R. Henke, *Cultivating Science, Harvesting Power: Science and Industrial Agriculture in California*

Helga Nowotny, *Insatiable Curiosity: Innovation in a Fragile Future*

Karin Bijsterveld, *Mechanical Sound: Technology, Culture, and Public Problems of Noise in the Twentieth Century*

Peter D. Norton, *Fighting Traffic: The Dawn of the Motor Age in the American City*

Joshua M. Greenberg, *From Betamax to Blockbuster: Video Stores tand the Invention of Movies on Video*

Mikael Hård and Thomas J. Misa, editors, *Urban Machinery: Inside Modern European Cities*

Christine Hine, *Systematics as Cyberscience: Computers, Change, and Continuity in Science*

Wesley Shrum, Joel Genuth, and Ivan Chompalov, *Structures of Scientific Collaboration*

Shobita Parthasarathy, *Building Genetic Medicine: Breast Cancer, Technology, and the Comparative Politics of Health Care*

Kristen Haring, *Ham Radio's Technical Culture*

Atsushi Akera, *Calculating a Natural World: Scientists, Engineers and Computers during the Rise of U.S. Cold War Research*

Donald MacKenzie, *An Engine, Not a Camera: How Financial Models Shape Markets*

Geoffrey C. Bowker, *Memory Practices in the Sciences*

Christophe Lécuyer, *Making Silicon Valley: Innovation and the Growth of High Tech, 1930–1970*

Anique Hommels, *Unbuilding Cities: Obduracy in Urban Sociotechnical Change*

David Kaiser, editor, *Pedagogy and the Practice of Science: Historical and Contemporary Perspectives*

Charis Thompson, *Making Parents: The Ontological Choreography of Reproductive Technology*

Pablo J. Boczkowski, *Digitizing the News: Innovation in Online Newspapers*

Dominique Vinck, editor, *Everyday Engineering: An Ethnography of Design and Innovation*

Nelly Oudshoorn and Trevor Pinch, editors, *How Users Matter: The Co-Construction of Users and Technology*

Peter Keating and Alberto Cambrosio, *Biomedical Platforms: Realigning the Normal and the Pathological in Late-Twentieth-Century Medicine*

Paul Rosen, *Framing Production: Technology, Culture, and Change in the British Bicycle Industry*

Maggie Mort, *Building the Trident Network: A Study of the Enrollment of People, Knowledge, and Machines*

Donald MacKenzie, *Mechanizing Proof: Computing, Risk, and Trust*

Geoffrey C. Bowker and Susan Leigh Star, *Sorting Things Out: Classification and Its Consequences*

Charles Bazerman, *The Languages of Edison's Light*

Janet Abbate, *Inventing the Internet*

Herbert Gottweis, *Governing Molecules: The Discursive Politics of Genetic Engineering in Europe and the United States*

Kathryn Henderson, *On Line and On Paper: Visual Representation, Visual Culture, and Computer Graphics in Design Engineering*

Susanne K. Schmidt and Raymund Werle, *Coordinating Technology: Studies in the International Standardization of Telecommunications*

Marc Berg, *Rationalizing Medical Work: Decision Support Techniques and Medical Practices*

Eda Kranakis, *Constructing a Bridge: An Exploration of Engineering Culture, Design, and Research in Nineteenth-Century France and America*

Paul N. Edwards, *The Closed World: Computers and the Politics of Discourse in Cold War America*

Donald MacKenzie, *Knowing Machines: Essays on Technical Change*

Wiebe E. Bijker, *Of Bicycles, Bakelites, and Bulbs: Toward a Theory of Sociotechnical Change*

Louis L. Bucciarelli, *Designing Engineers*

Geoffrey C. Bowker, *Science on the Run: Information Management and Industrial Geophysics at Schlumberger, 1920–1940*

Wiebe E. Bijker and John Law, editors, *Shaping Technology / Building Society: Studies in Sociotechnical Change*

Stuart Blume, *Insight and Industry: On the Dynamics of Technological Change in Medicine*

Donald MacKenzie, *Inventing Accuracy: A Historical Sociology of Nuclear Missile Guidance*

Pamela E. Mack, *Viewing the Earth: The Social Construction of the Landsat Satellite System*

H. M. Collins, *Artificial Experts: Social Knowledge and Intelligent Machines*

http://mitpress.mit.edu/books/series/inside-technology